INTRODUCTION TO PROBABILITY AND MEASURE

Introduction to Probability and Measure

K. R. Parthasarathy
Professor, Indian Statistical Institute, New Delhi

Springer-Verlag New York Inc.

First published in India 1977 by
The Macmillan Company of India Ltd

First published in the United Kingdom 1978 by
The Macmillan Press Ltd

Sole distributors for the United States
of America, its territories and possessions,
the Philippine Republic and Canada
Springer-Verlag New York Inc/New York

ISBN 0-387-91135-9

Printed in India by V. Varadarajan
at Macmillan India Press, Madras 600 002

To my grandfather

S Raghunathachari

PREFACE

In 1902 the French mathematician Henri Lebesgue wrote his famous dissertation *Intégrale, Longueur, Aire* (Integral, Length and Area). Since 1914 the theory of the Lebesgue measure has become a part of the undergraduate curriculum in analysis in all the technologically advanced countries of the world. In 1933 the Russian mathematician A. N. Kolmogorov wrote the famous book *Grundbegriffe der Wahrsche inlichkeitsrechnung* (Foundations of Probability) in which he gave the basic axioms of probability theory. The appearance of *Measure Theory* by P. R. Halmos and *An Introduction to Probability Theory and Its Applications* by W. Feller in 1950 made both subjects accessible to all undergraduate and graduate students in mathematics all over the world.

The present book has been written in the hope that it will provide the impetus to introduce in undergraduate and graduate programmes both measure theory and probability theory as a one-year course. Since the study of probability theory in its advanced stage depends on a knowledge of measure theory, special effort has been made to integrate the two subjects into a single volume.

The material of the book grew out of the lectures delivered by the author to M.Sc. students at the Centre of Advanced Study in Mathematics in the University of Bombay, M.Stat. students of the Indian Statistical Institute, Delhi, and M.Sc. students at the Indian Institute of Technology, Delhi.

The book is divided into eight chapters. Chapter I deals with combinatorial probability and the classical limit theorems of Poisson and Laplace–De Moivre. It provides a motivation for extending measures from boolean algebras to σ-algebras. Chapter II is devoted to extension of measures from boolean semi-algebras and classes of topologically important subsets to σ-algebras. Chapter III deals with properties of borel maps from a measure space into a separable metric space. In particular, Lusin's theorem and the isomorphism theorem are proved. Extension of measures to projective limits of borel spaces is also studied. Chapter IV deals with integration, Reisz's representation theorem that integration is the only linear operation on good function spaces, and properties of function spaces arising out of a measure space. Chapter V contains a discussion of measures and transition measures on product spaces. The Lebesgue measure in R^k, the change of variable formula for Lebesgue integrals and construction of infinitely differentiable functions are also considered. Chapter VI, which is the longest in the book, introduces the notion of conditional expectation through orthogonal projection and avoids the customary use of the Radon–Nikodym theorem. The Radon–Nikodym theorem and the Lebesgue decomposition are deduced as a corollary of a more general decomposition theorem due to von Neumann. Convergence of conditional expectations in various senses, the idea of regular conditional probability, ergodic theorems and ergodic decomposition are also treated in this chapter. Chapter VII gives a brief introduction to weak convergence of probability measures and characteristic functions. The last chapter introduces the construction of Haar measure on a locally compact group and invariant and quasi-invariant measures on homogeneous spaces. The Mackey–Weil theorem on groups with a quasi-invariant measure is also proved. For the benefit of the student a number of exercises are included. Connections between measure theory and probability theory on the one hand and various topics like functional analysis, statistics, ergodic theory, etc., on the other are indicated through Remarks, Examples and Exercises.

New Delhi, 1977 K.R.P.

ACKNOWLEDGEMENTS

Thanks to Professor S. S. Shrikhande whose enthusiasm brought me back from Manchester into a teaching career in India. It was in his department that I enjoyed my considerable freedom to violate the 'regular' syllabus and teach anything I wanted. My special thanks to Professor C. R. Rao who invited me to teach this subject to the students of the Indian Statistical Institute, which provided me with the opportunity to go much above the accepted levels of other degree-awarding institutions. Thanks to the authorities of the Indian Institute of Technology for providing me with a comfortable house in their pleasant campus. Thanks to Sri S. Ramasubramaniam who read the manuscript and made many corrections. Thanks to Sri Dev Raj Joshi for his efficient typing of the manuscript on his heavy and ancient mathematical typewriter. Finally, thanks to my wife Shyama who cheerfully exerted herself in no small measure to shield me from the children and provide the required solitude for writing this volume.

CONTENTS

CHAPTER ONE

Probability on Boolean Algebras

§1. Sets and Events

In probability theory we look at all possible basic outcomes of a statistical experiment and assume that they constitute a set X, called the *sample space*. The points or elements of X are called *elementary outcomes*. We shall illustrate by a few examples.

Example. 1.1. The simplest statistical experiment is one with two elementary outcomes. For example, tossing a coin where the outcome is a head or a tail; observing the sex of a new born baby where the outcome is male or female; examining whether a manufactured item is defective or not, etc.

In these cases we denote the basic outcomes by 0 and 1. It is customary to call them *failure* and *success* respectively. The sample space X contains exactly two points, namely, 0 and 1.

Example. 1.2. Throw a die and observe the score. The die has six faces and the possible scores form the set $X = \{1, 2, 3, 4, 5, 6\}$.

Example. 1.3. Go on tossing a coin till you get the first head and observe the outcome at every stage. If we denote head by H and tail by T, any elementary outcome of this experiment is a finite sequence of the form $TTT \ldots TH$. The sample space consists of all such sequences.

Example. 1.4. Shuffle a pack of cards and observe the order from top to bottom. The space X consists of $52!$ permutations.

Example. 1.5. Observe the atmospheric temperature at a specific place. The elementary outcomes are just real numbers. Thus the sample space is the real line.

Example. 1.6. Observe the pressure and temperature of a gas in a box. Here X may be assumed to be the plane R^2.

Example. 1.7. Observe the temperature graph of the atmosphere during a fixed hour. The sample space X may be identified with the set of all continuous curves in the interval $[0, 1]$.

Let $A \subset X$ be any subset of the sample space X of a statistical experiment. The performance of the experiment leads to the observation of an elementary outcome x which is an element of X. If $x \in A$, we say that the *event* A *has occurred.* If $x \notin A$, we say that the event A has not occurred or equivalently $X-A$ (the complement of A) has occurred. From a practical point of view not every event may be of interest. For example, in Example 1.5 above, consider the event ‘ the temperature measured is a transcendental number ’. Such an event is not of any practical significance. However, an event of the kind ‘ the temperature measured lies in the interval $[a, b]$ ’ is of value. We can sum up this discussion as follows: there is a collection $\mathscr{F}$ of subsets of the sample space, the events corresponding to the elements of which are of ‘ practical value ’. We assume that such a collection $\mathscr{F}$ of events or subsets of the sample space X is clearly specified. We simply say that *$\mathscr{F}$ is the collection of all events* concerning the statistical experiment whose sample space is X. By an *event* we mean an element of $\mathscr{F}$.

We shall now examine what are athe ntural conditions which the collection or family $\mathscr{F}$ of all events should satisfy. Let $A \subset B \subset X$ be such that $A, B \in \mathscr{F}$. If $x \in A$, then $x \in B$. In other words, whenever A occurs B also occurs. Thus set theoretic inclusion is equivalent to the logical notion of implication.

If $A, B \in \mathscr{F}$, consider the sets $A \cup B$, $A \cap B$ and $X-A$. Note that the occurrence of one of the events A, B is equivalent to saying that the experiment yields an observation x belonging to $A \cup B$. It is natural to expect that $A \cup B$ is also an event. The occurrence of both A and B means that the experimental observation x belongs to $A \cap B$. The non-occurrence of A means that x lies in $X-A$. So it is natural to demand that $\mathscr{F}$ is closed under finite union, finite intersection and complementation. Nothing is lost by assuming that the whole space X and hence its complement, the empty set $\emptyset$ also belong to $\mathscr{F}$. This leads us to the following.

Definition. 1.8. A collection $\mathscr{F}$ of subsets of a set X is called a *boolean algebra* if the following conditions are satisfied:

(1) If $A, B \in \mathscr{F}$, then $A \cup B \in \mathscr{F}$ and $A \cap B \in \mathscr{F}$;

(2) If $A \in \mathscr{F}$, the complement $X-A \in \mathscr{F}$;

(3) the empty set $\emptyset$ and whole space X belong to $\mathscr{F}$.

Remark. 1.9. Hereafter throughout the text we shall write A' (read A prime) for the complement $X-A$ of the set A. For any two subsets A, B of X we shall write AB for the intersection $A \cap B$, $A-B$ for the set AB' and $A \Delta B$ for the symmetric difference $(A-B) \cup (B-A)$.

Example. 1.10. Let X be any non-empty set and let $\mathcal{F}$ be the class of all subsets of X. Then $\mathcal{F}$ is a boolean algebra.

Example. 1.11. Let $X=R$ be the real line and let the family $\mathcal{I}$ be defined by

$$\mathcal{I} = \{\text{all intervals of the form } (-\infty, +\infty), (-\infty, a], (a, \infty), (a, b], \text{ where } a, b \in R\}.$$

Then the collection

$$\mathcal{F} = \{A : A \subset R, A = \bigcup_{i=1}^{n} A_i, A_i \in \mathcal{I}, A_i \cap A_j = \emptyset \text{ for } i \neq j \text{ for some positive integer } n\}$$

is a boolean algebra. (Here we consider the empty set as the interval $(a, b]$ when $b \leqslant a$.)

Example. 1.12. Let Y be any set and let X be the space of all sequences of elements from Y, i.e., any $x \in X$ can be written as $x = (y_1, y_2, \ldots)$ where $y_i \in Y$ for every $i = 1, 2, \ldots$. Let A be any subset of the cartesian product $Y \times Y \times \ldots \times Y$, taken k times. A subset $C \subset X$ of the form

$$C = \{x = (y_1, y_2, \ldots) : (y_{i_1}, y_{i_2}, \ldots, y_{i_k}) \in A\},$$

(where $i_1 < i_2 < \ldots < i_k$ are fixed positive integers) is called a k-*dimensional cylinder set*. Then the collection $\mathcal{F}$ of all finite dimensional cylinder sets is a boolean algebra.

Going back to the relation between the language of set theory and the language of events we summarise our conclusions in the form of a table. Let $\mathcal{F}$ be a boolean algebra of subsets of the sample space of a statistical experiment so that $\mathcal{F}$ is the collection of all events. Then we have the following dictionary:

Language of events	*Language of set theory*
A is an event	$A \in \mathcal{F}$
event A implies event B	$A \subset B$
event A does not occur	A'
one of the events A, B occurs	$A \cup B$
both the events A and B occur	AB
event which always occurs	X
event which never occurs	$\emptyset$
events A and B cannot occur at the same time	$A \cap B = \emptyset$

§2. Probability on a Boolean Algebra

Consider a statistical experiment whose elementary outcomes are described by a sample space X together with a boolean algebra $\mathcal{F}$ of subsets of X. Let the experiment be performed n times resulting in the elementary outcomes $x_1, x_2, \ldots, x_n \in X$. Let $A \subset X$ be an element of $\mathcal{F}$. Let

$$p_n(A) = m(A)/n$$

where $m(A)$ is the number of elementary outcomes x_i that lie in the set A. The number $p_n(A)$ may be called the frequency of occurrence of the event A in the given n trials. First of all we note that $A \rightarrow p_n(A)$ is a map from $\mathcal{F}$ into the unit interval $[0, 1]$. It is clear that

(i) $p_n(A \cup B) = p_n(A) + p_n(B)$ if $A \cap B = \emptyset$, $A, B \in \mathcal{F}$

(ii) $p_n(X) = 1$.

It follows from property (i) that

$$p_n(A_1 \cup A_2 \cup \ldots \cup A_k) = \sum_{i=1}^{k} p_n(A_i) \text{ if } A_i \cap A_j = \emptyset$$

for all $i \neq j$ and $A_1, A_2, \ldots, A_k \in \mathcal{F}$. We say that p_n is a non-negative finitely additive function on $\mathcal{F}$ such that $p_n(X) = 1$. If there is a 'statistical regularity' in the occurrence of the observations $x_1, x_2, \ldots,$ we expect that, for $A \in \mathcal{F}$, $p_n(A)$ will stabilise to a number $p(A)$. If it is indeed so, then the map $A \rightarrow p(A)$ will share the properties (i) and (ii). Motivated by these considerations we introduce the following definitions.

Definition. 2.1. Let $\{A_\alpha, \alpha \in I\}$ be a family of subsets of a set X, where I is some index set. Such a family is said to be *pairwise disjoint* if $A_\alpha \cap A_\beta = \emptyset$ whenever $\alpha \neq \beta$ and $\alpha, \beta \in I$.

Definition. 2.2. Let $\mathcal{F}$ be a boolean algebra of subsets of a set X. A map $m : \mathcal{F} \rightarrow [0, \infty]$ is said to be *finitely additive* if

$$m(A \cup B) = m(A) + m(B) \text{ whenever } A, B \in \mathcal{F} \text{ and } A \cap B = \emptyset$$

It is said to be *countably additive* if for any sequence $\{A_n\}$ of pairwise disjoint sets belonging to $\mathcal{F}$

$$m\left(\bigcup_{n=1}^{\infty} A_n\right) = \sum_{n=1}^{\infty} m(A_n), \text{ if } \bigcup_{n=1}^{\infty} A_n \in \mathcal{F}.$$

A map $p : \mathcal{F} \rightarrow [0, 1]$ is called a *probability distribution* on $\mathcal{F}$ if it is finitely additive and $p(X) = 1$.

We shall now introduce a few examples.

Example. 2.3. Let X be a finite or countable set and let $\mathcal{F}$ be the boolean algebra of all subsets of X. Let $\{x_1, x_2, \ldots\}$ be an enumeration of all the points of X. Let $\{p_1, p_2, \ldots\}$ be a sequence of non-negative numbers. For any $A \subset X$, let

$$m(A) = \sum_{j: x_j \in A} p_j.$$

Then it is clear that m is a countably additive function on $\mathcal{F}$. If $\sum_{j: x_j \in X} p_j = 1$, m is a probability distribution on $\mathcal{F}$.

Example. 2.4. Let F be a monotonically increasing function defined on the real line R. Let

$$m((a, b]) = F(b) - F(a) \text{ if } a < b \text{ and } a, b \in R.$$

Write $F(+\infty) = \lim_{a \to +\infty} F(a)$ and $F(-\infty) = \lim_{a \to -\infty} F(a)$. Put

$$\begin{aligned} m((-\infty, a]) &= F(a) - F(-\infty), \\ m((b, +\infty)) &= F(\infty) - F(b), \\ m((-\infty, +\infty)) &= F(+\infty) - F(-\infty). \end{aligned}$$

Then m is a finitely additive set function defined on the class $\mathcal{I}$ of intervals (Example 1.11), i.e.,

$$m(\bigcup_{j=1}^{k} I_j) = \sum_{j=1}^{k} m(I_j)$$

whenever $I_1, I_2, \ldots, I_k$ and $\bigcup_{j=1}^{k} I_j$ belong to $\mathcal{I}$ and the family $\{I_j, 1 \leqslant j \leqslant k\}$ is pairwise disjoint. Let now A be any set of the form

$$A = \bigcup_{r=1}^{k} I_r \tag{2.1}$$

where $I_1, I_2, \ldots, I_k$ belong to $\mathcal{I}$ and are pairwise disjoint. Define

$$\tilde{m}(A) = \sum_{r=1}^{k} m(I_r).$$

Now the question arises whether $\tilde{m}$ is well defined. For, it is quite possible that A is also of the form

$$A = \bigcup_{s=1}^{l} F_s \tag{2.2}$$

where $F_1, F_2, \ldots, F_l$ belong to $\mathcal{I}$ and are pairwise disjoint. Thus A has two representations (2.1) and (2.2). However,

$$\sum_{r=1}^{k} m(I_r) = \sum_{s=1}^{l} m(F_s) \tag{2.3}$$

Indeed, we have

$$I_r = I_r \cap A = \bigcup_{s=1}^{l} (I_r \cap F_s),$$

$$F_s = F_s \cap A = \bigcup_{r=1}^{k} (F_s \cap I_r).$$

We note that the family $\mathscr{I}$ is closed under finite intersection. Since m is additive on $\mathscr{I}$, it follows that

$$m(I_r) = \sum_{s=1}^{l} m(I_r \cap F_s),$$

$$m(F_s) = \sum_{r=1}^{k} m(F_s \cap I_r).$$

Now (2.3) is an immediate consequence of the above two equations. This argument implies that $\tilde{m}$ is a well defined finitely additive map on the boolean algebra $\mathscr{F}$ of all subsets which are finite disjoint unions of intervals from $\mathscr{I}$. In other words, corresponding to every monotonic increasing function F on R, one can construct a unique non-negative finitely additive function on the boolean algebra $\mathscr{F}$ of Example 1.11. This becomes a probability distribution if

$$\lim_{\substack{b \to +\infty \\ a \to -\infty}} F(b) - F(a) = 1$$

Proposition. 2.5. Let m be a non-negative finitely additive function on a boolean algebra $\mathscr{F}$ of subsets of X. If $A \subset B$ and $A, B \in \mathscr{F}$, then $m(A) \leqslant m(B)$. If $A_1, A_2, \ldots, A_k \in \mathscr{F}$, then

$$m\left(\bigcup_{i=1}^{k} A_i\right) \leqslant \sum_{i=1}^{k} m(A_i). \tag{2.4}$$

Proof. To prove the first part we observe that

$$B = A \cup (BA') \text{ if } A \subset B.$$

Since $\mathscr{F}$ is a boolean algebra A and BA' are disjoint subsets belonging to $\mathscr{F}$. Hence

$$m(B) = m(A) + m(BA') \geqslant m(A).$$

To prove the second part we note that

$$\bigcup_{i=1}^{k} A_i = \bigcup_{i=1}^{k} B_i,$$

where $B_1 = A_1$, $B_2 = A_2 A'_1$, ..., $B_i = A_i A'_{i-1} A'_{i-2} \ldots A'_1$, ... $B_k = A_k A'_{k-1}$

$A'_{k-2}, \ldots, A'_1$. Then $B_1, B_2, \ldots$ are disjoint sets belonging to $\mathscr{F}$ and $B_i \subset A_i$, for all $i=1, 2, \ldots, k$. Hence

$$\begin{aligned} m\left(\bigcup_{i=1}^{k} A_i\right) &= m\left(\bigcup_{i=1}^{k} B_i\right) \\ &= \sum_{i=1}^{k} m(B_i) \\ &\leqslant \sum_{i=1}^{k} m(A_i). \end{aligned}$$

This completes the proof.

Remark. 2.6. Property (2.4) is known as *finite subadditivity*. If m is countably additive (2.4) holds with $k = \infty$, provided $\bigcup_{i=1}^{\infty} A_i$ belongs to $\mathscr{F}$. The same proof goes through. When $k=\infty$, (2.4) is known as the property of *countable subadditivity*.

§3. Probability Distributions and Elementary Random Variables

Consider a statistical experiment, the performance of which leads to an observation x in the sample space X. Very often one is not interested in the exact observation but a function of the observation. We shall illustrate by means of a few examples.

Example. 3.1. Consider an individual performing an experiment with two elementary outcomes called ' success ' and ' failure ' (see Example 1.1). Suppose he gets a rupee if success occurs and loses a rupee if failure occurs. Then his gain can be expressed through the function f defined by

$$f(0) = -1, \quad f(1) = +1,$$

where 0 and 1 denote the outcomes failure and success respectively.

Example. 3.2. Suppose r objects are distributed in n cells. Assume the objects to be distinguishable from one another. Observe the configuration. It should be noted that more than one object can occupy a cell. Then the sample space X of all possible configurations contains n^r points. For each configuration $x \in X$, let $f(x)$ be the number of empty cells.

Example. 3.3. Let a bullet be shot from a gun and let the experiment consist of observing the trajectory of the bullet. For every trajectory x, let $f(x)$ be the coordinates of the point at which the bullet hits the ground.

From the above examples we understand that the value of the function depends on the outcome which is subject to chance. Thus the value of the function varies in a ' random ' manner. Till we make further progress in the subject we shall consider functions on X, which take only a finite number of values.

Let $f : X \to Y$ be a map from the sample space X into a set Y. Let $\mathcal{F}$ be a boolean algebra of subsets of X, on which we shall consider probability distributions. Suppose we wish to raise the following question: what is the probability that the experiment yields an elementary outcome $x \in X$ such that the function $f(x)$ takes a given value $y \in Y$? Consider the set

$$\{x : f(x) = y\} = f^{-1}(\{y\}).$$

If we wish to find the probability of the above event, it is necessary that $f^{-1}(\{y\}) \in \mathcal{F}$. For this reason we introduce the following definition.

Definition. 3.4. Let X be a sample space with a boolean algebra $\mathcal{F}$ of subsets of X. A map $f : X \to Y$ is called a Y-*valued simple random variable* if f takes only a finite number of values and, for every $y \in Y$,

$$f^{-1}(\{y\}) \in \mathcal{F}.$$

If Y is the real line we shall call f a *simple random variable.* We denote by $S(X, \mathcal{F})$ the set of all simple random variables.

For any set $A \subset X$, let

$$\begin{aligned} \chi_A(x) &= 1 \text{ if } x \in A, \\ &= 0 \text{ if } x \notin A. \end{aligned}$$

Then χ_A is called the *characteristic or indicator function of the set* A. If $A \in \mathcal{F}$, χ_A is a simple random variable assuming two values, namely 0 and 1. If $a_1, a_2, \ldots, a_k$ are real numbers and $A_1, A_2, \ldots, A_k \in \mathcal{F}$, then $\sum_{j=1}^{k} a_j \chi_{A_j}$ is a simple random variable. Conversely every simple random variable can be expressed in this form with A_i*'s* pairwise disjoint. It is clear that

$$\begin{aligned} \chi_{A \cup B} &= \chi_A + \chi_B - \chi_{AB}, \\ \chi_A \chi_B &= \chi_{AB}, \\ |\chi_A - \chi_B| &= \chi_{A \Delta B}, \text{ for all } A, B \subset X. \end{aligned}$$

In particular, it follows that the set $S(X, \mathcal{F})$ of all simple random variables is an algebra under the usual operations of addition, multiplication and scalar multiplication.

Definition. 3.5. By a *boolean space* we mean a pair $(X, \mathscr{F})$ where X is a set and $\mathscr{F}$ is a boolean algebra of subsets of X. By a *boolean probability space* we mean a triple $(X, \mathscr{F}, P)$ where $(X, \mathscr{F})$ is a boolean space and P is a probability distribution on $\mathscr{F}$. If s is a simple random variable on $(X, \mathscr{F})$ and P is a probability distribution on $\mathscr{F}$, we define the *integral* of s with respect to P as the number

$$\sum_i a_i P(s^{-1}(\{a_i\})),$$

where the summation is over all the values a_i which s can take. We denote this number by the symbol $\int s\, dP$ or simply $\mathbf{E}s$ when P is fixed. $\mathbf{E}s$ is also called the *expectation* of s with respect to P.

Proposition. 3.6. If $s = \sum_{i=1}^{k} a_i \chi_{A_i}$, where $A_1, A_2, \ldots, A_k$ are disjoint sects in $\mathscr{F}$ and $a_1, a_2, \ldots, a_k$ are real numbers then

$$\int s\, d P = \sum_{i=1}^{k} a_i P(A_i). \tag{3.1}$$

Further

(i) $\int (as_1+bs_2)dP = a \int s_1 dP + b \int s_2 dP$
for any two simple random variables s_1 and s_2 and any two real constants a and b;

(ii) the function Q on $\mathscr{F}$ defined by
$$Q(F) = \int s \chi_F\, dP, \; F \in \mathscr{F}$$
is finitely additive, i.e.,
$$Q\left(\bigcup_{i=1}^{j} F_i\right) = \sum_{i=1}^{j} Q(F_i)$$
whenever $F_1, F_2, \ldots, F_j$ are pairwise disjoint elements in $\mathscr{F}$;

(iii) $\int s dP \geqslant 0$ if $P(\{x : s(x) < 0\}) = 0$;

(iv) $\inf_{x \in X} s(x) \leqslant \int s dP \leqslant \sup_{x \in X} s(x)$

Proof. Without loss of generality we may assume that the a_i's are distinct and $\bigcup_{i=1}^{k} A_i = X$. Then $s^{-1}(\{a_i\}) = A_i$ and the range of s is the set $\{a_1, a_2, \ldots, a_k\}$. Hence Eq. (3.1) follows immediately from the definition of integral. To prove property (i) we may assume that

$$s_1 = \sum_{i=1}^{k} a_i \chi_{A_i}, \; s_2 = \sum_{j=1}^{l} \beta_j \chi_{B_j}$$

where $A_1, A_2, \ldots, A_k$ and $B_1, B_2, \ldots, B_l$ are two partitions of X into disjoint sets belonging to $\mathscr{F}$. Then

$$as_1 + bs_2 = \sum_{i=1}^{k} \sum_{j=1}^{l} (a\alpha_i + b\beta_j) \chi_{A_i B_j},$$

where the sets A_iB_j constitute another partition of X. Then

$$\begin{aligned}
\int (as_1 + bs_2)dP &= \sum_i \sum_j (a\alpha_i + b\beta_j) P(A_iB_j) \\
&= a \sum_i \alpha_i \left\{ \sum_j P(A_iB_j) \right\} + \\
&\quad b \sum_j \beta_j \left\{ \sum_i P(A_iB_j) \right\} \\
&= a \sum_i \alpha_i P(A_i) + b \sum_j \beta_j P(B_j) \\
&= a \int s_1 \, dP + b \int s_2 \, dP.
\end{aligned}$$

Here we have used the fact that P is finitely additive and

$$\bigcup_i A_iB_j = A_i \left(\bigcup_j B_j \right) = A_iX = A_i,$$

$$\bigcup_i A_iB_j = \left(\bigcup_i A_i \right) B_j = XB_j = B_j.$$

Property (ii) follows from property (i). Properties (iii) and (iv) follow from Eq. (3.1) immediately.

Remark. 3.7. It should be noted that property (ii) indicates a method of manufacturing new finitely additive functions on $\mathscr{F}$ from a given one by the process of integration.

We shall now prove some elementary results by using the notion of expectation and its properties described in Proposition 3.6.

Proposition. 3.8. Let $A_1, A_2, \ldots, A_n$ be subsets of X, and let $B = \bigcup_{i=1}^{n} A_i$. Then

$$\begin{aligned}
\chi_B = \sum_{i=1}^{n} \chi_{A_i} &- \sum_{1 \leqslant i < j \leqslant n} \chi_{A_i A_j} + \ldots \\
&+ (-1)^{r-1} \sum_{1 \leqslant i_1 < i_2 \ldots < i_r \leqslant n} \chi_{A_{i_1} A_{i_2} \ldots A_{i_r}} + \ldots \\
&+ (-1)^{n-1} \chi_{A_1 A_2 \ldots A_n},
\end{aligned} \tag{3.2}$$

where $A_{i_1} A_{i_2} \ldots, A_{i_r}$ stands for the intersection $\bigcap_{j=1}^{r} A_{i_j}$.

Proof. Let x belong to none of the A_i's. Then $\chi_B(x) = 0$ and every term on the right hand side of Eq. (3.2) is zero. If x belongs to exactly

k of the sets $A_1, A_2, \ldots, A_n$ then the rth term on the right hand side is $(-1)^{r-1}\binom{k}{r}$ if $r \leqslant k$ and 0 otherwise. Thus the right hand side of Eq. (3.2) is

$$\sum_{r=1}^{k} (-1)^{r-1}\binom{k}{r} = 1 - (1-1)^k = 1.$$

Then $\chi_B(x) = 1$ and Eq. (3.2) always holds. This completes the proof.

Corollary. 3.9. If P is a probability distribution on $(X, \mathcal{F})$, then for any $A_1, A_2, \ldots, A_n \in \mathcal{F}$.

$$P(\bigcup_{i=1}^{n} A_i) = \sum_{r=1}^{n} (-1)^{r-1} S_r, \tag{3.3}$$

where

$$S_r = \sum_{1 \leqslant i_1 < i_2 < \ldots i_r \leqslant n} P(A_{i_1} A_{i_2} \ldots A_{i_r}). \tag{3.4}$$

Proof. Equation (3.3) follows from Eq. (3.2) by integration of both sides and Proposition 3.6.

Proposition. 3.10. Let $A_1, A_2, \ldots, A_n$ be any subsets of X. Let B be the subset of all those points which belong to exactly k of the sets $A_1, A_2, \ldots, A_n$. Then

$$\chi_B = \sum_{r=k}^{n} (-1)^{r-k}\binom{r}{k} [\sum_{1 \leqslant i_1 < i_2 < \ldots < i_r \leqslant n} \chi_{A_{i_1} A_{i_2} \ldots A_{i_r}}]. \tag{3.5}$$

Proof. Let x be a point which belongs to exactly m of the sets $A_1, A_2, \ldots, A_n$. If $m < k$, then $\chi_B(x) = 0$ and every term on the right hand side of Equation (3.5) vanishes. If $m = k$, $\chi_B(x) = 1$. On the right hand side of Equation (3.5) the term within square brackets is one if $r = k$ and zero otherwise. Thus the right hand side is also unity. Now let $m > k$. Then $\chi_B(x) = 0$. The right hand side is

$$\sum_{r=k}^{m} (-1)^{r-k}\binom{r}{k}\binom{m}{r} = \binom{m}{k}(1-1)^{m-k} = 0.$$

Thus Equation (3.5) always holds and the proof is complete.

Corollary. 3.11. Let P be a probability distribution on $(X, \mathcal{F})$. Let $A_1, A_2, \ldots, A_n$ be n events belonging to $\mathcal{F}$, and let p_k be the probability that exactly k of the events $A_1, A_2, \ldots, A_n$ occur. Then

$$p_k = \sum_{r=k}^{n} (-1)^{r-k}\binom{r}{k} S_r \tag{3.6}$$

where S_r is defined by Equation (3.4).

Proof. This is obtained by integrating both sides of Equation (3.5) and using Proposition 3.6.

Example. 3.12. Let X be the set of all permutations of the integers $1, 2, 3, \ldots, N$ and let $\mathcal{F}$ be the boolean algebra of all subsets of X. Let P be the probability distribution which assigns the same probability $\frac{1}{N!}$ to every permutation (see Example 2.3). Let A_i be the set of all permutations which leave i fixed. In this case

$$P(A_{i_1} A_{i_2} \ldots A_{i_r}) = \frac{(N-r)!}{N!} \text{ if } i_1 < i_2 < \ldots < i_r.$$

Thus

$$S_r = \binom{N}{r} \frac{(N-r)!}{N!} = \frac{1}{r!}.$$

Corollary 3.9 implies that, the probability that a 'random' permutation leaves at least one of the i's fixed is given by the expression $1 - \frac{1}{2!} + \frac{1}{3!} - \ldots + \frac{(-1)^{N-1}}{N!}$. An application of Corollary 3.11 shows that the probability of exactly m of the elements $1, 2, \ldots, N$ being fixed by a random permutation is given by the expression

$$\begin{aligned} p_m &= \sum_{r=m}^{N} (-1)^{r-m} \binom{r}{m} \frac{1}{r!} \\ &= \frac{1}{m!} \left[1 - \frac{1}{1!} + \frac{1}{2!} - \frac{1}{3!} + \ldots + \frac{(-1)^{N-m}}{(N-m)!} \right]. \end{aligned} \tag{3.7}$$

We also observe that p_m converges to $\frac{e^{-1}}{m!}$ as $N \to \infty$. For every fixed N, consider the experiment of selecting a random permutation x from the set X of all permutations of $1, 2, \ldots, N$. Let $f(x)$ be the number of elements in the set $\{1, 2, \ldots, N\}$ left fixed by the permutation x. Then f is a simple random variable taking values $0, 1, 2, \ldots, N$. Then the probability that $f(x) = m$ is given by Eq. (3.7) for $m \leqslant N$. As $N \to \infty$, we get a distribution on the set of all non-negative integers such that the probability for m is $\frac{e^{-1}}{m!}$. This is a special example of many limit theorems to come in our subject. This particular limiting distribution is a special case of the Poisson distribution.

Example. 3.13. Suppose r objects (which are distinguishable from each other) are distributed randomly in n cells. The experiment consists in observing the configuration. Since any object can occupy any one of the n cells there are n^r possible configurations. Thus the sample space X consists of n^r points. Let $\mathcal{F}$ be the class of all subsets of X. Let A_i be the set of all configurations in which cell number i is empty. By random distribution of objects we mean that all configurations have equal probability n^{-r}. Now we can ask the question: what is the probability that exactly k cells are empty? Let it be p_k. Let $i_1 < i_2 < \ldots < i_j$. Then

$$P(A_{i_1} A_{i_2} \ldots A_{i_j}) = \frac{(n-j)^r}{n^r}.$$

In the notation of Eq. (3.4)

$$S_j = \binom{n}{j}\left(1 - \frac{j}{n}\right)^r.$$

By Corollary 3.11,

$$p_k = \sum_{j=k}^{n} (-1)^{j-k} \binom{j}{k}\binom{n}{j}\left(1 - \frac{j}{n}\right)^r. \tag{3.8}$$

As in the preceding example the number of empty cells in any observed configuration x can be thought of as a simple random variable assuming the values 0, 1, 2, ... $(n-1)$. The probability that this random variable assumes the value k is equal to p_k given by Equation (3.8). Its expectation is $\sum_{k=0}^{n-1} k\, p_k$.

We see that p_k is a function of r and n, namely, the number of objects and cells. We can ask the question: what happens when r and n tend to ∞ in some suitable manner? Does p_k converge to a limit for every fixed k in such a case?

Putting $j-k=s$ in Eq. (3.8), an elementary computation shows that

$$p_k = \frac{1}{k\,!} \sum_{s=0}^{n-k} \frac{(-1)^s}{s\,!}\left[\left(1 - \frac{s+k}{n}\right)^r n^{s+k}\left(1 - \frac{s+k-1}{n}\right) \times \right.$$
$$\left. \left(1 - \frac{s+k-2}{n}\right) \ldots 1\right]. \tag{3.9}$$

Since $1 - x \leqslant e^{-x}$ for $0 \leqslant x < 1$, we have

$$\left(1 - \frac{s+k}{n}\right)^r n^{s+k} \leqslant (ne^{-r/n})^{s+k}.$$

Thus the term within square brackets in Eq. (3.9) is dominated by $(ne^{-r/n})^{s+k}$. Let now r and n tend to ∞ in such a manner that $ne^{-r/n} \to \lambda > 0$. Thus every term in the summation (3.9) has absolute value dominated by $\frac{(\lambda + 1)^{s+k}}{s\,!}$ for all r and n sufficiently large.

Since the series $\sum_{s=0}^{\infty} \frac{(\lambda + 1)^s}{s\,!}$ is convergent we may take the limit in Eq. (3.9) under the summation sign. Since

$$\lim_{n \to \infty} \left[\left(1 - \frac{x}{n}\right)^n e^x\right]^{\log n} = 1 \text{ for all } x \geqslant 0,$$

and

$$\lim_{r, n \to \infty} \left(\log n - \frac{r}{n}\right) = \log \lambda,$$

we have

$$\lim_{r, n \to \infty} \left(1 - \frac{s + k}{n}\right)^r n^{s+k} =$$

$$\lim_{r, n \to \infty} \left[\left(1 - \frac{s + k}{n}\right)^n e^{s+k}\right]^{r/n} e^{(s+k)\left(\log n - \frac{r}{n}\right)} = \lambda^{s+k}.$$

Thus

$$\lim_{r, n \to \infty} p_k = e^{-\lambda} \frac{\lambda^k}{k\,!}.$$

The distribution on the set of all non-negative integers with probability for the single point set $\{k\}$ equal to $e^{-\lambda} \frac{\lambda^k}{k\,!}$ is called the *Poisson distribution* with parameter λ.

Example. 3.14. Consider now r indistinguishable objects being placed randomly in n cells. In this case a configuration corresponds to a vector $(x_1, x_2, \ldots, x_n)$ where x_i is the number of objects in the ith cell. If X is the sample space of all possible configurations we shall find out the number of points in X. We can look at a configuration as

$$\underbrace{00 \ldots 0}_{x_1} \mid \underbrace{00 \ldots 0}_{x_2} \mid \;\ldots\ldots\; \mid \underbrace{00 \ldots 0}_{x_n},$$

where 0 denotes the object and | denotes a wall of the cell. Since there are n cells the number of vertical bars in the above picture is $n-1$. The total number of positions occupied either by an object or a bar is $n-1+r$. Out of these, r positions are taken over by the objects. Thus the total number of configurations is $\binom{n-1+r}{r}$. If all configurations are equally likely, then each point of X has probability $\dfrac{1}{\binom{n-1+r}{r}}$. This distribution is known as *Bose-Einstein Statistics.* This distribution, discovered by the Physicist, S. N. Bose in the context of quantum statistical mechanics, is basic in the Theory of Bose-Einstein statistics.

In the above problem, let A_i be the event that the ith cell is empty. Then for $i_1 < i_2 < \ldots < i_j$, $A_{i_1} A_{i_2} \ldots A_{i_j}$ is the event that the j cells i_1, i_2, ..., i_j are empty. The number of such configurations is the number of ways in which r identical objects can be distributed in $n-j$ cells. Thus

$$P(A_{i_1} A_{i_2} \ldots A_{i_j}) = \binom{n-j+r-1}{r} \Big/ \binom{n+r-1}{r}.$$

If p_k denotes the probability of exactly k cells being empty, then

$$p_k = \sum_{j=k}^{n} (-1)^{j-k} \binom{j}{k} \binom{n}{j} \binom{n-j+r-1}{r} \Big/ \binom{n+r-1}{r}. \quad (3.10)$$

Exercise. 3.15. In (3.10), $p_k \to e^{-\lambda} \dfrac{\lambda^k}{k\,!}$ for $k=0, 1, 2, \ldots$, if $n, r \to \infty$ in such a manner that $\dfrac{n^2}{r} \to \lambda$.

Example. 3.16. Consider an urn with m white balls and n black balls. Choose a sample of k balls *at random without replacement.* This means that the probability for any sample of k balls out of the total of $m+n$ balls is the same and hence is equal to $\dfrac{1}{\binom{m+n}{k}}$. We now ask the question: what is the probability of r white balls occurring in the sample? If there are r white balls in the sample then there are $k-r$ black balls in the same sample. Such a choice can be made in $\binom{m}{r}\binom{n}{k-r}$ ways. If we denote the required probability by p_r then

$$p_r = \frac{\binom{m}{r}\binom{n}{k-r}}{\binom{m+n}{k}}, \quad 0 \leqslant r \leqslant \min(k, m).$$

The above distribution on the set of integers 0, 1, 2, ... min (k, m) is known as the *hypergeometric distribution.*

Exercise. 3.17. Consider an urn with balls of r different colours. Let there be n_1 balls of the first colour, n_2 of the second and so on. If k balls are drawn at random without replacement the probability of getting m_1 balls of the first colour, m_2 of the second,...and m_r of the rth colour is equal to

$$p_{m_1, m_2 \dots m_r} = \frac{\binom{n_1}{m_1}\binom{n_2}{m_2}\cdots\binom{n_r}{m_r}}{\binom{n_1+n_2+\dots n_r}{m_1+m_2+\dots m_r}}, \quad \begin{array}{l} 0 \leqslant m_j \leqslant \min(k, n_j), \\ m_1 + \dots + m_r = k. \end{array}$$

Exercise. 3.18. In the preceding exercise let $n_1, n_2, \dots, n_r \to \infty$ in such a manner that $\lim \dfrac{n_i}{n_1+n_2+\dots+n_r} = p_i$ for $i=1,2,\dots r$. Then

$$\lim p_{m_1, m_2, \dots, m_r} = \frac{k!}{m_1!\, m_2! \dots m_r!}\, p_1^{m_1} p_2^{m_2} \dots p_r^{m_r} \tag{3.11}$$

where $k=m_1+m_2+\dots+m_r$. (The probabilities specified by the right hand side of Eq. (3.11) determine the *multinomial distribution*).

Remark. 3.19. Examples 3.12 and 3.13 and Exercises 3.15 and 3.18 yield special cases of limit theorems in probability theory. The usefulness of such limit theorems lies in the fact that they make practical computation easier. We shall illustrate by means of examples later.

Definition. 3.20. Let s_1, s_2 be two simple random variables on a boolean probability space $(X, \mathscr{F}, P)$. Then the *covariance* between s_1 and s_2 is defined as the number $\mathbf{E}(s_1-\mathbf{E}s_1)\,(s_2-\mathbf{E}s_2)$. It is denoted by cov (s_1, s_2). If $s_1 = s_2 = s$, then **cov** (s, s) is called the *variance* of s and denoted by $\mathbf{V}(s)$. Then $\mathbf{V}(s) \geqslant 0$ and $\mathbf{V}(s)^{\frac{1}{2}}$ is called the *standard deviation* of s and denoted by $\sigma(s)$.

Exercise. 3.21. $\mathbf{V}(s) = 0$ if and only if s equals the constant $\mathbf{E}s$ on a set of probability one. **Cov** $(s_1, s_2) = \mathbf{E}s_1s_2 - \mathbf{E}s_1\mathbf{E}s_2$.

Proposition. 3.22. If $s_1, s_2, \dots, s_k$ are simple random variables on $(X, \mathscr{F}, P)$ the matrix Σ whose ijth term is $\sigma_{ij} = \mathbf{cov}\,(s_i, s_j)$ is positive semi-definite. If $a_1, a_2, \dots, a_k, b_1, b_2, \dots, b_k$ are real numbers, then

$$\mathbf{cov}\left(\sum_{i=1}^{k} a_i s_i, \sum_{j=1}^{k} b_j s_j\right) = a' \mathbf{S}\, b,$$

where **a** and **b** stand for the column vectors

$$\begin{pmatrix} a_1 \\ a_2 \\ \cdot \\ \cdot \\ \cdot \\ a_k \end{pmatrix} \text{ and } \begin{pmatrix} b_1 \\ b_2 \\ \cdot \\ \cdot \\ \cdot \\ b_k \end{pmatrix}$$

respectively and $\mathbf{a}'$ denotes the transpose of $\mathbf{a}$. The rank of $\mathbf{S}$ is not full if and only if there exist constants $a_1, a_2, \ldots, a_k$ such that $\sum_j a_j s_j$ is constant on a set of probability one. (The matrix $\mathbf{S}$ is known as the *covariance matrix* of $s_1, s_2, \ldots, s_k$.)

Proof. Since expectation is an additive operation on the algebra of simple random variables we have

$$\begin{aligned}\mathbf{Cov}\Big(\sum_i a_i s_i, \sum_j b_j s_j \Big) &= \sum_{i,j} a_i b_j \mathbf{E} s_i s_j - \sum_i a_i \mathbf{E} s_i \sum_j b_j \mathbf{E} s_j \\ &= \sum_{i,j} a_i b_j \, \mathbf{cov}(s_i, s_j) \\ &= \mathbf{a}' \mathbf{S} \mathbf{b}.\end{aligned}$$

Since $\mathbf{E}s \geqslant 0$ whenever $s \geqslant 0$, it follows that $V(s) \geqslant 0$ for all simple random variables s. Hence

$$0 \leqslant \mathbf{V}\Big(\sum_i a_i s_i\Big) = \mathbf{a}' \mathbf{S} \mathbf{a}$$

for all $\mathbf{a}$. Thus $\mathbf{S}$ is a positive semi-definite matrix. If $\mathbf{S}$ is not of full rank $\mathbf{a}' \mathbf{S} \mathbf{a} = 0$ for some $\mathbf{a} \neq 0$. In such a case $\mathbf{V}\{ \sum_i a_i s_i) = 0$. By Exercise 3.21, $\sum_i a_i s_i$ is constant on a set of probability one. This completes the proof.

Exercise. 3.23. Suppose

$$\mathbf{s} = \begin{pmatrix} s_1 \\ s_2 \\ \cdot \\ \cdot \\ \cdot \\ s_k \end{pmatrix}$$

is a column vector of simple random variables with variance-covariance matrix $\mathbf{S}$. For a real $k \times k$ constant matrix Γ, let

$$\Gamma \mathbf{s} = \begin{pmatrix} t_1 \\ t_2 \\ \cdot \\ \cdot \\ \cdot \\ t_k \end{pmatrix} = \mathbf{t}$$

Then the variance covariance matrix of $\mathbf{t}$ is $\Gamma \, \mathbf{S} \, \Gamma'$, where Γ' is the transpose of Γ. Hence any positive semi-definite real matrix can be represented as the covariance matrix of simple random variables on some boolean probability space $(X, \mathcal{F}, P)$.

Exercise. 3.24. If s is a simple random variable with expectation μ and variance σ^2, then $\frac{s-\mu}{\sigma}$ has mean zero and variance unity. $\left(\frac{s-\mu}{\sigma}\right.$ is called the *normalised* random variable.$\left.\right)$

Before proceeding further we shall introduce the following convention. If s is a simple random variable on a boolean probability space $(X, \mathcal{F}, P)$ and E is a subset of the real line such that the set $\{x : s(x) \in E\} = s^{-1}(E) \in \mathcal{F}$, we shall write

$$P(s^{-1}(E)) = P(s \in E).$$

In particular, $P(s \leqslant a)$ means the measure or probability of the set $\{x : s(x) \leqslant a\}$.

Proposition. 3.25. (Chebyshev's inequality). Let s be a simple random variable on a boolean probability space $(X, \mathcal{F}, P)$. Then for any $a \geqslant 0$,

$$P(|s| \geqslant a) \leqslant \frac{\mathbf{E}|s|}{a}.$$

Proof. Let $s = \Sigma\, a_i \chi_{E_i}$, where $\{E_i\}$ is a finite partition of X into sets belonging to $\mathcal{F}$. Then

$$\begin{aligned} \mathbf{E}|s| &= \Sigma\, |a_i|\, P(E_i) \\ &\geqslant \sum_{i:\,|a_i| \geqslant a} |a_i|\, P(E_i) \\ &\geqslant a \sum_{i:\,|a_i| \geqslant a} P(E_i) \\ &= a\, P(|s| \geqslant a). \end{aligned}$$

This completes the proof.

Corollary. 3.26. Under the same conditions as in Proposition 3.25,

$$P(|s| \geqslant a) \leqslant \frac{\mathbf{E}|s|^n}{a^n} \quad \text{if } n > 0,$$

$$P(|s - \mathbf{E}s| \geqslant a) \leqslant \frac{\mathbf{V}(s)}{a^2}.$$

§4. Repeated Trials and Statistical Independence

Suppose we perform two statistical experiments whose associated boolean spaces are $(X_1, \mathscr{F}_1)$ and $(X_2, \mathscr{F}_2)$ respectively. It is natural to ask how to describe them together as a single experiment. Performance of both leads to an observation (x_1, x_2) where $x_1 \in X_1$ and $x_2 \in X_2$. Thus the sample point for the joint experiment is the ordered pair (x_1, x_2). In other words the sample space is the cartesian product $X_1 \times X_2$. It is clear that events of the form '$x_1 \in F_1$, $x_2 \in F_2$', where $F_1 \in \mathscr{F}_1$, $F_2 \in \mathscr{F}_2$ are of some interest. This event is described by the subset $F_1 \times F_2 = \{(x_1, x_2): x_1 \in F_1, x_2 \in F_2\}$. Such events may be called ***boolean rectangles***. However, all boolean rectangles *do not* form a boolean algebra of subsets of the product space $X_1 \times X_2$. We shall prove soon that all finite unions of boolean rectangles constitute a boolean algebra. This is called the ***product*** of the two boolean algebras $\mathscr{F}_1$ and $\mathscr{F}_2$ and denoted by $\mathscr{F}_1 \times \mathscr{F}_2$. Thus we have a method of generating new boolean algebras out of given ones. The boolean space $(X_1 \times X_2, \mathscr{F}_1 \times \mathscr{F}_2)$ adequately describes the events of the joint experiment. Similarly one can define the product $(X_1 \times X_2 \times \ldots X_k, \mathscr{F}_1 \times \mathscr{F}_2 \times \ldots \times \mathscr{F}_k)$ of several boolean spaces $(X_i, \mathscr{F}_i)$, $i = 1, 2 \ldots, k$ by considering all finite unions of boolean rectangles of the form $F_1 \times F_2 \times \ldots \times F_k$, where $F_j \in \mathscr{F}_j$, $j = 1, 2 \ldots k$.

The class of all sets $F_1 \times F_2$, $F_1 \in \mathscr{F}_1$, $F_2 \in \mathscr{F}_2$ has certain features, which it is very useful to describe in the form of a definition.

Definition. 4.1. A collection $\mathscr{D}$ of subsets of X is called a ***boolean semi algebra*** if

(i) $A, B \in \mathscr{D}$ implies $AB \in \mathscr{D}$;

(ii) $A \in \mathscr{D}$ implies that A' can be expressed as a finite disjoint union of sets in $\mathscr{D}$;

(iii) $X \in \mathscr{D}$.

Proposition. 4.2. Let $\mathscr{D}$ be a boolean semi algebra. Then the family $\mathscr{F}$ of all finite disjoint unions of sets in $\mathscr{D}$ is a boolean algebra. ($\mathscr{F}$ is called the boolean algebra ***generated by*** $\mathscr{D}$. In particular, $\mathscr{F}$ is the family of all finite unions of sets in $\mathscr{D}$.)

Proof. Let A be any set of the form $\bigcup_{j=1}^{k} R_j$, $R_j \in \mathscr{D}$. We can write

$$A = R_1 \cup R_2 R'_1 \cup R_3 R'_2 R'_1 \cup \ldots \cup R_k R'_{k-1} \ldots R'_1.$$

By properties (i) and (ii) of the definition above it follows that A can be expressed as a finite disjoint union of elements from $\mathscr{D}$. In particular, $\mathcal{F}$ is closed under finite union. That $\mathcal{F}$ is closed under complementation follows trivially. This shows that $\mathcal{F}$ is a boolean algebra and completes the proof.

Exercise. 4.3. If $\mathcal{D}_1$ and $\mathcal{D}_2$ are boolean semi algebras of subsets of X_1 and X_2 respectively, then all rectangles of the form $R_1 \times R_2$, $R_1 \in \mathcal{D}_1$, $R_2 \in \mathcal{D}_2$ constitute a boolean semi algebra.

Exercise. 4.4. The class $\mathcal{I}$ of all intervals of the form $(-\infty, +\infty)$, $(-\infty, a]$, (a, ∞), $(a, b]$, a, b varying in R, constitute a boolean semi algebra.

Combining Exercises 4.3, 4.4, we have Exercise 4.5.

Exercise. 4.5. Let $\mathcal{I}$ be as in Exercise 4.4. Then the class $\mathcal{I}^k$ of all rectangles of the form $I_1 \times I_2 \times \ldots \times I_k$, $I_j \in \mathcal{I}$, $j=1, 2, \ldots, k$ is a boolean semi algebra of subests of the k-dimensional real Euclidean space R^k. All finite disjoint unions of sets belonging to $\mathcal{I}^k$ constitute a boolean algebra.

Exercise. 4.6. Let $(X_\alpha, \mathcal{F}_\alpha)$, $\alpha \in I$ be a family of boolean spaces. Let $X = \prod_{\alpha \in I} X_\alpha$ be the cartesian product of all the X_α's. Let $\mathcal{F}$ be the family of all subsets of the form $\{x : x \in X, (x(\alpha_1), x(\alpha_2), \ldots, x(\alpha_k)) \in F\}$, where $F \in \mathcal{F}_{\alpha_1} \times \mathcal{F}_{\alpha_2} \times \ldots \times \mathcal{F}_{\alpha k}$ and $\{\alpha_1, \alpha_2, \ldots, \alpha_k\}$ is a finite subset of T. Then $\mathcal{F}$ is a boolean algebra. The space $(X, \mathcal{F})$ is called the *product boolean space.*

Proposition. 4.7. Let $\mathcal{D}$ be a boolean semi algebra of subsets of X. Let $p : \mathcal{D} \to [0, 1]$ be a map such that $p(X) = 1$ and

$$p\left(\bigcup_{i=1}^{k} E_i\right) = \sum_{i=1}^{k} p(E_i),$$

whenever E_i's are pairwise disjoint elements of $\mathcal{D}$ and $\bigcup_{i=1}^{k} E_i \in \mathcal{D}$. Then there exists a unique probability distribution P on the boolean algebra generated by $\mathcal{D}$ such that

$$P(E) = p(E) \text{ for } E \in \mathcal{D}.$$

Proof. This is proved exactly in the same way as Example 2.4. Nevertheless we repeat. Let $\mathcal{F}$ be the boolean algebra generated by $\mathcal{D}$. By Proposition 4.2 any $A \in \mathcal{F}$ can be expressed as

$$A = \bigcup_{i=1}^{k} E_i,\ E_i \in \mathcal{D},\ E_i \cap E_j = \emptyset \text{ for } i \neq j.$$

Define

$$P(A) = \sum_{i=1}^{k} p(E_i).$$

If

$$A = \bigcup_{j=1}^{m} F_j,\ F_j \in \mathscr{D} \text{ and } F_i \cap F_j = \emptyset \text{ for } i \neq j,$$

we shall show that $P(A) = \sum_{j=1}^{m} p(F_j)$. Indeed, we have

$$F_j = \bigcup_{i=1}^{k} F_j E_i.$$

since $\mathscr{D}$ is closed under intersection and p is additive on $\mathscr{D}$,

$$p(F_j) = \sum_{i=1}^{k} p(F_j E_i).$$

Hence

$$\sum_{j=1}^{m} p(F_j) = \sum_{j=1}^{m} \sum_{i=1}^{k} p(F_j E_i).$$

By symmetry the right hand side above is also equal to $\sum_{i=1}^{k} p(E_i)$.

It also follows from the definition that P is finitely additive on $\mathscr{F}$. Uniqueness is trivial and the proof is complete.

Remark. 4.8. The above proposition holds if p is a map from $\mathscr{D}$ into $[0, \infty]$ with the modification that P is replaced by a finitely additive function.

Let $(X_i, \mathscr{F}_i)$, $i = 1, 2, \ldots, k$ be boolean spaces. Then by Exercise 4.6 we can construct the product boolean space $(X_1 \times X_2 \times \ldots \times X_k, \mathscr{F}_1 \times \mathscr{F}_2 \times \ldots \times \mathscr{F}_k)$. If P_i is a probability distribution on $\mathscr{F}_i$ we can ask the question: Is there a probability distribution P on $\mathscr{F}_1 \times \mathscr{F}_2 \times \ldots \times \mathscr{F}_k$ such that

$$P(F_1 \times F_2 \times \ldots \times F_k) = P_1(F_1)\, P_2(F_2) \ldots P_k(F_k),$$
$$F_i \in \mathscr{F}_i,\ i = 1, 2, \ldots k? \qquad (4.1)$$

Our next result answers the question in the affirmative and gives a method of generating new probability spaces out of given ones.

Proposition. 4.9. Let P be defined by Eq. (4.1) on the boolean semi algebra of all rectangles. Then P is finitely additive and can be extended uniquely to a probability distribution on $\mathscr{F}_1 \times \mathscr{F}_2 \times \ldots \times \mathscr{F}_k$.

Proof. Let $F_i \in \mathcal{F}_i$, $i = 1, 2, \ldots, k$ be such that

$$F_1 \times F_2 \times \ldots \times F_k = \bigcup_{i=1}^{n} (F_{1i} \times F_{2i} \times \ldots \times F_{ki}),$$

where $F_{ri} \in \mathcal{F}_r$, $r = 1, 2, \ldots, n$. Suppose the rectangles appearing within the union sign on the right hand side are disjoint. Then for any $(x_1, x_2, \ldots, x_k)$, $x_j \in X_j$, we have

$$\prod_{j=1}^{k} \chi_{F_j}(x_j) = \sum_{i=1}^{n} \chi_{F_{1i}}(x_1)\, \chi_{F_{2i}}(x_2) \ldots \chi_{F_{ki}}(x_k). \tag{4.2}$$

When $k-1$ of the variables are fixed, both sides of Eq. (4.2) happen to be simple random variables as functions of the remaining variable. If we integrate both sides with respect to P_k, we obtain

$$\left[\prod_{j=1}^{k-1} \chi_{F_j}(x_j)\right] P_k(F_k) = \sum_{i=1}^{n} \left[\prod_{i=1}^{k-1} \chi_{F_{ji}}(x_j)\right] P_k(F_{ki}).$$

In each of the remaining variables $x_1, x_2, \ldots, x_{k-1}$, both sides are simple random variables. Thus integrating successively with respect to P_{k-1}, $P_{k-2}, \ldots, P_1$ we obtain

$$P_1(F_1)\, P_2(F_2) \ldots P_k(F_k) = \sum_{i=1}^{n} P_1(F_{1i})\, P_2(F_{2i}) \ldots P_k(F_{ki}).$$

This shows that P is finitely additive on the boolean semi algebra of all boolean rectangles. By proposition 4.7, P extends uniquely to the product boolean algebra $\mathcal{F}_1 \times \mathcal{F}_2 \times \ldots \times \mathcal{F}_k$.

Definition. 4.10. If $(X_i, \mathcal{F}_i, P_i)$, $i = 1, 2, \ldots, k$ are boolean probability spaces, the probability distribution P constructed in Proposition 4.9 is called the *product of the distributions* $P_1, P_2, \ldots, P_k$ and denoted by $P_1 \times P_2 \times \ldots \times P_k$. The boolean probability space $(X_1 \times X_2 \times \ldots \times X_k, \mathcal{F}_1 \times \ldots \times \mathcal{F}_k, P_1 \times P_2 \times \ldots \times P_k)$ is known as the *product boolean probability space.*

Remark. 4.11. Let now $(X_\alpha, \mathcal{F}_\alpha, P_\alpha)$, $\alpha \in \Gamma$ be a family of boolean probability spaces. Consider the cartesian product $\widetilde{X} = \prod_{\alpha \in \Gamma} X_\alpha$. Any point $x \in \widetilde{X}$ is a function on Γ with values in $\bigcup_{\alpha \in \Gamma} X_\alpha$ such that $x(\alpha) \in X_\alpha$ for all α. In words we say that x has the property that its α-th coordinate lies in X_α. For any finite set $A \subset \Gamma$, $A = \{\alpha_1, \alpha_2, \ldots, \alpha_k\}$, consider all sets of the form

$$\{x : (x(\alpha_1), x(\alpha_2), \ldots, x(\alpha_k)) \in E\}, \tag{4.3}$$

where $E \in \mathscr{F}_{a_1} \times \mathscr{F}_{a_2} \times \ldots \times \mathscr{F}_{a_k}$. We may call such sets as k-dimensional boolean cylinders in $\tilde{X}$. We introduce the map

$$\pi_A : \quad \tilde{X} \to X_{a_1} \times X_{a_2} \times \ldots \times X_{a_k}$$

defined by

$$\pi_A(x) = (x(a_1),\ x(a_2),\ \ldots,\ x(a_k)).$$

Then the set (4.3) is nothing but $\pi_A{}^{-1}(E)$. In $\tilde{X}$ it may be considered as an event concerning the coordinates (or observations) at ' times ' $a_1, a_2, \ldots, a_k$. It is quite easily seen that the class

$$\tilde{\mathscr{F}}_A = \{\pi_A{}^{-1}(E),\ E \in \mathscr{F}_{a_1} \times \mathscr{F}_{a_2} \times \ldots \times \mathscr{F}_{a_k}\}$$

is a boolean algebra in $\tilde{X}$. $\tilde{\mathscr{F}}_A$ may be considered as the collection of all events concerning observations at times $a_1, a_2, \ldots, a_k$. If $A \subset B \subset \Gamma$, and A and B are finite then $\tilde{\mathscr{F}}_A \subset \tilde{\mathscr{F}}_B$. Indeed, if $A = \{a_1, a_2, \ldots, a_k\}$ $B = \{a_1, a_2, \ldots, a_k, a_{k+1}, \ldots, a_j\}$ then an event concerning the observations at times $a_1, a_2, \ldots, a_k$ is also an event concerning the observations at times $a_1, a_2, \ldots, a_k, a_{k+1}, \ldots, a_j$. Hence it follows that the family

$$\tilde{\mathscr{F}} = \bigcup_{A \subset \Gamma,\ A \text{ finite}} \tilde{\mathscr{F}}_A$$

is also a boolean algebra. By the preceding proposition we can construct the distribution

$$P_A = P_{a_1} \times P_{a_2} \times \ldots \times P_{a_k}$$

on the boolean algebra $\mathscr{F}_{a_1} \times \mathscr{F}_{a_2} \times \ldots \times \mathscr{F}_{a_k}$. Using this we can construct a distribution $\tilde{P}_A$ on $\tilde{\mathscr{F}}_A$ by defining

$$\tilde{P}_A(\pi_A{}^{-1}(E)) = P_A(E),\ E \in \mathscr{F}_{a_1} \times \mathscr{F}_{a_2} \times \ldots \times \mathscr{F}_{a_k},$$

whenever $A = \{a_1, a_2, \ldots, a_k\}$. A bit more effort shows that $\tilde{P}_A$ is well defined on $\tilde{\mathscr{F}}_A$. If $A \subset B \subset \Gamma$ and both A and B are finite subsets then

$$\tilde{P}_A(F) = \tilde{P}_B(F) \text{ for all } F \in \tilde{\mathscr{F}}_A \subset \tilde{\mathscr{F}}_B. \tag{4.4}$$

This enables us to construct a single probability distribution $\tilde{P}$ on $\tilde{\mathscr{F}}$ by putting

$$\tilde{P}(F) = \tilde{P}_A(F) \text{ if } F \in \tilde{\mathscr{F}}_A, \tag{4.5}$$

for all finite sets $A \subset \Gamma$. Thus we have a boolean probability space $(\tilde{X}, \tilde{\mathscr{F}}, \tilde{P})$ with the property

$$\tilde{P}\,(\pi_A{}^{-1}\,(E)) = (P_{a_1} \times P_{a_2} \times \ldots \times P_{a_k})\,(E) \tag{4.6}$$

for all $E \in \mathscr{F}_{a_1} \times \mathscr{F}_{a_2} \times \ldots \times \mathscr{F}_{a_k}$.

Condition (4.4) is of considerable interest in the theory of stochastic processes. It is known as the *consistency condition.*

Remark. 4.12. Suppose two statistical experiments with boolean spaces $(X_1, \mathcal{F}_1)$ and $(X_2, \mathcal{F}_2)$ are performed n_1 and n_2 times respectively. Let $A_i \in \mathcal{F}_i$, $i = 1, 2$ be two events. Let m_1 and m_2 be the number of times A_1 and A_2 occur respectively in each of the experiments. Suppose the outcome of the second experiment has nothing to do with the first and vice versa. Consider the ith trial of the first experiment and jth trial of the second experiment as the ij th trial of a 'joint experiment' with sample space $X_1 \times X_2$ and collection of events $\mathcal{F}_1 \times \mathcal{F}_2$. Then we have $n_1 n_2$ trials of the joint experiment. The event$A_1 \times A_2$ has occurred $m_1 m_2$ times. Then the frequency of $A_1 \times A_2$ is $m_1 m_2 / n_1 n_2$, which is the product of the individual frequencies m_1/n_1 and m_2/n_2 of A_1 and A_2 respectively. If frequency is assumed to stabilise as the numbers of trials n_1 and n_2 increase to infinity, it is natural to impose the condition that the probability of $A_1 \times A_2$ in the joint experiment is the product of the probabilities of the individual events A_1 and A_2. This intuitive criterion is derived under the assumption that the outcomes of one experiment do not have any influence on those of the other. In view of this heuristic discussion we introduce the following definition.

Definition. 4.13. A statistical experiment described by the product boolean probability space $(X_1 \times X_2 \times \ldots \times X_k, \mathcal{F}_1 \times \mathcal{F}_2 \times \ldots \times \mathcal{F}_k, P_1 \times P_2 \times \ldots \times P_k)$ is called a *series of independent experiments* $(X_i, \mathcal{F}_i, P_i)$, $i=1, 2, \ldots, k$. If $X_i = X$, $\mathcal{F}_i = \mathcal{F}$, $P_i = P$ for all $i=1, 2, \ldots, k$ then we say that $(X \times X \times \ldots \times X, \mathcal{F} \times \ldots \times \mathcal{F}, P \times P \times \ldots \times P)$ is a *series of k independent trials* of the experiment $(X, \mathcal{F}, P)$.

Definition. 4.14. Two events A and B in a boolean probability space $(X, \mathcal{F}, P)$ are said to be *statistically independent* or simply *independent* if $P(AB) = P(A)\, P(B)$. A family $\{A_\alpha\}$, $\alpha \in \Gamma$ of events is said to be *mutually independent* if

$$P(A_{\alpha_1} A_{\alpha_2} \ldots A_{\alpha_j}) = P(A_{\alpha_1})\, P(A_{\alpha_2}) \ldots P(A_{\alpha_j})$$

for any finite set $\{\alpha_1, \alpha_2, \ldots, \alpha_j\}$ contained in Γ.

Definition. 4.15. A collection $\{s_\alpha\}$, $\alpha \in \Gamma$ of simple random variables on a boolean probability space $(X, \mathcal{F}, P)$ is said to be *mutually independent* if for any finite set $\{\alpha_1, \alpha_2, \ldots, \alpha_j\} \subset \Gamma$ and arbitrary subsets $E_1, E_2, \ldots, E_j$ of the real line

$$P\{s_{\alpha_1}^{-1}(E_1) \cap s_{\alpha_2}^{-1}(E_2) \ldots \cap s_{\alpha_j}^{-1}(E_j)\}$$

$$= \prod_{i=1}^{j} P\{s_{\alpha_i}^{-1}(E_i)\}.$$

Example. 4.16. Let X be the finite set consisting of the integers 1, 2, ..., N and let $\mathscr{F}$ be the class of all subsets of X. Let P be a distribution on $\mathscr{F}$ such that

$$P(\{i\}) = p_i,\ i = 1, 2, \dots, N.$$

Then $p_i \geqslant 0$, $\sum_i p_i = 1$. Consider n independent trials of the experiment $(X, \mathscr{F}, P)$. We can now ask the question: what is the probability that in n independent trials of $(X, \mathscr{F}, P)$ the observed sequence $(i_1, i_2, \dots, i_n)$ contains r_1 1's, r_2 2's, ..., r_N N's? If a particular sequence $(i_1, i_2, \dots, i_n)$ has r_j j's for $j = 1, 2, \dots, N$ its probability of occurrence is

$$p_{i_1} p_{i_2} \cdots p_{i_n} = p_1^{r_1} p_2^{r_2} \dots p_N^{r_N}.$$

Thus the required probability is $k\, p_1^{r_1} p_2^{r_2} \dots p_N^{r_N}$, where k is the number of n long sequences of elements from X with r_1 1's r_2 2's, ..., r_N N's. If there are n positions we can choose r_1 positions in $\binom{n}{r_1}$ ways and fill them by 1. Out of the remaining $(n-r_1)$ positions we can choose r_2 positions in $\binom{n-r_1}{r_2}$ ways and fill them by 2. After filling r_1 positions with 1, r_2 with 2, ..., r_j with j, we can select r_{j+1} positions in

$$\binom{n - r_1 - r_2 - \dots - r_j}{r_{j+1}}$$

ways and fill them by $j+1$. Thus the number of sequences with r_1 1's, r_2 2's, ... r_N N's is

$$\binom{n}{r_1}\binom{n-r_1}{r_2} \cdots \binom{n - r_1 - r_2 - \dots - r_{N-1}}{r_N} = \frac{n\,!}{r_1\,!\, r_2\,! \dots r_N\,!},$$

where $r_1 + r_2 + \dots + r_N = n$. Thus the required probability is

$$\frac{n\,!}{r_1\,!\, r_2\,! \dots r_N\,!}\, p_1^{r_1}\, p_2^{r_2} \dots p_N^{r_N}. \tag{4.7}$$

Thus we get the multinomial distribution (see Exercise 3.18).

We can translate what we have done into the language of random variables. Perform n independent trials of $(X, \mathscr{F}, P)$. To any observed sequence $i_1, i_2, \dots, i_n$ define

$$\mathbf{r}\,(i_1, i_2, \dots, i_n) = (r_1, r_2, \dots, r_N)$$

where r_j is the number of j's in the sequence for every $1 \leqslant j \leqslant N$. Thus $\mathbf{r}$ is a map from $X \times X \times \dots \times X$ into the set of N dimensional vectors with non negative integral coordinates whose total is n. In other words $\mathbf{r}$ is a vector valued simple random variable. Further the probability that $\mathbf{r}$ takes the value $(r_1, r_2, \dots, r_N)$ is given by the expression (4.7).

Let X be the space of all integral vectors $\mathbf{r} = (r_1, \ldots, r_N)$ with $0 \leqslant r_j \leqslant N$, $\sum_{j=1}^{N} r_j = n$, and let $\mathscr{F}$ be the class of all subsets of X. Let P be the distribution on $(X, \mathscr{F})$ such that $P(\{\mathbf{r}\})$ is equal to the expression (4.7). Then P is called the *multinomial distribution* with parameters $p_1, p_2, \ldots, p_N$, where $p_i \geqslant 0$ for all i and $\sum_{i=1}^{N} p_i = 1$.

The case $N = 2$ is of fundamental importance in our subject. In this case we write $p_1 = p$, $p_2 = 1 - p = q$, call 1 as *success* and 2 as *failure*. Then the probability for r successes in n independent trials is given by the expression

$$b(n, r, p) = \binom{n}{r} p^r q^{n-r}, \; r = 0, 1, 2, \ldots, n. \tag{4.8}$$

If X is the set of integers $\{0, 1, 2, \ldots, n\}$, $\mathscr{F}$ is the class of all subsets of X and P is the distribution defined on $(X, \mathscr{F})$ where $P(\{r\}) = b(n, r, p)$ for $r = 0, 1, 2, \ldots, n$ then P is called the *binomial distribution* with probability for success equal to p and number of trials equal to n. It describes the probability for r successes in n independent trials of an experiment with two elementary outcomes, one called success and the other called failure.

Exercise 3.18 shows that the hypergeometric distribution converges to the binomial distribution if the number of balls in the urn increases to infinity in such a manner that the proportion of white balls tends to a value p.

§5. The Poisson Approximation to the Binomial Distribution

Consider n independent binomial trials with the same probability p for success. Let S_n be the number of successes. S_n is a random variable and

$$P\{S_n = r\} = \binom{n}{r} p^r q^{n-r}.$$

A simple calculation shows that

$$\mathbf{E}\, S_n = \sum_{0}^{n} r \binom{n}{r} p^r q^{n-r} = np, \tag{5.1}$$

$$\mathbf{V}(S_n) = \sum_{0}^{n} (r - np)^2 \binom{n}{r} p^r q^{n-r} = npq. \tag{5.2}$$

These can be obtained from the identity

$$(p+q)^n = \sum_0^n \binom{n}{r} p^r q^{n-r}$$

and differentiating both sides with respect to the variable p twice successively.

Suppose now that p depends on n in such a manner that $np \to \lambda$ as $n \to \infty$, i.e., the expected number of successes converges to a fixed number λ. Then $\mathbf{V}(S_n)$ also tends to λ. We shall now see what happens to $b(n, r, p)$ in the limit.

Proposition. 5.1. Let $p_n > 0$, $n = 1, 2, \ldots$ be such that $\lim_{n \to \infty} np_n = \lambda$. Then

$$\lim_{n\to\infty} b(n, r, p_n) = e^{-\lambda} \frac{\lambda^r}{r!}, \quad r = 0, 1, 2, \ldots \tag{5.3}$$

Proof. We have

$$b(n, r, p_n) = \frac{n(n-1)(n-2)\ldots(n-r+1)}{r!} p^r{}_n (1-p_n)^{n-r}$$

$$= \frac{1}{r!} \left\{ \frac{\left(1-\frac{1}{n}\right)\left(1-\frac{2}{n}\right)\cdots\left(1-\frac{r-1}{n}\right)}{(1-p_n)^r} \right\} \times (np_n)^r \left(1 - \frac{np_n}{n}\right)^n.$$

We observe that the expression within braces on the right hand side tends to unity as $n \to \infty$. It is a well known result in real analysis that $\left(1 - \frac{x_n}{n}\right)^n \to e^{-x}$ as $n \to \infty$ if $x_n \to x$ as $n \to \infty$. Hence Eq. (5.3) holds and the proof is complete.

Remark. 5.2. The limit theorem of Proposition 5.1 is of considerable importance. The evaluation of expression (4.8) is difficult for large n because it involves factorials of large numbers. If np is stable around a value λ, one can use the approximation (5.3). Further there are many examples in nature where the probability of a certain phenomenon is very small but the number of observations is quite large and the expectation of the number of times the phenomenon takes place in a large number of trials is a certain fixed quantity. We shall illustrate by a few examples.

Consider an individual walking on the road. The probability p of his being involved in an accident is small. Now consider a large number of individuals walking on the road. The average number of accidents may be taken to be a fixed number λ. If the number Z of accidents in a day is the random quantity that is being observed and it is assumed that the individuals walk independently of each other, then we may use the model:

$$P(Z=r)=e^{-\lambda}\frac{\lambda^r}{r!}. \tag{5.4}$$

Consider an individual making a telephone call during a certain busy hour. The probability p of a particular individual making a call is small. Since the total number of subscribers is very large and different individuals make calls independently of each other we may say that the number X of telephone calls being made during the busy hour is distributed according to Eq. (5.4).

For more examples the reader may consult the well known book by W. Feller [4].

Exercise. 5.3. Let $M(n;\ r_1, r_2,\ldots, r_N;\ p_1, p_2,\ldots, p_N)$ denote the expression (4.7) of the multinomial distribution. Suppose $p_1, p_2,\ldots, p_N$ depend on n in such a manner that as $n\to\infty$, $np_j\to\lambda_j$ for $j=1, 2,\ldots, N-1$. Then

$$\lim_{n\to\infty} M(n;\ r_1, r_2,\ldots, r_N;\ p_1, p_2,\ldots, p_N)$$
$$=\frac{e^{-(\lambda_1+\lambda_2+\ldots+\lambda_{N-1})}}{r_1!\ r_2!\ \ldots\ r_{N-1}!}\lambda_1^{r_1}\lambda_2^{r_2}\ldots\lambda_{N-1}^{r_{N-1}}.$$

§6. The Normal Approximation to the Binomial Distribution

As in Sec. 5 let S_n be the number of successes in n independent binomial trials with probability p for success. Let $0 < p < 1$. Consider the normalised random variable

$$Z_n=\frac{S_n-np}{\sqrt{npq}}. \tag{6.1}$$

It follows from Eqs. (5.1) and (5.2) that $\boldsymbol{E}Z_n=0$, $\boldsymbol{E}Z_n^2=1$ Then Chebyshev's inequality implies that

$$P\{|Z_n|\geqslant a\}\leqslant\frac{1}{a^2}.$$

This shows that the random variable Z_n lies in the interval $[-a, a]$ with probability at least $1-a^{-2}$ for all n. In classical analysis Bolzano-Weierstrass' theorem tells us that every bounded sequence has a convergent subsequence. Instead of a sequence of numbers we now have a sequence of random variables Z_n which lie in a bounded interval with a large probability. We can, therefore, naturally ask whether there is a subsequence Z_{n_k} with the property that $P\{\alpha < Z_{n_k} \leqslant \beta\}$ converges as $k \to \infty$ for every α, β, i.e., the probability laws of Z_n have convergent subsequences in some sense. In this connection we shall prove the following theorem.

Proposition. 6.1. (Laplace-DeMoivre Theorem). Let the random variables Z_n be described by Eq. (6.1). Then

$$\lim_{n\to\infty} P\{\alpha < Z_n \leqslant \beta\} = \frac{1}{\sqrt{2\pi}} \int_\alpha^\beta e^{-x^2/2}\, dx \tag{6.2}$$

for all α, β.

Proof. We shall make use of the well known *Stirling's formula* (see[4]).

$$\lim_{n\to\infty} \frac{n!}{\sqrt{2\pi}\, e^{-n}\, n^{n+\frac{1}{2}}} = 1. \tag{6.3}$$

For simplicity we write $Z_n = x$. Then the number S_n of successes and the number $n-S_n$ of failures can be written as

$$S_n = np + x\sqrt{npq},$$
$$n-S_n = nq - x\sqrt{npq}.$$

Suppose x lies in a bounded interval $[\alpha, \beta]$. Then S_n and $n-S_n$ tend to ∞ as $0(n)$. Using Eq. (6.3) we can write

$$\frac{n!}{S_n!\,(n-S_n)!} = \left[\frac{n}{2\pi\, S_n\,(n-S_n)}\right]^{\frac{1}{2}} \left(\frac{n}{S_n}\right)^{S_n} \times \left(\frac{n}{n-S_n}\right)^{n-S_n} (1+o(1)),$$

where $o(1)$ is a term depending on n and x, but whose absolute value is dominated by a sequence ε_n which tends to 0 as $n \to \infty$ and ε_n depends only on α, β, p and n. Then

$$\begin{aligned} b(n, S_n, p) &= \left[\frac{n}{2\pi\, S_n\,(n-S_n)}\right]^{\frac{1}{2}} \left(\frac{np}{S_n}\right)^{S_n} \left(\frac{nq}{n-S_n}\right)^{n-S_n} (1+o(1)) \\ &= \left[\frac{n}{2\pi\,(np + x\sqrt{npq})\,(nq - x\sqrt{npq})}\right]^{\frac{1}{2}} \times \\ &\qquad (1 + x\sqrt{q/np})^{-(np+x\sqrt{npq})} \times \\ &\qquad (1 - x\sqrt{p/nq})^{-(nq-x\sqrt{npq})} (1+o(1)) \end{aligned}$$

Since $\alpha \leqslant x \leqslant \beta$, we have

$$\frac{n}{2\pi\,(np + x\sqrt{npq})\,(nq - x\sqrt{npq})} = \frac{1}{2\pi\, npq}\,(1+o(1)).$$

Thus

$$b(n, S_n, p) = \frac{1}{\sqrt{2\pi\, npq}}\, f_n(x)\,(1+o(1)), \tag{6.4}$$

where

$$\begin{aligned} -\log f_n(x) = {} & (np + x\sqrt{npq})\log(1 + x\sqrt{q/np}) + \\ & (nq - x\sqrt{npq})\log(1 - x\sqrt{p/nq}). \end{aligned} \tag{6.5}$$

We have

$$\log(1 + x\sqrt{q/np}) = x\sqrt{q/np} - \tfrac{1}{2}x^2\,\frac{q}{np} + 0\,(n^{-3/2}),$$

$$\log(1 - x\sqrt{p/nq}) = -x\sqrt{p/nq} - \tfrac{1}{2}x^2\,\frac{p}{nq} + 0\,(n^{-3/2}),$$

if $\alpha \leqslant x \leqslant \beta$. Substituting these in Eq. (6.5) we have

$$-\log f_n(x) = \frac{x^2}{2} + 0\left(\frac{1}{\sqrt{n}}\right). \tag{6.6}$$

Combining Eqs. (6.4) to (6.6) we have

$$b(n, S_n, p) = \frac{1}{\sqrt{2\pi\, npq}}\, e^{-x^2/2}\,(1+o(1)), \tag{6.7}$$

where $x = \dfrac{S_n - np}{\sqrt{npq}}$, provided $\alpha \leqslant x \leqslant \beta$. Hence

$$P(\alpha < Z_n \leqslant \beta) = \frac{1}{\sqrt{2\pi\, npq}} \left\{ \sum_r e^{-(r-np)^2/2npq} \right\} (1+o(1)), \tag{6.8}$$

where the summation within wavy brackets ranges over all non-negative integers r such that

$$\alpha\sqrt{npq} + np < r \leqslant \beta\sqrt{npq} + np.$$

Consider the interval $[\alpha, \beta]$ and the points

$$x_{rn} = \frac{r - np}{\sqrt{npq}}$$

in it. We have

$$x_{(r+1)n} - x_{rn} = \frac{1}{\sqrt{npq}}.$$

Thus, for fixed n, the points $\{x_{rn}\}$ constitute a partition of $[\alpha, \beta]$ into sub-intervals of length $\leqslant \frac{1}{\sqrt{npq}}$. All except the two end intervals have length $\frac{1}{\sqrt{npq}}$. Hence by the definition of Riemann integral we have

$$\lim_{n\to\infty} \frac{1}{\sqrt{2\pi\, npq}} \sum_r e^{-\frac{1}{2}x^2_{rn}} = \frac{1}{\sqrt{2\pi}} \int_\alpha^\beta e^{-x^2/2}\, dx.$$

Now Eq. (6.8) implies

$$\lim_{n\to\infty} P(\alpha < Z_n \leqslant \beta) = \frac{1}{\sqrt{2\pi}} \int_\alpha^\beta e^{-x^2/2}\, dx.$$

This completes the proof.

Corollary. 6.2. When $\frac{r-np}{\sqrt{npq}}$ is bounded,

$$b(n, r, p) = \frac{1}{\sqrt{2\pi npq}}\, e^{-(r-np)^2/2npq}\, (1+o(1)).$$

Proof. This is just a restatement of Eq. (6.7).

§7. The Multivariate Normal Approximation to the Multinomial Distribution

Consider n independent multinomial trials with N elementary outcomes 1, 2, ..., N having probabilities $p_1, p_2, \ldots, p_N$ where $0 < p_i < 1$, for all $i=1, 2, \ldots, N$ and $\Sigma p_i = 1$. If S_{nj} is the number of times the outcome j occurs in n trials, then from Example 4.16 we have

$$\begin{aligned} &P\{S_{n1} = r_1, S_{n2} = r_2, \ldots, S_{nN} = r_N\} \\ &= \frac{n!}{r_1!\, r_2! \ldots r_N!}\, p_1^{r_1} p_2^{r_2} \ldots p_N^{r_N} \\ &= M(n, r_1, r_2, \ldots, r_N; p_1, p_2, \ldots, p_N), \text{ say.} \end{aligned} \tag{7.1}$$

Note that

$$\mathbf{E}S_{nj} = np_j,\ \mathbf{V}(S_{nj}) = np_j(1-p_j).$$

Exactly as in Section 6, we shall make an asymptotic analysis of the expression

$$P\left\{\alpha_j < \frac{S_{nj} - np_j}{\sqrt{n}} \leqslant \beta_j,\ j = 1, 2, \ldots, N-1\right\}.$$

Indeed, we have

Proposition. 7.1. $\lim_{n \to \infty} P\left\{ \alpha_j < \frac{S_{nj} - np_j}{\sqrt{n}} \leqslant \beta_j, j=1, 2, \ldots, N-1 \right\}$

$$= \int_{\alpha_j < x_j \leqslant \beta_j,\, j=1, 2, \ldots, N-1} \frac{1}{[(2\pi)^{N-1} p_1 p_2 \ldots p_N]^{1/2}} \times$$

$$\left\{ \exp - \frac{1}{2} \sum_{1}^{N} \frac{x_j^2}{p_j} \right\} dx_1\, dx_2 \ldots dx_{N-1},$$

where x_N is defined by $\sum_{i=1}^{N} x_i = 0$.

Proof. Let

$$x_j = \frac{S_{nj} - np_j}{\sqrt{n}}, \quad j = 1, 2, \ldots, N. \tag{7.2}$$

Then

$$\sum_{j=1}^{N} x_j = 0. \tag{7.3}$$

If $\alpha_j < x_j \leqslant \beta_j$ for $j=1, 2, \ldots, N-1$, then S_{nj} tends to infinity as $0(n)$ for every $j=1, 2, \ldots, N$. So we can apply Stirling's formula to the multinomial probabilities (7.1). We have from Eq. (6.3).

$M(n;\, S_{n1}, S_{n2}, \ldots, S_{nN};\, p_1, p_2, \ldots, p_N)$

$$= \left\{ \frac{n}{(2\pi)^{N-1} S_{n1} S_{n2} \ldots S_{nN}} \right\}^{1/2} \prod_{j=1}^{N} \left(\frac{np_j}{S_{nj}} \right)^{S_{nj}} (1 + o(1))$$

$$= \frac{n^{1/2}}{(2\pi)^{\frac{N-1}{2}} \prod_{j=1}^{N} (x_j \sqrt{n} + np_j)^{1/2}} f_n(x_1, x_2, \ldots, x_N)\, (1 + o(1))$$

$$= \frac{1}{(2\pi n)^{\frac{N-1}{2}} (p_1, p_2 \ldots p_N)^{1/2}} f_n(x_1, x_2, \ldots, x_N)\, (1 + o(1)),$$

where, by Eqs. (7.2) and (7.3) we have

$$- \log f_n(x_1, x_2, \ldots, x_N) = \sum_{j=1}^{N} (x_j \sqrt{n} + np_j) \log \left(1 + \frac{x_j}{p_j \sqrt{n}} \right)$$

$$= \sum_{j=1}^{N} (x_j \sqrt{n} + np_j) \left[\frac{x_j}{p_j \sqrt{n}} - \frac{x_j^2}{2np_j^2} + 0 \left(\frac{1}{n^{3/2}} \right) \right]$$

$$= \frac{1}{2}\sum_{j=1}^{N}\frac{x_j^2}{p_j} + \sqrt{n}\sum_{=1}^{N} x_j + 0\left(\frac{1}{\sqrt{n}}\right)$$

$$= \frac{1}{2}\sum_{j=1}^{N}\frac{x_j^2}{p_j} + 0\left(\frac{1}{\sqrt{n}}\right),$$

where $0\left(\frac{1}{\sqrt{n}}\right)$ is a term whose absolute value is less than or equal to $c/\sqrt{n}$, c being a constant depending only on $p_1, p_2, \ldots, p_N$ and $\alpha_1, \beta_1, \alpha_2, \beta_2, \ldots \alpha_{N-1}, \beta_{N-1}$. Thus

$$M(n; S_{n1}, S_{n2}, \ldots, S_{nN}; p_1, p_2, \ldots, p_N)$$

$$= \frac{1}{(2\pi n)^{\frac{(N-1)}{2}}(p_1, p_2 \ldots p_N)^{1/2}} \exp\left(-\tfrac{1}{2}\sum_{j=1}^{N}\frac{x_j^2}{p_j}\right\}(1+o(1)), \tag{7.4}$$

whenever $x_j = \frac{S_{nj}-np_j}{\sqrt{n}}$ lies in the bounded interval $[\alpha_j, \beta_j]$ for $j=1, 2,\ldots, N-1$. Now exactly as in the proof of Proposition 6.1, it follows from the definition of multidimensional Riemann integral that

$$\lim_{n\to\infty} \sum_{\substack{np_j-\alpha_j\sqrt{n}<S_{nj} \\ <np_j+\beta_j\sqrt{n}, \\ j=1,2,\ldots,N-1}} M(n; S_{n1}, S_{n2},\ldots, S_{nN}; p_1, p_2, \ldots p_N)$$

$$= \frac{1}{[(2\pi)^{N-1}p_1, p_2 \ldots p_N]^{1/2}} \int\limits_{\substack{\alpha_j<x_j\leqslant\beta_j \\ j=1,2,\ldots,N-1}} \exp\left(-\tfrac{1}{2}\sum_{j=1}^{N}\frac{x_j^2}{p_j}\right) \times dx_1, dx_2 \ldots dx_{N-1}$$

where x_N denotes the quantity $-\sum_{j=1}^{N-1} x_j$. This completes the proof.

§8. Some Applications of the Normal Approximation

Consider the monotonic increasing function

$$\Phi(x) = \frac{1}{\sqrt{2\pi}}\int_{-\infty}^{x} e^{-u^2/2}\, du.$$

This is known as the *standard normal distribution function* on the real line. From the theory of gamma integrals it is known that

$$\lim_{x \to +\infty} \Phi(x) = \frac{1}{\sqrt{2\pi}} \int_{-\infty}^{+\infty} e^{-u^2/2}\, du = 1.$$

Obviously

$$\lim_{x \to -\infty} \Phi(x) = 0.$$

From Example 2.4 it follows that Φ determines a probability distribution on the boolean algebra $\mathcal{F}$ described therein. The function

$$\phi(x) = \frac{1}{\sqrt{2\pi}} e^{-x^2/2}$$

is known as the *standard normal density function.*

We shall now illustrate the usefulness of the limit theorem described in Proposition 6.1.

Example. 8.1. (Telephone trunking problem). Suppose there are N subscribers who wish to have telephone connections between exchanges A and B. We wish to decide the number of lines to be provided between A and B. Suppose that in any busy hour a subscriber is likely to require a line for k minutes. We can say that at any particular moment the probability of a subscriber likely to require a line is $p=k/60$. We can compare the various subscribers requiring a line to N binomial trials with probability of success equal to p. The probability that r subscribers require a line is $b(N, r, p) = \binom{N}{r} p^r q^{N-r}$. Suppose we wish to lay n lines between A and B so that the probability that any subscriber is unable to get a connection is less than or equal to α. (α is usually chosen to be .01 or .05). If all the n lines are engaged at a moment it means more than n subscribers need a line. The probability for such an event is given by

$$\sum_{r=n}^{N} b(N, r, p) \leqslant \alpha. \tag{8.1}$$

Since it is rather difficult to determine n from Eq. (8.1) we can make use of the normal approximation given by Proposition 6.1. The left hand side of (8.1) is $P(S_N \geqslant n)$, where S_N is the number of successes in N independent binomial trials. It is also equal to

$P\left(\frac{S_N - Np}{\sqrt{Npq}} \geqslant \frac{n-Np}{\sqrt{Npq}}\right)$. By Proposition 6.1, this is approximately equal to

$$1-\Phi\left(\frac{n-Np}{\sqrt{Npq}}\right),$$

where Φ is the standard normal distribution function. There exists a unique point x_α on the real line such that

$$1-\Phi(x_\alpha) = \alpha. \tag{8.2}$$

As an approximate value of n we may put

$$\frac{n-Np}{\sqrt{Npq}} = x_\alpha$$

or $n=[Np+x_\alpha\sqrt{Npq}]+1$, where the bracketed part denotes the integral part. For any given α the value of x_α may be read off from standard statistical tables for the normal distribution function.

Example. 8.2. Suppose a factory has to run n identical machines to maintain a schedule of production. Suppose the probability of any one of the machines to break down is p. We wish to find the number N of machines to be operated so that the probability of less than n machines running on a certain day is less than or equal to a preassigned value α. We may compare the running of N machines to N independent binomial trials with probability of success (i.e., break down) equal to p.

$$P\{\text{number of breakdowns} > N-n\}$$
$$= \sum_{N-n+1}^{N} \binom{N}{r} p^r q^{N-r}.$$

If we use the normal approximation the above expression is approximately $1-\Phi\left(\frac{N-n-Np}{\sqrt{Npq}}\right)=1-\Phi\left(\frac{Nq-n}{\sqrt{Npq}}\right)$. We choose x_α (as in the preceding example) such that Eq. (8.2) is fulfilled. Now set

$$\frac{Nq-n}{\sqrt{Npq}} = x_\alpha.$$

Solving this quadratic equation in $\sqrt{N}$ we obtain

$$N=\frac{n}{q}+\frac{x_\alpha}{2q}(p^2x_\alpha{}^2+4np)^{1/2}+\frac{px_\alpha{}^2}{2q}.$$

When n is large, N is approximately equal to $\frac{n}{q}+\frac{x_\alpha\sqrt{np}}{q}$.

Example. 8.3. Now we shall consider a generalisation of Example 8.1. Consider the case when telephone lines have to be laid between exchanges A, B and C, D. Suppose the number of lines to be laid between A, B is n_1 and C, D is n_2. Let, as before, N be the number of subscribers and let p_1 and p_2 be the probabilities that a subscriber requires the line AB and CD respectively at any moment. Let p_3 be chosen such that $p_1+p_2+p_3=1$. Suppose we wish that the probability of a subscriber not being able to get his required line at any particular moment is less than or equal to a preassigned α. This means we have to select n_1 and n_2 such that

$$\sum_{\substack{r_1 \leqslant n_1,\, r_2 \leqslant n_2 \\ r_1 + r_2 + r_3 = N}} \frac{N!}{r_1!\, r_2!\, r_3!} p_1^{r_1} p_2^{r_2} p_3^{r_3} \geqslant 1-\alpha.$$

To select the numbers n_1 and n_2 satisfying the above inequality is a rather difficult problem. So we may use the multivariate normal approximation of Proposition 7.1. If r_1 and r_2 are the numbers of subscribers requiring lines AB and CD respectively at a moment, then

$$\left(\frac{r_1 - Np_1}{\sqrt{Np_1 q_1}}, \frac{r_2 - Np_2}{\sqrt{Np_2 q_2}}\right)$$

is a random variable such that the probability

$$P\left(a_i < \frac{r_i - Np_i}{\sqrt{Np_i q_i}} \leqslant b_i, i=1, 2\right)$$

is approximately equal to

$$\frac{1}{2\pi (1-\rho^2)} \int_{a_2}^{b_2} \int_{a_1}^{b_1} \left[\exp - \frac{1}{2 (1-\rho)^2} (x^2+y^2-2\rho xy)\right] dx\, dy,$$

where

$$\rho = -\left(\frac{p_1 p_2}{q_1 q_2}\right)^{\frac{1}{2}}, \quad q_1=1-p_1, \; q_2=1-p_2.$$

The function

$$\Phi_\rho (x, y) = \frac{1}{2\pi (1-\rho^2)} \int_{-\infty}^{y} \int_{-\infty}^{x} \left[\exp\left\{-\frac{1}{2(1-\rho^2)} (t^2+u^2-2\rho tu)\right\}\right]$$

$$\times\, dt\, du$$

is known as the bivariate normal distribution function with covariance matrix

$$\begin{pmatrix} 1 & \rho \\ \rho & 1 \end{pmatrix}.$$

Now select (x_α, y_α) such that

$$1 - \Phi_\rho (x_\alpha, y_\alpha) = \alpha.$$

Then we may put

$$n_1 = [Np_1 + x_\alpha \sqrt{Np_1 q_1}] + 1,$$
$$n_2 = [Np_2 + y_\alpha \sqrt{Np_2 q_2}] + 1,$$

where the bracketed part denotes integral part. However, there is no unique choice for (x_α, y_α). We may solve the problem as follows. Suppose the lines between A and B are somewhat more important than those between C and D. We wish to ensure that the probability of a subscriber not being able to get the line AB is $\leqslant \beta$, where $\beta < \alpha$. Then we choose x_β such that

$$1 - \Phi (x_\beta) = \beta,$$

where Φ is the standard normal distribution function. Put

$$n_1 = [Np_1 + x_\beta \sqrt{Np_1 q_1}] + 1.$$

Now choose y such that

$$\Phi_\rho (x_\beta, y) = 1 - \alpha.$$

Note that such a $y = y(\alpha, \beta)$ exists. Then choose

$$n_2 = [Np_2 + y \sqrt{Np_2 q_2}] + 1.$$

To select x_β and $y(\alpha, \beta)$ we use the normal and bivariate normal tables.

Remark. 8.4. In Example 8.1, 8.2 one may use the Poisson approximation to the binomial distribution. However, the Poisson distribution is close to the normal distribution when it is suitably normalised and the parameter becomes large. Indeed, one can show that

$$\lim_{\lambda \to \infty} \sum_{\lambda + \alpha \sqrt{\lambda} < r \leqslant \lambda + \beta \sqrt{\lambda}} e^{-\lambda} \frac{\lambda^r}{r!} = \frac{1}{\sqrt{2\pi}} \int_\alpha^\beta e^{-x^2/2}\, dx.$$

Such an approximation enables us to completely do away with evaluation of factorials of large numbers.

§9. Independent Simple Random Variables and Central Limit Theorem

We recall (from Definition 4.15) that a family $\{s_\alpha, \alpha \in I\}$ of Υ valued simple random variables on a boolean probability space $(X, \mathscr{F}, P)$ is mutually independent if for any distinct $\alpha_1, \alpha_2, \ldots, \alpha_k$ in I and positive integer k,

$$P\{s_{\alpha_i} = a_i, \ i = 1, 2, \ldots, k\} = \prod_{i=1}^{k} P\{s_{\alpha_i} = a_i\},$$

for every $a_1, a_2, \ldots, a_k$ in Υ.

Proposition. 9.1. If $s_1, s_2, \ldots, s_k$ are mutually independent simple random variables then

$$\mathbf{E}\, s_i s_j = \mathbf{E}\, s_i\, \mathbf{E}\, s_j \text{ [or } \mathbf{Cov}\,(s_i, s_j) = 0] \text{ for } i \neq j,$$

$$\mathbf{V}\Big(\sum_{i=1}^{k} s_i\Big) = \sum_{i=1}^{k} \mathbf{V}(s_i).$$

Proof. If s_1, s_2 take values $a_1, a_2, \ldots, a_m$, $b_1, b_2, \ldots, b_n$ respectively on sets $E_1, E_2, \ldots, E_m$; $F_1, F_2, \ldots, F_n$ respectively, then $s_1 s_2$ takes the value $a_i b_j$ on $E_i F_j$. Hence

$$\begin{aligned}\mathbf{E}\, s_1 s_2 &= \sum_{i,j} a_i b_j P(E_i F_j) \\ &= \sum_{i,j} a_i b_j P(E_i) P(F_j) \\ &= \Big[\sum_i a_i P(E_i)\Big] \Big[\sum_j b_j P(F_j)\Big] \\ &= \mathbf{E}\, s_1\, \mathbf{E}\, s_2.\end{aligned}$$

This proves the first part. The second part follows from Proposition 3.22.

Exercise. 9.2. Let $\mathbf{s}$ and $\mathbf{t}$ be k-dimensional vector valued simple random variables whose ith components are s_i and t_i respectively. Suppose $\mathbf{s}$ and $\mathbf{t}$ are independent and

$$\mathbf{E}\, s_i = \mathbf{E}\, t_i = 0 \text{ for } i = 1, 2, \ldots, k.$$

Let $\mathbf{u}$ be the vector whose ith component is $s_i t_i$ for all $i = 1, 2, \ldots, k$. If $A = ((a_{ij}))$, $B = ((b_{ij}))$ are the covariance matrices of $\mathbf{s}$ and $\mathbf{t}$ respectively then the covariance matrix of $\mathbf{u}$ is $\mathbf{C} = ((c_{ij}))$, where $c_{ij} = ((a_{ij}\, b_{ij}))$. Hence, for any two positive semi-definite matrices $((a_{ij}))$ and $((b_{ij}))$, the matrix whose ijth element is $a_{ij}\, b_{ij}$ is also positive semi-definite. (see Exercise 3.23).

Let $s_1, s_2, \ldots, s_k$ be k simple random variables on a boolean probability space $(X, \mathscr{F}, P)$. Each of these random variables takes a finite number of values. Consider events of the form

$$\mathbf{E} = \{s_1 = a,\ s_2 = b,\ \ldots,\ s_k = c\}.$$

Such events constitute a partition of the sample space into sets E_1, E_2, ..., E_N say. Then each s_i can be written as

$$s_i = \sum_{j=1}^{N} a_{ij}\, \chi_{E_j},\ i = 1, 2, \ldots, k.$$

Consider n independent trials of the experiment $(X, \mathscr{F}, P)$. Let $x_1, x_2, \ldots, x_n$ be the outcomes. Let

$$\xi_{ni} = s_i(x_1) + s_i(x_2) + \ldots + s_i(x_n).$$

Then $(\xi_{n1}, \xi_{n2}, \ldots, \xi_{nk})$ is a k-dimensional vector valued random variable on the product probability space $(X \times X \times \ldots \times X, \mathscr{F} \times \mathscr{F} \times \ldots \times \mathscr{F}, P \times P \times \ldots \times P)$, where the product is taken n-fold. If we write

$$S_{nj} = \chi_{E_j}(x_1) + \chi_{E_j}(x_2) + \ldots + \chi_{E_j}(x_n),\ j = 1, 2, \ldots, N,$$

then S_{nj} is the number of times the event E_j occurs in n independent trials. Thus $(S_{n1}, S_{n2}, \ldots, S_{nN})$ has the multinomial distribution. Let $P(E_j) = p_j$, and let

$$Z_{nj} = \frac{S_{nj} - np_j}{\sqrt{n}},\ j = 1, 2, \ldots, N. \tag{9.1}$$

We note that

$$\frac{\xi_{ni} - \mathbf{E}\xi_{ni}}{\sqrt{n}} = \sum_{j=1}^{N} a_{ij}\, Z_{nj},\ i = 1, 2, \ldots, k. \tag{9.2}$$

We shall now make an asymptotic analysis of the probability

$$P\left\{u_i < \frac{\xi_{ni} - \mathbf{E}\xi_{ni}}{\sqrt{n}} \leqslant v_i,\ i = 1, 2, \ldots, k\right\},$$

as $n \to \infty$, where u_i, v_i, $i=1, 2, \ldots, k$ are constants. Indeed, we have

Proposition. 9.3. $\displaystyle\lim_{n \to \infty} P\left\{u_i < \frac{\xi_{ni} - \mathbf{E}\xi_{ni}}{\sqrt{n}} \leqslant v_i,\ i=1, 2, \ldots, k\right\},$

$$= \frac{1}{[(2\pi)^{N-1} p_1 p_2 \ldots p_N]^{1/2}} \times$$

$$\int \left[\exp\left(-\frac{1}{2} \sum_{j=1}^{N} \frac{x_j^2}{p_j}\right)\right] dx_1\, dx_2 \ldots dx_N, \tag{9.3}$$

$$\sum_{j=1}^{N} x_j = 0$$

$$u_i < \sum_{j=1}^{N} a_{ij}\, x_j \leqslant v_i,$$

$i = 1, 2, \ldots, k$

where u_i, v_i, $i=1, 2, \ldots, k$ are arbitrary but fixed real numbers.

Proof. Let

$$A_n = \left\{u_i < \frac{\xi_{ni} - \mathbf{E}\xi_{ni}}{\sqrt{n}} \leqslant v_i,\ i=1, 2, \ldots, N\right\}, \tag{9.4}$$

$$B_{n\alpha} = \{|Z_{nj}| \leqslant \alpha \text{ for all } j = 1, 2, \ldots, N-1\}, \tag{9.5}$$

$$E = \left\{\mathbf{x} : \mathbf{x} = (x_1, x_2, \ldots, x_N),\ \sum_{j=1}^{N} x_j = 0,\right.$$

$$\left. u_i < \sum_{j=1}^{N} a_{ij} x_j \leqslant v_i,\ i=1, 2, \ldots, k\right\} \tag{9.6}$$

$$F_\alpha = \{\mathbf{x} : |x_j| \leqslant \alpha, j = 1, 2, \ldots, N-1\}. \tag{9.7}$$

Then by the argument of Proposition 7.1,

$$\lim_{n\to\infty} P(A_n \cap B_{n\alpha}) = \int_{E\cap F_\alpha} f(\mathbf{x})\, dx_1\, dx_2 \ldots dx_{N-1}, \tag{9.8}$$

where

$$f(\mathbf{x}) = \frac{1}{[(2\pi)^{N-1} p_1 p_2 \ldots p_N]^{1/2}} \exp\left(-\tfrac{1}{2} \sum_{j=1}^{N} \frac{x_j^2}{p_j}\right).$$

By Proposition 2.5 and Chebyshev's inequality,

$$P(B'_{n\alpha}) \leqslant \sum_{j=1}^{N} P\{|Z_{nj}| > \alpha\}$$

$$\leqslant \frac{1}{\alpha^2} \sum_{j=1}^{N} p_j(1-p_j)$$

$$\leqslant \frac{1}{\alpha^2}, \text{ for all } n.$$

Since $f(\mathbf{x})$ is Riemann integrable over R^{N-1}, for any given $\epsilon > 0$, we can choose α such that

$$\int_{F'_\alpha} f(\mathbf{x})\, dx_1\, dx_2 \ldots dx_{N-1} < \frac{\epsilon}{2},$$

and

$$P(B'_{n\alpha}) < \frac{\epsilon}{2} \text{ for all } n.$$

Hence we have

$$\Big| P(A_n) - \int_E f(\mathbf{x}) dx_1\, dx_2 \ldots dx_{N-1} \Big|$$

$$\leqslant P(A_n B'_{n\alpha}) + \Big| P(A_n B_{n\alpha}) - \int_{EF_\alpha} f(\mathbf{x}) dx_1\, dx_2 \ldots dx_{N-1} \Big| + \int_{EF'_\alpha} f(\mathbf{x}) dx_1\, dx_2 \ldots dx_{N-1}$$

$$\leqslant \epsilon + \Big| P(A_n B_{n\alpha}) - \int_{EF_\alpha} f(\mathbf{x}) dx_1\, dx_2 \ldots dx_{N-1} \Big|.$$

Now by Eq. (9.8) the second term on the right hand side tends to zero as $n \to \infty$. Since ϵ is arbitrary, we have

$$\lim_{n\to\infty} P(A_n) = \int_E f(\mathbf{x}) dx_1\, dx_2 \ldots dx_{N-1}$$

This completes the proof.

Remark. 9.9. From the theory of gamma integrals, it follows that

$$\lim_{n\to\infty} P\left\{u_1 < \frac{\xi_{n1} - \mathbf{E}\xi_{n1}}{\sqrt{n}} \leqslant v_1\right\}$$

$$= \frac{1}{\sqrt{2\pi}\,\sigma} \int_{u_1}^{v_1} \left[\exp\left(-\frac{x^2}{2\sigma^2}\right)\right] dx,$$

where $\sigma^2 = \mathbf{V}(s_1)$.

§10. Conditional Probability

The following example leads us in a natural manner to the idea of conditional probability. Suppose there are n horses taking part in a race and let their chances of winning the race be $p_1, p_2, \ldots, p_n$ respectively. Suppose it happens that horse number n is unable to participate in the race for some reason. The relative chances of the first $n-1$ horses to win are in the ratio $p_1 : p_2 : \ldots : p_{n-1}$. In other words the probability of horse number r winning, given that the first $n-1$ horses take part in the race, is given by $p_r/(1-p_n)$. We formalise this simple idea into a definition. Before this we remark that any general probability statement is with respect to a fixed boolean probability space $(X, \mathcal{F}, P)$.

Definition. 10.1. For any two events A and B such that $P(B) > 0$, *the conditional probability of the event A given the hypothesis B* is defined as the number $\dfrac{P(AB)}{P(B)}$. It is denoted by $P(A|B)$.

Remark. 10.2. Events A and B are independent if and only if $P(A \mid B) = P(B)$, whenever $P(B) > 0$. Whenever one of the events A, B have probability zero, they are always independent.

In section 4, we saw how to construct new probability spaces out of given probability spaces $(X_i, \mathscr{F}_i, P_i)$, $i = 1, 2, \ldots, k$ by taking products. Now we shall illustrate how one can construct a new probability space from a given one by ' conditioning '. We have

Proposition. 10.3. Let $(X, \mathscr{F}, P)$ be a boolean probability space and let $\Upsilon \in \mathscr{F}, \Upsilon \subset X, P(\Upsilon) > 0$. Let

$$P_\Upsilon(A) = P(A \mid \Upsilon) = \frac{P(A\Upsilon)}{P(\Upsilon)}.$$

Then $(X, \mathscr{F}, P_\Upsilon)$ is a boolean probability space.

Proof. Obvious.

Remark. 10.4. In the proposition above let $\mathscr{F} \cap \Upsilon$ denote the family $\{A \cap \Upsilon,\ A \in \mathscr{F}\}$. If $E \in \mathscr{F} \cap \Upsilon$ and $E = A \cap \Upsilon = B \cap \Upsilon$, where $A, B \in \mathscr{F}$, it is clear that $P(A \cap \Upsilon) = P(A \cap B \cap \Upsilon) = P(B \cap \Upsilon)$. Define $P^\Upsilon(E) = P(A \mid \Upsilon)$. Then $(\Upsilon, \mathscr{F} \cap \Upsilon, P^\Upsilon)$ is a boolean probability space.

Proposition. 10.5. Let $X = \bigcup_{i=1}^{k} H_i$ be a partition of X into disjoint sets $H_i \in \mathscr{F}$ where $P(H_i) > 0$ for all i. For any $A \in \mathscr{F}$, the following holds:

$$P(A) = \sum_{i=1}^{k} P(A \mid H_i)\, P(H_i).$$

Proof. It is clear that $A = \bigcup_{i=1}^{k} AH_i$. Since H_i are disjoint AH_i are also disjoint. Hence $P(A) = \sum_i P(AH_i) = \sum_i P(A \mid H_i)\, P(H_i)$. This completes the proof.

Proposition. 10.6. (Bayes' theorem) In the notation of proposition 10.5.

$$P(H_i \mid A) = \frac{P(AH_i)}{P(A)} = \frac{P(A \mid H_i)\, P(H_i)}{\sum_{j=1}^{k} P(A \mid H_j)\, P(H_j)}. \tag{10.1}$$

Remark. 10.7. The above proposition is usually given the following interpretation. The H_i's are considered as probable hypotheses. $P(H_i)$, $i = 1, 2, \ldots, k$ are considered as *prior probabilities* for the different hypotheses to be true. Suppose the performance of an experiment leads to the occurrence of the event A. Then the probabilities $P(H_i|A)$, $i=1, 2, \ldots, k$ given by Eq. (10.1) are interpreted as *posterior probabilities* for the different hypotheses $H_1, H_2, \ldots, H_k$ to be true (in the light of the experience that A has occurred).

Example. 10.8. (Polya's urn scheme). Consider an urn containing a white balls and b black balls. Select a ball at random, replace it and add c more balls of the same colour drawn. Repeat the experiment n times. Note that the outcome of the rth draw depends on the outcomes of the preceding $r-1$ draws. Let W_i, B_i denote respectively the events that the ith draw results in a white or black ball. Let

$$P(W_1 W_2 \ldots W_j) = p_j.$$

Then

$$\begin{aligned} p_j &= P(W_j \mid W_1 W_2 \cdots W_{j-1})\, P(W_1 W_2 \cdots W_{j-1}) \\ &= \frac{a + \overline{j-1}\, c}{a + b + \overline{j-1}\, c}\, p_{j-1}. \end{aligned}$$

In general, the probability of having first j_1 draws resulting in white, next j_2 in black, next j_3 white, etc., and the last j_r black (where $j_1 + j_2 + \ldots + j_r = n$) is given by

$$\frac{a(a+c)\ldots(a+\overline{j_1-1}\,c).b(b+c)\ldots(b+\overline{j_2-1}\,c).(a+j_1c)\ldots(a+\overline{j_1+j_3-1}\,c)\ldots}{(a+b)\,(a+b+c)\,(a+b+2c)\,\ldots\,(a+b+\overline{n-1}\,c)}$$

$$= \frac{a(a+c)\,(a+2c)\,\ldots\,(a+\overline{n_1-1}\,c)\; b(b+c)\,\ldots\,(b+\overline{n_2-1}\,c)}{(a+b)\,(a+b+c)\,\ldots\,(a+b+\overline{n-1}\,c)}$$

where $n_1 = j_1+j_3+j_5+\ldots+j_{r-1}$, $n_2 = j_2+j_4+\ldots+j_r$, $n_1+n_2 = n$. In other words the required probability is also the probability that the first n_1 draws result in white and the next n_2 in black.

Exercise. 10.9. In the preceding example

$$\begin{aligned} P(W_i \mid W_j) &= P(W_j \mid W_i), \\ P(B_i \mid B_j) &= P(B_j \mid B_i), \\ P(W_i \mid B_j) &= P(W_j \mid B_i), \\ P(B_i \mid W_j) &= P(B_j \mid W_i). \end{aligned}$$

Exercise. 10.10. In Example 10.8, let $p_k\,(n)$ be the probability that k white balls result in n draws. Then

$$p_k(n+1) = p_k(n)\,\frac{b+(n-k)c}{a+b+nc} + p_{k-1}(n)\,\frac{a+(k-1)c}{a+b+nc}.$$

Example. 10.11. Example 10.8 can be thought of as a sequence of binomial trials. If the first m trials resulted in k white balls and l

black balls, the $(m+1)$th trial has probability $(a+kc)/(a+b+mc)$ for white and $b+lc/(a+b+mc)$ for black. We may call white as success and black as failure. The probability p_{m+1} for success at the $(m+1)$th trial depends on the outcome of the last m trials. Thus we have a sequence of dependent binomial trials. We can now generalise this idea as follows: let S denote success and F denote failure. Suppose the first m trials result in k successes and l failures. Then the probability for success in the $(m+1)$th trial is assumed to be $(p+k\alpha)/(1+m\alpha)$. Hence the probability for failure at $(m+1)$th trial given k successes in the first m trials is $[q+(m-k)\alpha]/(1+m\alpha)$. Let $\pi(k, n)$ be the probability of k successes in the first n trials. Let S_{kn} denote the event 'k successes in the first n trials' and let S_n, F_n denote the events 'success at the nth trial' and 'failure at the nth trial' respectively. Then

$$S_{k,\,n+1} = S_{n+1}\, S_{k-1,\,n} \cup F_{n+1}\, S_{kn}$$

and the two events on the right hand side are disjoint. Thus

$$P(S_{k,\,n+1}) = P(S_{n+1} \mid S_{k-1,\,n})\, P(S_{k-1,\,n}) + P(F_{n+1} \mid S_{kn})\, P(S_{kn}),$$

or equivalently,

$$\pi(k, n+1) = \frac{p+(k-1)\alpha}{1+n\alpha}\,\pi(k-1, n) + \frac{q+(n-k)\alpha}{1+n\alpha}\,\pi(k, n).$$

Exercise. 10.12. In Example 10.11,

$$\pi(k, n) = \binom{n}{k} \frac{p(p+\alpha)\,(p+2\alpha)\ldots(p+\overline{k-1}\;\alpha)\; q(q+\alpha)\ldots q+\overline{n-k-1}\,\alpha)}{(\alpha+1)\,(2\alpha+1)\,\ldots\,(\overline{n-1}\;\alpha+1)}.$$

If $n\to\infty$, $p\to 0$ so that $np\to\lambda$, $n\alpha\to\rho^{-1}$, then

$$\lim \pi(k, n) = \binom{\lambda\rho+k-1}{k}\left(\frac{\rho}{1+\rho}\right)^{\lambda\rho}\left(\frac{1}{1+\rho}\right)^{k}.$$

As $\rho\to\infty$, the last expression converges to $e^{-\lambda}\,\lambda^k/k!$. (The distribution $\pi(k, n)$ for k successes in the first n trials is known as *Polya's distribution*).

Remark. 10.13. Polya's distribution is used as a model for studying the spread of an infectious disease. If already k people in a population of n have the infection and an $(n+1)$th individual enters the population then the probability of his catching the infection is dependent on k and n. If getting the infection is considered as a success then the probability of success for the $(n+1)$th individual depends on k and n. Thus we may compare the situation with a sequence of dependent binomial trials.

Example. 10.14. It may be noted that Proposition 10·5 holds with $k=\infty$, provided P is countably additive. As an example we may consider the following application. Suppose a hen lays n eggs with probability $e^{-\lambda}\frac{\lambda^n}{n!}$. Suppose these eggs behave independently and the probability that an egg hatches is p. Let A_r be the event that r eggs from the hen hatch. Let B_n be the event that the hen lays n eggs. Then

$$\begin{aligned}
P(A_r) &= \sum_{n=0}^{\infty} P(A_r|B_n)\,P(B_n) \\
&= \sum_{m=r}^{\infty} P(A_r|B_n)\,P(B_n) \\
&= \sum_{n=r}^{\infty} \binom{n}{r} p^r q^{n-r} e^{-\lambda} \frac{\lambda^n}{n!} \\
&= e^{-\lambda p} \frac{(\lambda p)^r}{r!}.
\end{aligned}$$

This shows that the 'mixture' of a Poisson distribution with parameter λ and a binomial distribution with probability p for success is again a Poisson distribution with parameter λp.

Example. 10.15. (Laplace) Consider $N+1$ urns where urn number k contains k white balls and $N-k$ black balls. Here k assumes the values 0, 1, 2, ..., N. Choose an urn at random and make n drawings with replacement. We have

$$\begin{aligned}
p_n &= P \text{ (all draws are white)} \\
&= \sum_{k=1}^{N} \left(\frac{k}{N}\right)^n \frac{1}{N+1},
\end{aligned}$$

$$P(n+1\text{th draw is white} \mid \text{the first } n \text{ draws are white}) =$$

$$\frac{p_{n+1}}{p_n} = \frac{\dfrac{1}{N}\displaystyle\sum_{k=1}^{N}\left(\frac{k}{N}\right)^{n+1}}{\dfrac{1}{N}\displaystyle\sum_{k=1}^{N}\left(\frac{k}{N}\right)^{n}}.$$

If N is large, we can approximate the above quantity by

$$\frac{\int_0^1 x^{n+1}\,dx}{\int_0^1 x^n\,dx} = \frac{n+1}{n+2}.$$

Remark. 10.16. Before the development of modern probability theory the notion of equal probabilities was usually interpreted as 'no prior knowledge'. Laplace used the above Example 10.15 to compute the probability that the sun will rise tomorrow given that it has arisen daily for $n = 5000$ years! For further historical details see Feller [4].

§11. Laws of Large Numbers

Suppose $s_1, s_2, \ldots, s_n$ are independent simple random variables with the same mean and variance. Let

$$\mathbf{E}s_i = m; \quad \mathbf{V}(s_i) = \sigma^2, \quad i = 1, 2, \ldots$$

Let

$$\bar{s}_n = \frac{s_1 + s_2 + \ldots + s_n}{n}.$$

Then $\mathbf{E}(\bar{s}_n - m)^2 = \sigma^2/n$. Hence by Chebyshev's inequality we have

$$P(|\bar{s}_n - m| > \varepsilon) \leqslant \frac{\sigma^2}{n\varepsilon^2} \tag{11.1}$$

Thus, for every $\varepsilon > 0$, it follows that

$$\lim_{n\to\infty} P(|\bar{s}_n - m| > \varepsilon) = 0.$$

This is known as the *weak law of large numbers*. We state it as a proposition.

Proposition. 11.1. Let $s_1, s_2, \ldots$ be a sequence of independent simple random variables on a boolean probability space $(X, \mathscr{F}, P)$ with the same mean m and same variance σ^2. Let

$$\bar{s}_n = \frac{1}{n}(s_1 + s_2 + \ldots + s_n).$$

Then for every $\varepsilon > 0$,

$$\lim_{n\to\infty} P(|\bar{s}_n - m| > \varepsilon) = 0$$

Remark. 11.2. Consider any boolean probability space $(X, \mathscr{F}, P)$. Let $E \in \mathscr{F}$. Repeat the experiment $(X, \mathscr{F}, P)$ n times independently. Let $x_1, x_2, \ldots, x_n$ be the outcomes. Then $\chi_E(x_1), \chi_E(x_2), \ldots, \chi_E(x_n)$ are independent random variables taking values 1 and 0 with probability $p(E)$ and $(1 - p(E))$ respectively. Hence they have expectation $p(E)$ and variance $p(E)(1 - p(E))$. It follows as a consequence that

$$\lim_{n\to\infty} P\left\{\left|\frac{\chi_E(x_1) + \chi_E(x_2) + \ldots + \chi_E(x_n)}{n} - p(E)\right| > \varepsilon\right\} = 0.$$

In the above sense the frequency of occurrence of the event E 'converges' to the probability $p(E)$ of E as the number of trials tends to infinity.

Now we shall prove a much stronger version of the inequality (11.1).

Proposition. 11.3. (Kolmogorov's inequality). Let $s_1, s_2, \ldots, s_n$ be independent simple random variables on a boolean probability space $(X, \mathscr{F}, P)$. Let

$$\begin{aligned} \mathbf{E}s_i &= m_i, \mathbf{V}(s_i) = \sigma_i^2, i = 1, 2, \ldots, n, \\ S_k &= s_1 + s_2 + \ldots + s_k, \\ M_k &= m_1 + m_2 + \ldots + m_k, \\ v_k^2 &= \sigma_1^2 + \sigma_2^2 + \ldots + \sigma_k^2, \; k = 1, 2, \ldots, n. \end{aligned}$$

Then

$$P\left\{\frac{|S_k - M_k|}{v_n} \leqslant t, \; k = 1, 2, \ldots, n\right\} \geqslant 1 - t^{-2} \tag{11.2}$$

Proof. Consider the events

$$E = \left\{\frac{|S_k - M_k|}{v_n} > t \text{ for some } k = 1, 2, \ldots, n\right\},$$

$$E_j = \left\{\frac{|S_i - M_i|}{v_n} \leqslant t \text{ for } i = 1, 2, \ldots, j-1; \right.$$
$$\left. \frac{|S_j - M_j|}{v_n} > t\right\}.$$

Then E_j's are disjoint events and

$$\bigcup_{j=1}^{n} E_j = E.$$

Hence

$$\chi_E = \sum_{j=1}^{n} \chi_{E_j}.$$

Now we have

$$\begin{aligned} v_n^2 &= \mathbf{V}(S_n) = \mathbf{E}(S_n - M_n)^2 \\ &\geqslant \mathbf{E}(S_n - M_n)^2 \chi_E \\ &= \sum_j \mathbf{E}(S_n - M_n)^2 \chi_{E_j} \\ &= \sum_j \mathbf{E}(S_n - S_j + M_j - M_n + S_j - M_j)^2 \chi_{E_j} \\ &\geqslant \sum_j \{\mathbf{E}(S_j - M_j)^2 \chi_{E_j} + 2\mathbf{E}[(S_n - S_j + M_j - M_n) \times \\ &\qquad (S_j - M_j) \chi_{E_j}]\}. \end{aligned} \tag{11.3}$$

From the definition of E_j and the independence of the random variables s_i, it is clear that $(S_j - M_j)\,\chi_{E_j}$ is independent of $S_n - S_j$. Indeed, the first one is a function of $s_1, s_2, \ldots, s_j$ and the second is a function of $s_{j+1}, s_{j+2}, \ldots, s_n$. Hence

$$\mathbf{E}[(S_n - S_j + M_j - M_n)\,(S_j - M_j)\,\chi_{E_j}]$$
$$= \mathbf{E}[S_n - S_j + M_j - M_n]\,\mathbf{E}[(S_j - M_j)\,\chi_{E_j}] = 0. \tag{11.4}$$

Again from the definition of E_j, we have

$$(S_j - M_j)^2\,\chi_{E_j} \geqslant t^2\,\chi_{E_j}\,v_n^2. \tag{11.5}$$

Hence Eqs. (11.3) to (11.5) imply

$$v_n^2 \geqslant \mathbf{E} \sum_j (t^2\,\chi_{E_j}\,v_n^2)$$
$$= t^2\,v_n^2\,P(E).$$

Thus

$$P(E) \leqslant t^{-2}.$$

This completes the proof.

As an application of Kolmogorov's inequality we shall prove the following proposition and deduce an important corollary.

Proposition. 11.4. Let $s_1, s_2, \ldots, s_n, \ldots$ be independent simple random variables on a boolean probability space $(X, \mathscr{F}, P)$. Let $\mathbf{E}s_i = m_i$, $\mathbf{V}(s_i) = \sigma_i^2$, $i = 1, 2, \ldots, n, \ldots$ Let $A_j(\varepsilon)$ be the event

$$\frac{|S_n - M_n|}{n} \geqslant \varepsilon \text{ for at least one } n \text{ in } (2^{j-1}, 2^j],$$

where $M_n = m_1 + m_2 + \ldots + m_n$ and $S_n = s_1 + s_2 + \ldots + s_n$. Then

$$\sum_{j=1}^{\infty} P(A_j(\varepsilon)) \leqslant \frac{16\varepsilon^{-2}}{3} \sum_{k=1}^{\infty} k^{-2}\,\sigma_k^2 \tag{11.6}$$

Proof. Let $v_n^2 = \sigma_1^2 + \sigma_2^2 + \ldots + \sigma_n^2$. Then by Kolmogorov's inequality (11.2), we have

$$P(A_j(\varepsilon)) = P\left\{\frac{|S_n - M_n|}{v_{2^j}} \geqslant \frac{n\,\varepsilon}{v_{2^j}} \text{ for some } n \text{ in } (2^{j-1}, 2^j]\right\}$$

$$\leqslant P\left\{\frac{|S_n - M_n|}{v_{2^j}} \geqslant \frac{2^{j-1}\varepsilon}{v_{2^j}} \text{ for some } n \text{ in } (2^{j-1}, 2^j]\right\}$$

$$\leqslant \left(\frac{2^{j-1}\,\varepsilon}{v_{2^j}}\right)^{-2} = 4\,\varepsilon^{-2}\left(\frac{v_{2^j}^2}{2^{2j}}\right).$$

Hence

$$\sum_{j=1}^{\infty} P(A_j(\varepsilon)) = 4\,\varepsilon^{-2} \sum_{j=1}^{\infty} \left(\frac{\sigma_1^2+\sigma_2^2+\ldots+\sigma_{2^j}^2}{2^{2j}}\right)$$

$$\leqslant 4\,\varepsilon^{-2} \sum_{k=1}^{\infty} \sigma_k^2 \left\{ \sum_{j:\,2^j \geqslant k} \frac{1}{2^{2j}} \right\}$$

$$\leqslant 4\,\varepsilon^{-2} \sum_{k=1}^{\infty} k^{-2}\sigma_k^2 \left(1+\frac{1}{4}+\frac{1}{4^2}+\ldots\right)$$

$$\leqslant \frac{16}{3}\,\varepsilon^{-2} \sum_{k=1}^{\infty} k^{-2}\sigma_k^2.$$

This completes the proof.

Corollary. 11.5. Let $s_1, s_2, \ldots, s_n, \ldots$ be independent simple random variables on $(X, \mathcal{F}, P)$. Let $\mathbf{E}s_i = m_i$, $\mathbf{V}(s_i) = \sigma_i^2$, $i = 1, 2, \ldots$. Suppose $\sum_{k=1}^{\infty} k^{-2}\sigma_k^2 < \infty$. Then, for any $\varepsilon > 0$, $\delta > 0$ there exists an integer N such that

$$P\left\{\frac{|S_n - M_n|}{n} < \varepsilon \text{ for every } n \in [N, N+r]\right\} \geqslant 1-\delta,$$

for every $r = 1, 2, \ldots$, where $S_n = s_1 + s_2 + \ldots + s_n$ and $M_n = m_1 + m_2 + \ldots + m_n$.

Proof. Since the infinite series $\sum_k k^{-2}\sigma_k^2$ is convergent it follows from Proposition 11.4 that the left hand side of Eq. (11.6) is convergent. In particular, for any $\delta > 0$, there exists a j_0 such that

$$P\Big(\bigcup_{j=j_0}^{j_0+k} A_j(\varepsilon)\Big) \leqslant \delta \text{ for all } k.$$

If we put $N = 2^{j_0-1}$, the proof is complete.

Corollary. 11.6. Let $s_1, s_2, \ldots, s_n, \ldots$ be independent simple random variables with $\mathbf{E}s_i = m$ and $\mathbf{V}(s_i) = \sigma^2$ for every i. Then, for any $\varepsilon > 0$, $\delta > 0$, there exists an integer N such that

$$P\left\{\left|\frac{s_1+s_2+\ldots+s_n}{n} - m\right| < \varepsilon \text{ for every } n \text{ such that } N \leqslant n \leqslant N+r\right\} \geqslant 1-\delta,$$

for all $r = 1, 2, 3, \ldots$.

Proof. This is an immediate consequence of corollary 11.5.

Remark. 11.7. Let $(X, \mathcal{F}, P)$ be a boolean probability space and let s be a simple random variable defined on X. Consider an infinite sequence of independent trials of the experiment $(X, \mathcal{F}, P)$. Then any outcome may be represented by the sequence $\mathbf{x}=(x_1, x_2,\ldots, x_n, \ldots)$. Let X^∞ be the space of all such sequences $\mathbf{x}$. The outcome of the nth trial is x_n. Let $(X^\infty, \mathcal{F}^\infty, P^\infty)$ be the infinite product of the boolean probability space $(X, \mathcal{F}, P)$. (See Remark 4.11). Let

$$\bar{s}_n = \frac{s(x_1)+s(x_2)+\ldots+s(x_n)}{n}.$$

Then $\bar{s}_1, \bar{s}_2,\ldots$ are simple random variables on the boolean probability space $(X^\infty, \mathcal{F}^\infty, P^\infty)$. Let $\mathbf{E}s=\mu$, $\mathbf{V}(s)=\sigma^2$. Let

$$E_{n,\,\varepsilon} = \left\{ \mathbf{x} : \left| \frac{s(x_1)+s(x_2)+\ldots+s(x_n)}{n} - \mu \right| < \varepsilon \right\}.$$

Now we define

$$s_n(\mathbf{x})=s(x_n),\ n=1, 2,\ldots.$$

Then $s_1, s_2,\ldots, s_n,\ldots$ are independent simple random variables on $(X^\infty, \mathcal{F}^\infty, P^\infty)$. The weak law of large numbers implies that

$$\lim_{n\to\infty} P^\infty(E_{n,\varepsilon})=1 \text{ for every } \varepsilon>0.$$

Corollary 11.6 implies that

$$\lim_{N\to\infty} \inf_r P^\infty \Big(\bigcap_{n=N}^{N+r} E_{n,\varepsilon}\Big)=1 \text{ for every } \varepsilon>0. \tag{11.7}$$

Now consider the set

$$E=\left\{ \mathbf{x} : \frac{s(x_1)+s(x_2)+\ldots+s(x_n)}{n} \text{ converges to } \mu \text{ as } n\to\infty \right\}.$$

We can express the set E in set theoretic language as

$$E = \bigcap_{k=1}^{\infty} \bigcup_{N=1}^{\infty} \bigcap_{n=N}^{\infty} E_{n,\frac{1}{k}}.$$

Equation (11.7) tempts us to think that the 'probability' of E is also unity, i.e., 'with probability one' the average value $\bar{s}_n$ based on n independent trials converges to the true expectation μ. However, there is one difficulty in making such a statement. Since $\mathcal{F}^\infty$ is a boolean algebra, every event $E_{n,\,\varepsilon}$ which is a statement only about the first n observations belongs to E^∞. However, the event E is obtained by making a *countable* number of operations on the sets $E_{n,\frac{1}{k}}$.

Hence E lies outside $\mathcal{F}^\infty$. Since P^∞ is defined only on $\mathcal{F}^\infty$, we cannot make the statement $P^\infty(E)=1$. In order to circumvent this difficulty we can try to *extend* the definition of the distribution P^∞ on $\mathcal{F}^\infty$ to a larger collection of events which includes sets of the type E. In the next chapter we shall demonstrate the existence of a *unique* extension of P^∞ to a collection of events which includes $\mathcal{F}^\infty$ and is closed under all countable set operations.

Corollary 11.6 is a version of what is known as the *strong law of large numbers*.

§12. An Application of the Law of Large Numbers to a Problem in Analysis

In the last section we saw that the weak law of large numbers is an immediate consequence of Chebyshev's inequality. The power of the very elementary inequality due to Chebyshev was demonstrated in analysis by S.N. Bernstein who gave a very beautiful proof of Weierstrass' theorem on approximation of continuous functions by means of polynomials. We shall present this proof here.

Let f be a real valued continuous function defined on the unit interval $[0, 1]$. For any $\delta>0$, let

$$\omega(f;\delta) = \sup_{\substack{x,y\in[0,1]\\ |x-y|\leqslant\delta}} |f(x)-f(y)|.$$

For every positive integer n, we define the nth degree *Bernstein polynomial* $B_n(f, x)$ by

$$B_n(f, x) = \sum_{r=0}^{n} \binom{n}{r} x^r (1-x)^{n-r} f\left(\frac{r}{n}\right). \tag{12.1}$$

If S_n is the number of successes in n independent binomial trials with probability of success equal to x, then

$$B_n(f, x) = \mathbf{E} f\left(\frac{S_n}{n}\right).$$

We now have the following proposition.

Proposition. 12.1. For any real continuous function f on $[0, 1]$, and any $\delta>0$,

$$\sup_{x\in[0,1]} |B_n(f, x) - f(x)| \leqslant \omega(f;\delta) + \frac{M}{2n\delta^2}, \tag{12.2}$$

where

$$M = \sup_{x\in[0,1]} |f(x)|.$$

Proof. We have

$$
\begin{aligned}
&| B_n(f, x) - f(x) | \\
&= \left| \sum_{r=0}^{n} \binom{n}{r} x^r (1-x)^{n-r} \left[f\left(\frac{r}{n}\right) - f(x) \right] \right| \\
&\leqslant \left\{ \sum_{\left|\frac{r}{n} - x\right| \leqslant \delta} \binom{n}{r} x^r (1-x)^{n-r} \right\} \omega(f; \delta) + \\
&\quad \left\{ \sum_{\left|\frac{r}{n} - x\right| > \delta} \binom{n}{r} x^r (1-x)^{n-r} \right\} (2M) \\
&\leqslant \omega(f; \delta) + 2M\, P\left(\left| \frac{S_n}{n} - x \right| > \delta \right),
\end{aligned} \tag{12.3}
$$

where P stands for the binomial distribution of the number of successes S_n in n independent trials with probability of success equal to x. By Chebyshev's inequality,

$$
\begin{aligned}
P\left(\left| \frac{S_n}{n} - x \right| > \delta \right) &\leqslant \frac{x(1-x)}{n\delta^2} \\
&\leqslant \frac{1}{4n\delta^2},
\end{aligned} \tag{12.4}
$$

because the maximum value of $x(1-x)$ is attained at $x = \frac{1}{2}$. Now Eqs. (12.3) and (12.4) imply (12.2). This completes the proof.

Corollary. 12.2.
$$
\sup_{x \in [0, 1]} | B_n(f, x) - f(x) | \leqslant \omega(f, n^{-1/4}) + \frac{M}{2\sqrt{n}}.
$$
In particular $B_n(f, x)$ converges uniformly to $f(x)$ in $[0, 1]$.

Proof. This is obtained by putting $\delta = n^{-1/4}$.

Remark. 12.3. It may be noted that even when f is not continuous the Bernstein polynomial converges to f at all continuity points of f as $n \to \infty$.

Exercise. 12.4. Let $f(x_1, x_2, \ldots, x_k)$ be a real continuous function of k variables $x_1, x_2, \ldots, x_k$ in the domain $x_i \geqslant 0$ for all i, $\sum_{i=1}^{k} x_i = 1$.

Let $B_{n,k}$ be the kth order Bernstein polynomial defined by

$$B_{n,k}(f, x_1, x_2, \ldots, x_k) = \sum_{r_i \geqslant 0\; r_1+\ldots+r_k=n} \frac{n!}{r_1!\, r_2! \ldots r_k!} \times$$

$$x_1^{r_1} x_2^{r_2} \ldots x_k^{r_k}\; f\left(\frac{r_1}{n}, \frac{r_2}{n}, \ldots, \frac{r_k}{n}\right)$$

Then $B_{n,k}(f, x_1, \ldots, x_k)$ converges uniformly to $f(x_1, x_2, \ldots, x_k)$ in the set $\{(x_1, x_2, \ldots, x_k) : x_i \geqslant 0, \sum_i x_i = 1\}$ as $n \to \infty$. Hence deduce that every real continuous function f defined on a compact subset of R^n can be approximated uniformly by a sequence of polynomials.

CHAPTER TWO

Extension of Measures

§13. σ-algebras and Borel Spaces

In Section 11 we mentioned about the fruitfulness of introducing the idea of a collection of sets closed under countable set operations and introducing probability distributions on such a collection. To this end we introduce the following definition.

Definition. 13.1. A collection $\mathscr{B}$ of subsets of a set X is called a *σ-algebra* if

(i) $\emptyset \in \mathscr{B}$;
(ii) if $A \in \mathscr{B}$, then A' (complement of A) $\in \mathscr{B}$;
(iii) if $A_1, A_2, \ldots, A_n, \ldots, \in \mathscr{B}$ then $\bigcup_{i=1}^{\infty} A_i \in \mathscr{B}$.

Since $\bigcap_i A_i = (\bigcup_i A_i')'$, it follows that $\mathscr{B}$ is closed under countable intersection. Thus a σ-algebra is closed under all countable set operations, namely, union, intersection, complementation and symmetric difference. A σ-algebra is, in particular, a boolean algebra.

If $\mathscr{R}_1$ and $\mathscr{R}_2$ are two collections of subsets of a set X, we denote by $\mathscr{R}_1 \cap \mathscr{R}_2$, the collection of all subsets which belong to both $\mathscr{R}_1$ and $\mathscr{R}_2$. If $\{\mathscr{R}_\alpha\}$, $\alpha \in \Gamma$ is a family of collections of subsets of X, then we write $\bigcap_{\alpha \in \Gamma} \mathscr{R}_\alpha$ for the collection of all those subsets which belong to every $\mathscr{R}_\alpha$, $\alpha \in \Gamma$. If $A \subset X$, and $\mathscr{R}$ is a collection of subsets of X, then we write $\mathscr{R} \cap A$ for the family $\{B \cap A, B \in \mathscr{R}\}$.

Proposition. 13.2. If $\{\mathscr{B}_\alpha\}$, $\alpha \in \Gamma$ is a family of σ-algebras of subsets of a set X, then $\bigcap_{\alpha \in \Gamma} \mathscr{B}_\alpha$ is also a σ-algebra.

Proof. This follows immediately from the definitions.

Definition. 13.3. Let $\mathscr{R}$ be any collection of subsets of a set X.

Consider the family $\Gamma = \{\mathscr{B} : \mathscr{B}$ is a σ-algebra of subsets of X, $\mathscr{B} \supset \mathscr{R}\}$. Then the σ-algebra

$$\mathscr{B}(\mathscr{R}) = \bigcap_{\mathscr{B} \in \Gamma} \mathscr{B} \tag{13.1}$$

is called the σ-algebra *generated* by $\mathscr{R}$. It is also called the *smallest σ-algebra containing* $\mathscr{R}$. (It may be noted that Γ is nonempty because the σ-algebra of all subsets of X belongs to Γ.)

Proposition. 13.4. Let $\mathscr{R}$ be any collection of subsets of a set X. If $\mathscr{B}$ is any σ-algebra containing $\mathscr{R}$, then $\mathscr{B} \supset \mathscr{B}(\mathscr{R}) \supset \mathscr{R}$.

Proof. This is an immediate consequence of the preceding definition and Eq. (13.1).

Proposition. 13.5. For any collection $\mathscr{R}$ of subsets of a set X and any subset A of X,

$$\mathscr{B}(\mathscr{R}) \cap A = \mathscr{B}_A(\mathscr{R} \cap A),$$

where $\mathscr{B}_A(\mathscr{R} \cap A)$ denotes the σ-algebra generated by $\mathscr{R} \cap A$ in the set A.

Proof. First of all, we observe that $\mathscr{B}(\mathscr{R}) \cap A$ is, indeed, a σ-algebra of subsets in the space A. Hence by Proposition 13.4,

$$\mathscr{B}(\mathscr{R}) \cap A \supset \mathscr{B}_A(\mathscr{R} \cap A). \tag{13.2}$$

Let now $\mathscr{C}$ denote the class of all subsets of X of the form $B \cup (CA')$, where $B \in \mathscr{B}_A(\mathscr{R} \cap A)$ and $C \in \mathscr{B}(\mathscr{R})$. Then $\mathscr{C}$ is a σ-algebra in X. Indeed, it is easily seen that $\mathscr{C}$ is closed under countable union. It is enough to show that $\mathscr{C}$ is closed under complementation. Let $B \in \mathscr{B}_A(\mathscr{R} \cap A)$ and $C \in \mathscr{B}(\mathscr{R})$. We have

$$\begin{aligned}[B \cup (CA')]' &= B' \cap (C' \cup A) \\ &= B'C' \cup B'A \\ &= AB'C' \cup A'B'C' \cup AB' \\ &= AB' \cup C'A',\end{aligned}$$

because $B \subset A$. Thus $\mathscr{C}$ is a σ-algebra. Further, if $E \in \mathscr{R}$,

$$E = EA \cup EA'$$

and $EA \in \mathscr{R} \cap A \subset \mathscr{B}_A(\mathscr{R} \cap A)$. Thus $\mathscr{R} \subset \mathscr{C}$. By Proposition 13.4, $\mathscr{B}(\mathscr{R}) \subset \mathscr{C}$. Hence

$$\mathscr{B}(\mathscr{R}) \cap A \subset \mathscr{C} \cap A = \mathscr{B}_A(\mathscr{R} \cap A).$$

This together with Eq. (13.2) completes the proof.

Definition. 13.6. By a *borel space* we mean a pair $(X, \mathscr{B})$, where X is a set and $\mathscr{B}$ is a σ-algebra of subsets of X. (A borel space is, in particular, a boolean space). Any element of $\mathscr{B}$ is called a *measurable set*.

Remark. 13.7. Proposition 13.5 implies that whenever $(X, \mathscr{B})$ is a borel space and $Y \subset X$, then $(Y, \mathscr{B} \cap Y)$ is a new borel space. This is one method of constructing new borel spaces from given ones. Later we shall see other methods of constructing new borel spaces from given ones.

The term borel space is coined in honour of the French mathematician E. Borel, who first investigated measures on abstract σ-algebras.

Definition. 13.8. A σ-algebra $\mathscr{B}$ of subsets of a set X is said to be *countably generated* if there exists a countable family $\mathscr{R}$ of subsets of X such that $\mathscr{B} = \mathscr{B}(\mathscr{R})$.

Proposition. 13.9. Let $\mathscr{R}$ be any class of subsets of X. Then for any set $A \in \mathscr{B}(\mathscr{R})$ there exists a countable family $\mathscr{R}_1 \subset \mathscr{R}$ such that $A \in \mathscr{B}(\mathscr{R}_1)$.

Proof. Let the class $\mathscr{A}$ be defined by

$$\mathscr{A} = \{A \in \mathscr{B}(\mathscr{R}),\ A \in \mathscr{B}(\mathscr{R}_1) \text{ for some countable family } \mathscr{R}_1 \subset \mathscr{R}\}.$$

It is easy to check that $\mathscr{A}$ is a σ-algebra and $\mathscr{A} \supset \mathscr{R}$. Hence $\mathscr{A} \supset \mathscr{B}(\mathscr{R})$. Thus $\mathscr{A} = \mathscr{B}(\mathscr{R})$.

Remark. 13.10. The above mentioned argument is the commonest technique of proof used in measure theory. If it is necessary to prove a property p for every element of a σ-algebra $\mathscr{B}$ one tries to show that the class of all sets with property p is a σ-algebra which includes a family of sets generating $\mathscr{B}$. We shall illustrate by an example.

Proposition. 13.11. Let $\mathscr{E}_i$ be any family of subset of the set X_i such that $X_i \in \mathscr{E}_i$, $i=1, 2$. Let $\mathscr{E} = \{E_1 \times E_2,\ E_i \in \mathscr{E}_i,\ i=1, 2\}$. Let $\mathscr{B}_i$ be the σ-algebra generated by $\mathscr{E}_i$ and let $\mathscr{B}$ be the σ-algebra generated by $\mathscr{E}$. Then $\mathscr{B} = \widetilde{\mathscr{B}}$, where $\widetilde{\mathscr{B}}$ is the σ-algebra generated by $\{B_1 \times B_2,\ B_i \in \mathscr{B}_i,\ i=1, 2\}$.

Proof. From definitions we have $\mathscr{B} \subset \widetilde{\mathscr{B}}$. Let $E_2 \in \mathscr{E}_2$ and let

$$\mathscr{L}_1 = \{A : A \subset X_1,\ A \times E_2 \in \mathscr{B}\}.$$

If $A \in \mathscr{L}_1$, we have

$$A' \times E_2 = (A \times E_2)' \cap (X_1 \times E_2) \in \mathscr{B}.$$

Hence $A' \in \mathscr{L}_1$. If $A_1, A_2, \ldots \in \mathscr{L}_1$, then

$$(\bigcup_i A_i) \times E_2 = \bigcup_i (A_i \times E_2) \in \mathscr{B}.$$

Hence $\bigcup_i A_i \in \mathscr{L}_1$. Thus $\mathscr{L}_1$ is a σ-algebra containing $\mathscr{E}_1$. Thus

$$A \times E_2 \in \mathscr{B} \text{ if } A \in \mathscr{B}_1, E_2 \in \mathscr{E}_2.$$

Now, for $A \in \mathscr{B}_1$, let

$$\mathscr{L}_2 = \{B : B \subset X_2, A \times B \in \mathscr{B}\}.$$

As before $\mathscr{L}_2$ is a σ-algebra containing $\mathscr{E}_2$ and hence $\mathscr{B}_2$. Thus $A \times B \in \mathscr{B}$ whenever $A \in \mathscr{B}_1$, $B \in \mathscr{B}_2$. Hence $\tilde{\mathscr{B}} \subset \mathscr{B}$ and the proof is complete.

§14. Monotone Classes

It is impossible to give a constructive procedure for obtaining the σ-algebra generated by a class of subsets of a set X. This is the reason why σ-algebras are invariably specified by identifying a generating family of subsets. Of course, there are a few exceptions as in the following cases.

Example. 14.1. The most obvious example of a σ-algebra is the σ-algebra of all subsets of any set X. If X is uncountable, then the class $\{A : A \subset X, \text{ either } A \text{ or } A' \text{ is countable}\}$ is a σ-algebra. (However, σ-algebras of this kind are seldom useful from the point of view of probability theory).

It is easy to describe constructively the boolean algebra generated by a class $\mathscr{R}$ of subsets of a set X. Indeed, for any class $\mathscr{C}$ of subsets of X, let $\mathscr{C}^*$ denote the class of all finite unions of differences of sets in $\mathscr{C}$, i.e., any element $B \in \mathscr{C}^*$ is of the form

$$B = (C_1 D_1') \cup (C_2 D_2') \ldots \cup (C_k D_k'),$$

where $C_1, C_2, \ldots, C_k, D_1, D_2, \ldots, D_k \in \mathscr{C}$ and k is some positive integer. Let $\mathscr{R}_1$ be the class obtained by adding the whole space X to $\mathscr{R}$. Define the classes $\mathscr{R}_2, \mathscr{R}_3, \ldots$, by

$$\mathscr{R}_n = \mathscr{R}_{n-1}{}^*, \; n = 2, 3, \ldots$$

Let $\mathscr{F} = \bigcup_1^\infty \mathscr{R}_n$. It is clear that $\mathscr{F}$ is a boolean algebra containing $\mathscr{R}$. Further, if $\mathscr{F}_1$ is a boolean algebra containing $\mathscr{R}$ then $\mathscr{F}_1 \supset \mathscr{F}$, i.e., $\mathscr{F}$ is the smallest boolean algebra containing $\mathscr{R}$. Unfortunately, there is no such precise description of the smallest σ-algebra containing $\mathscr{R}$ without appealing to transfinite induction. However, there is another type of class, less restricted than a σ-algebra which is used often in the proofs of all basic theorems of measure theory.

Definition. 14.2. A collection $\mathcal{M}$ of subsets of a set X is called a *monotone class* if

(i) $E_1 \subset E_2 \subset \ldots, E_n \in \mathcal{M}, n=1, 2, \ldots$ implies that $\bigcup_i E_i \in \mathcal{M}$;

(ii) $E_1 \supset E_2 \supset \ldots, E_n \in \mathcal{M}, n=1, 2, \ldots$ implies that $\bigcap_i E_i \in \mathcal{M}$.

Remark. 14.3. Let $\mathcal{R}$ be any class of subsets of a set X. The class of all subsets of X is a monotone class. Intersection of any family of monotone classes is a monotone class. Hence intersection of *all* monotone classes of subsets of X, which include $\mathcal{R}$ is a monotone class. This is the *smallest monotone class* containing $\mathcal{R}$. It is denoted by $\mathcal{M}(\mathcal{R})$.

The following proposition is one of the most useful technical results of our subject.

Proposition. 14.4. Let $\mathcal{F}$ be a boolean algebra of subsets of X. Then $\mathcal{M}(\mathcal{F}) = \mathcal{B}(\mathcal{F})$.

Proof. First of all, we observe that a σ-algebra is also a monotone class. Hence $\mathcal{B}(\mathcal{F}) \supset \mathcal{M}(\mathcal{F})$. Since $\mathcal{M}(\mathcal{F}) \supset \mathcal{F}$, it follows that ϕ, $\emptyset$ belong to $\mathcal{M}(\mathcal{F})$. It is enough to show that $\mathcal{M}(\mathcal{F})$ is closed under complementation and countable union. Then $\mathcal{M}(\mathcal{F})$ would be a σ-algebra containing the smallest one, namely $\mathcal{B}(\mathcal{F})$. To this end, for any $E \subset X$, we introduce the collection

$$\mathcal{L}(E) = \{F : F \subset X, E \cup F, EF', FE' \in \mathcal{M}(\mathcal{F})\}.$$

Then $\mathcal{L}(E)$ is a monotone class and $F \in \mathcal{L}(E)$ if and only if $E \in \mathcal{L}(F)$. Let now $E \in \mathcal{F}$. Since $\mathcal{F}$ is a boolean algebra and $\mathcal{M}(\mathcal{F}) \supset \mathcal{F}$, it follows that $\mathcal{L}(E) \supset \mathcal{F}$ and hence $\mathcal{L}(E) \supset \mathcal{M}(\mathcal{F})$. Thus, for any $A \in \mathcal{M}(\mathcal{F})$, $E \in \mathcal{F}$, we have $E \in \mathcal{L}(A)$. This implies $\mathcal{F} \subset \mathcal{L}(A)$. Thus $\mathcal{M}(\mathcal{F}) \subset \mathcal{L}(A)$ for every $A \in \mathcal{M}(\mathcal{F})$. In other words $\mathcal{M}(\mathcal{F})$ is a boolean algebra. Since it is a monotone class, it is a σ-algebra. This completes the proof.

§15. Measures on Boolean Semi-Algebras and Algebras

Let $\mathcal{D}$ be a boolean semi-algebra. A map $\mu : \mathcal{D} \to [0, \infty]$ is called a *measure* on $\mathcal{D}$ if the following conditions are satisfied:

(i) if $A_1, A_2, \ldots,$ are disjoint elements of $\mathcal{D}$ and $\bigcup A_i \in \mathcal{D}$, then
$$\mu(\bigcup_i A_i) = \sum_i \mu(A_i);$$

(ii) $\mu(\emptyset) = 0$.

Since boolean algebras are also boolean semi-algebras the notion of measure is defined on them, too.

Proposition. 15.1. Let $\mu : \mathscr{D} \to [0, \infty]$ be a measure on the boolean semi-algebra $\mathscr{D}$. Let $\mathscr{F}$ be the boolean algebra of all finite unions of pairwise disjoint sets in $\mathscr{D}$. For any $E \in \mathscr{F}$, $E = \bigcup_{i=1}^{k} A_i$ and $A_i \cap A_j = \emptyset$ for $i \neq j$, define

$$\mu(E) = \sum_{i=1}^{k} \mu(A_i).$$

Then μ is well defined on $\mathscr{F}$ and μ is a measure on $\mathscr{F}$.

Proof. That μ is well defined and finitely additive is already proved in Proposition 4.7 and Remark 4.8. We shall establish countable additivity. Let $A, A_1, A_2, \ldots, \in \mathscr{F}$, $A = \bigcup_{1}^{\infty} A_n$, where the A_n's are disjoint. Let $A = \bigcup_{i=1}^{k} B_i$, $A_n = \bigcup_{j=1}^{k_n} B_{nj}$ be partitions of A and A_n respectively into disjoint elements of $\mathscr{D}$. We have

$$\mu(A) = \sum_{i=1}^{k} \mu(B_i),$$

$$\mu(A_n) = \sum_{j=1}^{k_n} \mu(B_{nj}).$$

Since

$$B_i = \bigcup_{n=1}^{\infty} \bigcup_{j=1}^{k_n} (B_i \cap B_{nj}),$$

and $B_i, B_i \cap B_{nj} \in \mathscr{D}$, the countable additivity of μ on $\mathscr{D}$ implies that

$$\mu(B_i) = \sum_{n=1}^{\infty} \sum_{j=1}^{k_n} \mu(B_i \cap B_{nj}).$$

Adding over i and noting that an infinite series of non-negative terms can be added in any order, we have

$$\mu(A) = \sum_{i} \mu(B_i) = \sum_{n=1}^{\infty} \left\{ \sum_{i=1}^{k} \sum_{j=1}^{k_n} \mu(B_i \cap B_{nj}) \right\}$$

$$= \sum_{n=1}^{\infty} \mu(A_n).$$

This completes the proof.

Proposition. 15.2. Let $\mathscr{D}$ be the boolean semi-algebra of all intervals of the form $(-\infty, +\infty)$ $(-\infty, a]$, $(a, b]$, $(b, +\infty)$, where a, b take all values in the real line R. Let μ be a measure on $\mathscr{D}$ such that $\mu((a, b]) < \infty$ for all $a, b \in R$. Then there exists a monotonic increasing right continuous function F on R such that

$$\mu((a, b]) = F(b) - F(a) \text{ for all } a, b \in R. \tag{15.1}$$

Conversely, if F is a real valued monotonic increasing right continuous function on R, then there exists a measure μ on $\mathscr{D}$ such that Eq. (15.1) is fulfilled.

Proof. Let μ be a measure on $\mathscr{D}$ such that $\mu((a, b]) < \infty$ for all a, b. Define $F(x)$ by

$$\begin{aligned} F(x) &= \mu((a, x]) && \text{if } x > a, \\ &= 0 && \text{if } x = a, \\ &= -\mu((x, a]) && \text{if } x < a, \end{aligned}$$

where a is any fixed real number. Let $x \geqslant a$ and let x_n be a sequence decreasing to the limit x. Then

$$(a, x_1] = (a, x] \cup (x_2, x_1] \cup (x_3, x_2] \cup \ldots \cup (x_n, x_{n-1}] \cup \ldots$$

Since μ is a measure on $\mathscr{D}$,

$$\begin{aligned} F(x_1) &= F(x) + [F(x_1) - F(x_2)] + \ldots + [F(x_{n-1}) - F(x_n)] + \ldots \\ &= \lim_{n \to \infty} [F(x) + F(x_1) - F(x_n)]. \end{aligned}$$

Hence $F(x) = \lim_{n \to \infty} F(x_n)$. Now let $x < a$ and let x_n descend monotonically to x. Without loss of generality we may assume that $x < x_1 < a$. We have

$$(x, a] = (x_1, a] \cup (x_2, x_1] \cup \ldots \cup (x_n, x_{n-1}] \cup \ldots$$

Hence

$$\begin{aligned} \mu((x, a]) &= \mu((x_1, a]) + \mu((x_2, x_1]) + \ldots + \mu((x_n, x_{n-1}]) + \ldots \\ &= \lim_{n \to \infty} \mu((x_n, a]) \\ &= \lim_{n \to \infty} F(a) - F(x_n). \end{aligned}$$

Since $\mu((x, a]) = F(a) - F(x)$, it follows that $\lim_{n \to \infty} F(x_n) = F(x)$. Thus F is a right continuous function. That F is monotonic increasing is obvious.

Conversely, let F be a monotonic increasing right continuous funtion. For any interval of the form $(a, b]$, where a, b are finite, define μ by the Equation (15.1). Define

$$\mu((a, \infty)) = \lim_{x \to +\infty} [F(x) - F(a)],$$

$$\mu((-\infty, a]) = \lim_{x \to -\infty} [F(a) - F(x)], \text{ for all } a \in R,$$

$$\mu((-\infty, \infty)) = \lim_{x \to +\infty} [F(x) - F(-x)].$$

From Example 2.4 it is clear that μ is finitely additive. We shall now prove countable additivity. Let $(a, b] = \bigcup_{n=1}^{\infty} (a_n, b_n]$ be a countable disjoint union of left open right closed intervals. Let $\varepsilon > 0$, $\delta > 0$ be arbitrary. Since F is right continuous we can choose $\varepsilon_k > 0$ such that

$$F(b_k + \varepsilon_k) - F(b_k) < \frac{\varepsilon}{2^k} \text{ for every } k. \tag{15.2}$$

The closed interval $[a+\delta, b]$ is covered by the open intervals $(a_k, b_k+\varepsilon_k)$, $k = 1, 2, \ldots$ Hence, by Heine-Borel theorem, there exists an integer N such that

$$(a+\delta, b] \subset [a+\delta, b] \subset \bigcup_{k=1}^{N} (a_k, b_k+\varepsilon_k) \subset \bigcup_{k=1}^{N} (a_k, b_k+\varepsilon_k].$$

Since finite additivity implies finite sub-additivity we have by Eq. (15.2),

$$\begin{aligned} F(b) - F(a+\delta) &\leqslant \sum_{k=1}^{N} [F(b_k+\varepsilon_k) - F(a_k)] \\ &\leqslant \sum_{k=1}^{\infty} [F(b_k) - F(a_k)] + \sum_{k=1}^{\infty} \frac{\varepsilon}{2^k} \\ &\leqslant \sum_{k=1}^{\infty} [F(b_k) - F(a_k)] + \varepsilon. \end{aligned}$$

Letting $\delta \to 0$ and $\varepsilon \to 0$ and using right continuity of F, we have

$$F(b) - F(a) \leqslant \sum_{k=1}^{\infty} [F(b_k) - F(a_k)]. \tag{15.3}$$

Proposition 2.5, Remark 4.8, and the fact that

$$(a, b] \supset \bigcup_{k=1}^{n} (a_k, b_k]$$

imply that

$$F(b)-F(a) \geqslant \sum_{1}^{n} [F(b_k)-F(a_k)] \text{ for every } n.$$

Hence

$$F(b)-F(a) \geqslant \sum_{1}^{\infty} [F(b_k)-F(a_k)]. \tag{15.4}$$

Now (15.3) and (15.4) imply that μ is countably additive if

$$\bigcup_{i=1}^{\infty} (a_i, b_i] = (a, b], \text{ where } -\infty < a < b < +\infty.$$

It is fairly easy to show that for any $I \in \mathscr{D}$,

$$\mu(I) = \sum_{n=-\infty}^{+\infty} \mu(I \cap (n, n+1]).$$

If now $I = \bigcup_{1}^{\infty} I_j$, where $I, I_1, I_2, \ldots \in \mathscr{D}$ and I_j's are disjoint, then by what has already been proved we have

$$\begin{aligned} \Sigma\, \mu(I_j) &= \sum_{j=1}^{\infty} \sum_{n=-\infty}^{+\infty} \mu(I_j \cap (n, n+1]) \\ &= \sum_{n=-\infty}^{+\infty} \mu(I \cap (n, n+1]) \\ &= \mu(I). \end{aligned}$$

This shows that μ is a measure on $\mathscr{D}$ and completes the proof.

Remark. 15.3. Since there are monotonic increasing functions which are not right continuous it follows that there exist finitely additive non-negative functions which are not measures on $\mathscr{D}$. If F_1 and F_2 are two monotonic increasing and right continuous functions on R such that

$$\mu((a, b]) = F_1(b)-F_1(a) = F_2(b)-F_2(a)$$

for all a and b, it follows that there exists a constant α such that

$$F_2(x) = F_1(x)+\alpha \text{ for all } x.$$

If μ is a measure on $\mathscr{D}$ such that $\mu(R) = \rho < \infty$, then we can choose a monotonic increasing right continuous function F on R such that

$$\lim_{x \to -\infty} F(x) = 0,$$

$$\lim_{x \to +\infty} F(x) = \rho,$$

$$F(b) - F(a) = \mu((a, b]) \text{ for all } a, b \in R.$$

Such a function F is unique. If $\rho = 1$, such a function is called a *probability distribution function.*

Combining Proposition 15.1 and Proposition 15.2 we obtain a large class of measures on the boolean algebra generated by intervals of the form $(a, b]$, $(-\infty, a]$, (a, ∞) and $(-\infty, \infty)$ on the real line. Indeed, they are determined by monotonic increasing right continuous functions. We shall now describe another method of constructing boolean algebras and measures on them.

Proposition. 15.4. Let X be any compact topological space and let $\mathscr{F}$ be the class of all subsets of X which are both open and closed. Then $\mathscr{F}$ is a boolean algebra and any finitely additive measure on $\mathscr{F}$ is countably additive.

Proof. It is obvious that $\mathscr{F}$ is a boolean algebra. Let now $A \in \mathscr{F}$ and let $B_1, B_2, \ldots$ be a sequence of disjoint elements of $\mathscr{F}$ such that $A = \bigcup_{i=1}^{\infty} B_i$. Since A is closed, it is compact. Since B_i's are open, a finite number of them cover A. Since $B_i \cap B_j = \emptyset$ for $i \neq j$ there exists n_0 such that $B_n = \emptyset$ for all $n \geqslant n_0$. Thus $A = \bigcup_{i=1}^{n_0} B_i$. Hence for any finitely additive measure μ, we have

$$\mu(A) = \sum_{i=1}^{n_0} \mu(B_i) = \sum_{i=1}^{\infty} \mu(B_i).$$

In other words μ is countably additive and the proof is complete.

Example. 15.5. Let X_i be the finite set $\{1, 2, \ldots, N_i\}$ for $i = 1, 2, \ldots$ and let X^∞ be the Cartesian product of the X_i's, i.e., $X^\infty = \prod_{i=1}^{\infty} X_i$. Let each X_i be assigned the discrete topology (i.e., when every subset is declared open) and X^∞ the product topology. Then X^∞ is a compact topological space. For any $a_j \in X_j$, $j = 1, 2, \ldots, k$, let

$$C_{a_1, a_2, \ldots, a_k} = \{\mathbf{x} : \mathbf{x} = (x_1, x_2, \ldots), x_i \in X_i, i = 1, 2, \ldots;$$
$$x_j = a_j, j = 1, 2, \ldots, k\}.$$

Such sets may be called *elementary cylinder sets.* Together with the empty set and whole space they constitute a boolean semi-algebra. Finite unions of such elementary cylinder sets are the finite dimensional cylinder sets. This is also the boolean algebra of all sets which are both open and closed in the topological space X^{∞}. (It may be noted that open sets in X^{∞} could depend on countably many coordinates.)

For every finite sequence $(a_1, a_2, \ldots, a_k)$, $a_j \in X_j$, $j=1, 2, \ldots, k$ and $k=1, 2, 3, \ldots$ let $p(a_1, a_2, \ldots, a_k)$ be non-negative and let the family $\{p(a_1, a_2, \ldots, a_k)\}$ satisfy the following conditions:

$$p(a_1, a_2, \ldots, a_{k-1}) = \sum_{n=1}^{N_k} p(a_1, a_2, \ldots, a_{k-1}, n),$$

$$\sum_{n=1}^{N_1} p(n)=1. \tag{15.5}$$

Then there exists a countably additive probability measure μ on the boolean algebra $\mathscr{F}$ of all finite dimensional cylinder sets in X^{∞}, such that

$$\mu(C_{a_1, a_2, \ldots, a_k})=p(a_1, a_2, \ldots, a_k),$$
$$a_j \in X_j, j = 1, 2, \ldots, k; k = 1, 2, 3,\ldots \tag{15.6}$$

Indeed, if we define μ for elementary cylinder sets by Eq. (15.6) and put $\mu(\varnothing)=0$ and $\mu(X)=1$, μ is finitely additive on the boolean semi-algebra of cylinder sets. Thus μ extends naturally to $\mathscr{F}$ and by Proposition 15.4, μ is countably additive on $\mathscr{F}$.

As special examples of Equation (15.5) we may consider the following cases.

Example. 15.6. Let $q_{r1}, q_{r2}, \ldots, q_{rN_r}$ be non-negative numbers such that $\sum_{j=1}^{N_r} q_{rj} = 1$, for $r = 1, 2, \ldots,$ Let

$$p(a_1, a_2, \ldots, a_k)= \prod_{i=1}^{k} q_{ia_i} \tag{15.7}$$

Then p satisfies the equations (15.5) and hence there exists a countably additive measure μ on $\mathscr{F}$, satisfying Eq. (15.6). Measures defined in this manner describe a 'sequence' of independent simple random variables. The case when each X_i is the two point space $\{0, 1\}$ and $q_{r0}=q$, $q_{r1}=p$, $p+q=1$ is said to describe a sequence of independent *Bernoulli trials.*

Example. 15.7. For every $r = 1, 2, \ldots$, let P_r be an $N_r \times N_{r+1}$ matrix given by

$$P_r = \begin{pmatrix} p^{(r)}_{11} & p^{(r)}_{12} & \cdots & p^{(r)}_{1N_{r+1}} \\ p^{(r)}_{21} & p^{(r)}_{22} & \cdots & p^{(r)}_{2N_{r+1}} \\ \cdots & \cdots & \cdots & \cdots \\ p^{(r)}_{N_r1} & p^{(r)}_{N_r2} & \cdots & p^{(r)}_{N_rN_{r+1}} \end{pmatrix} \tag{15.8}$$

such that every element of P_r is non-negative and all the row totals are equal to unity. Let p_i, $i=1, 2, \ldots, N_1$ be non-negative numbers which add upto unity. Let

$$p(a_1, a_2, \ldots, a_k) = p_{a_1} p^{(1)}_{a_1 a_2} p^{(2)}_{a_2 a_3} \ldots p^{(k-1)}_{a_{k-1} a_k} \tag{15.9}$$

Such a function satisfies Eq. (15.5). Hence there exists a probability measure μ on $\mathscr{F}$ such that $\mu(C_{a_1, a_2, \ldots, a_k})$ is given by Eq. (15.6). Probability measures defined in this manner are used to describe a sequence of simple random variables forming a *Markov chain.* The matrix P_r is called the *transition probability matrix* at time r.

Definition. 15.8. A triple $(X, \mathscr{F}, \mu)$ where $(X, \mathscr{F})$ is a boolean space and μ is a measure on $\mathscr{F}$ is called a boolean measure space. The measure μ is said to be *σ-finite* if there exists a sequence $\{A_n\}$ of sets such that $X = \bigcup_{n=1}^{\infty} A_n$, $A_n \in \mathscr{F}$ and $\mu(A_n) < \infty$ for $n=1, 2, 3, \ldots$. It is said to be *totally finite* if $\mu(X) < \infty$. It is called a *probability measure* if $\mu(X)=1$.

Proposition. 15.9. Let $(X, \mathscr{F})$ be a boolean space and let μ be a finitely additive and countably subadditive map on $\mathscr{F}$ with values in $[0, \infty]$. Then μ is a measure. Conversely, every measure is countably sub-additive.

Proof. Let $A_1, A_2, \ldots$ be disjoint elements of $\mathscr{F}$ such that $B = \bigcup_i A_i \in \mathscr{F}$. Then for any positive integer n

$$\mu(B) \geqslant \mu(\bigcup_{i=1}^{n} A_i) = \sum_{i=1}^{n} \mu(A_i).$$

Letting $n \to \infty$, we have

$$\mu(B) \geqslant \sum_{i=1}^{\infty} \mu(A_i).$$

The reverse inequality is ensured by countable sub-additivity. The converse follows from Remark 2.6. This completes the proof.

Proposition. 15.10. Let $(X, \mathscr{F}, \mu)$ be a boolean measure space. If $A_1 \subset A_2 \subset \ldots$ is an increasing sequence of sets in $\mathscr{F}$ and $\bigcup_{n=1}^{\infty} A_n \in \mathscr{F}$, then

$$\mu(\bigcup_n A_n) = \lim_{n \to \infty} \mu(A_n). \tag{15.10}$$

If $A_1 \supset A_2 \supset \ldots$ is a decreasing sequence of sets in $\mathscr{F}$, $\bigcap_n A_n \in \mathscr{F}$ and $\mu(A_k) < \infty$ for some k, then

$$\mu(\bigcap_n A_n) = \lim_{n \to \infty} \mu(A_n). \tag{15.11}$$

Proof. Let $\{A_n\}$ be increasing. If $\mu(A_n) = \infty$ for some $n = n_0$, then $\mu(A_n) = \infty$ for all $n \geqslant n_0$ and hence $\mu(\bigcup_n A_n) = \lim_{n \to \infty} \mu(A_n) = \infty$. Let $\mu(A_n) < \infty$ for all n. Since

$$\bigcup_n A_n = A_1 \cup (A_2 - A_1) \cup \ldots \cup (A_n - A_{n-1}) \cup \ldots,$$

and $\mu(A_n) - \mu(A_{n-1}) = \mu(A_n - A_{n-1})$, we have from the disjointness of $A_1, A_2 - A_1, \ldots, A_n - A_{n-1}, \ldots,$

$$\begin{aligned} \mu(\bigcup_n A_n) &= \sum_{n=1}^{\infty} [\mu(A_n) - \mu(A_{n-1})] \\ &= \lim_{n \to \infty} \mu(A_n), \end{aligned}$$

where A_0 is defined as the empty set. This proves Equation (15.10).

To prove Eq. (15.11) we may assume, without loss of generality, that $\mu(A_1) < \infty$. Put $B_n = A_1 - A_n$. Then B_n increases to the set $A_1 - \bigcap_n A_n$, i.e., $\bigcup_n B_n = A_1 - \bigcap_n A_n$. Hence by the first part of the proposition we have

$$\begin{aligned} \mu(A_1 - \bigcap_n A_n) &= \mu(A_1) - \mu(\bigcap_n A_n) = \\ \mu(\bigcup_n B_n) &= \lim_{n \to \infty} \mu(B_n) = \\ \lim_{n \to \infty} &[\mu(A_1) - \mu(A_n)]. \end{aligned}$$

Thus

$$\mu(\bigcap_n A_n) = \lim_{n \to \infty} \mu(A_n).$$

This completes the proof.

Remark. 15.11. Let $F(x) \equiv x$ in Proposition 15.2 and let μ be the measure determined by this monotonic increasing function (which is continuous). Suppose $A_n = (n, \infty)$, $n = 1, 2, \ldots$ Then $\mu(A_n) = \infty$ for every n and $\bigcap_n A_n = \varnothing$. This shows that we cannot remove the condition that $\mu(A_k) < \infty$ for some k in the second part of the above proposition.

§16. Extension of Measures to σ-Algebras

In the preceding section we constructed examples of boolean measure spaces where the boolean algebra itself was not a σ-algebra. As mentioned earlier in Remark 11.7, one of our aims is to extend the idea of measures on boolean algebras to larger classes of sets, namely σ-algebras. Since σ-algebras are closed under countable set operations it is natural to study countably additive functions on them. The aim of the present section is to construct examples of measures on σ-algebras.

Throughout this section let $(X, \mathscr{F}, \mu)$ be a fixed boolean measure space. For any $E \subset X$, let

$$\mu^*(E) = \inf \left\{ \sum_{i=1}^{\infty} \mu(F_i), F_i \in \mathscr{F}, \bigcup_i F_i \supset E \right\}. \tag{16.1}$$

Thus μ^* is defined for every subset E of X. The set function μ^* is called the *outer measure* determined by μ. $\mu^*(E)$ is called the outer measure of the set E. In a sense we have attempted to measure the size of any set E given the measures of sets in $\mathscr{F}$. However, μ^* does not always turn out to be countably additive on the class of all subsets of X. Thus we shall try to look for a large class of sets including $\mathscr{F}$, where μ^* happens to be countably additive.

Proposition. 16.1. For any $E \subset X$,

$$\mu^*(E) = \inf \left\{ \sum_{i=1}^{\infty} \mu(F_i), F_i \in \mathscr{F}, F_i\text{'s disjoint}, \bigcup_i F_i \supset E \right\}$$

and

$$\mu^*(F) = \mu(F) \text{ for every } F \in \mathscr{F}.$$

Proof. Let $F_i \in \mathscr{F}$, $i = 1, 2, \ldots$, be such that $\bigcup_{i=1}^{\infty} F_i \supset E$. Define $\tilde{F}_1 = F_1$, $\tilde{F}_i = F_i - \bigcup_{j<i} F_j$. Then $\tilde{F}_1, \tilde{F}_2, \ldots$, are disjoint and $\bigcup_i \tilde{F}_i = \bigcup_i F_i \supset E$. Since $\mathscr{F}$ is a boolean algebra each $\tilde{F}_i \in \mathscr{F}$. Further $\tilde{F}_i \subset F_i$ and hence

$$\sum_{i=1}^{\infty} \mu(\tilde{F}_i) \leqslant \sum_{i=1}^{\infty} \mu(F_i).$$

By the definition of infimum the proof of the first part is complete. Since μ is countably additive on $\mathscr{F}$, the second part follows from the first part.

Proposition. 16.2. The outer measure μ^* is countably sub-additive, i.e., for any sequence $\{E_n\}$ of subsets of X,

$$\mu^*(\bigcup_1^\infty E_n) \leqslant \sum_{n=1}^\infty \mu^*(E_n).$$

Proof. Note that when the right hand side is ∞, there is nothing to prove. Assume the contrary and for every n, choose a sequence $F_{n1}, F_{n2},\ldots$ of sets in $\mathscr{F}$ such that

$$\bigcup_{j=1}^\infty F_{nj} \supset E_n,$$

$$\sum_{j=1}^\infty \mu(F_{nj}) \leqslant \mu^*(E_n) + \frac{\varepsilon}{2^n},$$

where ε is any arbitrarily chosen and fixed positive number. Hence

$$\bigcup_{n=1}^\infty \bigcup_{j=1}^\infty F_{nj} \supset \bigcup_{n=1}^\infty E_n$$

and

$$\sum_n \sum_j \mu(F_{nj}) \leqslant [\sum_n \mu^*(E_n)] + \varepsilon.$$

Since the family $\{F_{nj}, j=1, 2,\ldots; n=1, 2\ldots\}$ is a countable covering of $\bigcup_n E_n$, we have

$$\mu^*(\bigcup_n E_n) \leqslant \sum_n \mu^*(E_n) + \varepsilon$$

Since ε is arbitrary, the proof is complete.

Definition. 16.3. A set $E \subset X$ is said to be *μ^*-measurable* if, for every $A \subset X$,

$$\mu^*(A) = \mu^*(AE) + \mu^*(AE'). \tag{16.2}$$

Remark. 16.4. Since μ^* is sub-additive it follows that a set E is μ^*-measurable if and only if, for every $A \subset X$,

$$\mu^*(A) \geqslant \mu^*(AE) + \mu^*(AE').$$

Exercise. 16.5. If $A \subset B$, then $\mu^*(A) \leqslant \mu^*(B)$.

Exercise. 16.6. If $\mu^*(E)=0$, then E is μ^*-measurable.

Proposition. 16.7. Let $\mathscr{B}^*$ denote the class of all μ^*-measurable sets. Then $\mathscr{B}^*$ is a boolean algebra. Further $\mathscr{B}^* \supset \mathscr{F}$.

Proof. Since Eq. (16.2) is symmetric in E and E', it follows that $\mathscr{B}^*$ is closed under complementation. Let now $E, F \in \mathscr{B}^*$. Replacing A by $A \cap (E \cup F)$ in (16.2), we have

$$\mu^*(A \cap (E \cup F)) = \mu^*(AE) + \mu^*(AE'F). \tag{16.3}$$

Substituting AE' for A and F for E in Eq. (16.2), we get

$$\mu^*(AE') = \mu^*(AE'F) + \mu^*(AE'F'). \tag{16.4}$$

Combining Eqs. (16.3) and (16.4) we have

$$\mu^*(A \cap (E \cup F)) + \mu^*(A \cap (E \cup F)') =$$
$$\mu^*(AE) + \mu^*(AE') = \mu^*(A).$$

Thus $E \cup F \in \mathscr{B}^*$. This shows that $\mathscr{B}^*$ is a boolean algebra.

Now let $A \subset X$, and let $F \in \mathscr{F}$. Then, for any $\varepsilon > 0$ there exists a sequence $\{F_n\}$ of disjoint sets in $\mathscr{F}$ such that $\bigcup_n F_n \supset A$ and

$$\begin{aligned} \varepsilon + \mu^*(A) &\geqslant \sum_n \mu(F_n) \\ &= \sum_n \mu(F_n F) + \sum_n \mu(F_n F'). \end{aligned}$$

Since $\bigcup_n F_n F \supset AF$ and $\bigcup_n F_n F' \supset AF'$, we have

$$\varepsilon + \mu^*(A) \geqslant \mu^*(AF) + \mu^*(AF').$$

The arbitrariness of ε implies that $F \in \mathscr{B}^*$ and completes the proof.

Proposition. 16.8. $\mathscr{B}^*$ is a σ-algebra which includes the smallest σ-algebra $\mathscr{B}(\mathscr{F})$ generated by $\mathscr{F}$. Further μ^* is countably additive on $\mathscr{B}^*$.

Proof. Let $E_1, E_2, \ldots, E_n, \ldots$, be a sequence of disjoint sets in $\mathscr{B}^*$. For any $A \subset X$, we have

$$\begin{aligned} \mu^*(A) &= \mu^*(AE_1) + \mu^*(AE'_1) \\ &= \mu^*(AE_1) + \mu^*(AE_2) + \mu^*(AE'_2 E'_1) \\ &= \ldots\ldots\ldots\ldots\ldots\ldots\ldots\ldots \\ &= \left[\sum_1^n \mu^*(AE_i)\right] + \mu^*(A'E_n E'_{n-1} \ldots E'_1) \\ &\geqslant \left[\sum_1^n \mu^*(AE_i)\right] + \mu^*(A \cap (\bigcup_1^\infty E_i)'). \end{aligned}$$

Since n is arbitrary and μ^* is countably sub-additive we have

$$\mu^*(A) = \mu^*(A \cap (\bigcup_1^\infty E_i)) + \mu^*(A \cap (\bigcup_1^\infty E_i)')$$

$$= [\sum_1^\infty \mu^*(AE_i)] + \mu^*(A \cap (\bigcup_1^\infty E_i)').$$

The first equation implies that $\bigcup_1^\infty E_i \in \mathscr{B}^*$. Thus $\mathscr{B}^*$ is closed under countable disjoint union. Since $\mathscr{B}^*$ is a boolean algebra it follows that $\mathscr{B}^*$ is a σ-algebra. The second equation implies (after putting $A = \bigcup_1^\infty E_i$) that μ^* is countably additive on $\mathscr{B}^*$. Proposition 16.7 implies that $\mathscr{B}^* \supset \mathscr{B}(\mathscr{F})$. The proof is complete.

Propositions 16.7 and 16.8 imply that any measure μ defined on a boolean algebra $\mathscr{F}$ can be extended to the σ-algebra $\mathscr{B}(\mathscr{F})$ generated by $\mathscr{F}$. Coupling this with Proposition 15.1, we get the following corollary.

Corollary. 16.9. (Extension theorem) Let $\mu : \mathscr{D} \to [0, \infty]$ be a measure on a boolean semi-algebra $\mathscr{D}$ of subsets of a set X. Then μ can be extended to a measure on the σ-algebra generated by $\mathscr{D}$, i.e., there exists a measure ν on the σ-algebra $\mathscr{B}(\mathscr{D})$ such that

$$\nu(E) = \mu(E) \text{ for all } E \in \mathscr{D}.$$

Now that we know the existence of measures on σ-algebras we introduce the following definition.

Definition. 16.10. Let $(X, \mathscr{B})$ be a borel space and let μ be a measure on $\mathscr{B}$. Then the triple $(X, \mathscr{B}, \mu)$ is called a *measure space*. It is called a *σ-finite measure space* if μ is σ-finite on $\mathscr{B}$. It is called a *totally finite measure space* if $\mu(X) < \infty$. It is called a *probability space* if $\mu(X) = 1$.

Throughout our text we shall consider only σ-finite measure spaces.

§17. Uniqueness of Extensions of Measures

Having convinced ourselves that every measure on a boolean algebra $\mathscr{F}$ of sets has an extension to the σ-algebra $\mathscr{B}(\mathscr{F})$ generated by $\mathscr{F}$, it is only natural to ask whether such an extension is unique. Our next proposition answers this question in the affirmative.

Proposition. 17.1. Let $(X, \mathscr{F}, \mu)$ be a σ-finite boolean measure space. If μ_1 and μ_2 are two measures on $\mathscr{B}(\mathscr{F})$ such that $\mu_1(F) = \mu_2(F) = \mu(F)$ for all $F \in \mathscr{F}$, then $\mu_1(E) = \mu_2(E)$ for all $E \in \mathscr{B}(\mathscr{F})$.

Proof. First of all, we shall prove the proposition when μ is totally finite. To this end, let

$$\mathscr{M} = \{E : E \in \mathscr{B}(\mathscr{F}),\ \mu_1(E) = \mu_2(E)\}.$$

By hypothesis, $\mathscr{B}(\mathscr{F}) \supset \mathscr{M} \supset \mathscr{F}$. By Proposition 15.10, $\mathscr{M}$ is a monotone class. By Proposition 14.4, $\mathscr{M} \supset \mathscr{B}(\mathscr{F})$. Thus $\mathscr{M} = \mathscr{B}(\mathscr{F})$. Now let μ be σ-finite. Then we can write $X = \bigcup_{i=1}^{\infty} X_i$, where X_i are disjoint elements of $\mathscr{F}$ with $\mu(X_i) < \infty$ for each i. Then μ_1 and μ_2 agree on the σ-algebra $\mathscr{B}(\mathscr{F}) \cap X_i = \mathscr{B}_{X_i}(\mathscr{F} \cap X_i)$ in the space X_i. (See Proposition 13.5). If $A \in \mathscr{B}(\mathscr{F})$, we have $A = \bigcup_i (A \cap X_i)$ and $\mu_1(A) = \sum_i \mu_1(A \cap X_i) = \sum_i \mu_2(A \cap X_i) = \mu_2(A)$. This completes the proof.

Corollary. 17.2. Let $(X, \mathscr{B})$ be a borel space. If μ_1, μ_2 are two measures on $\mathscr{B}$, $\mathscr{D}$ is a boolean semi-algebra such that $\mathscr{B} = \mathscr{B}(\mathscr{D})$ and the restrictions of μ_1 and μ_2 to $\mathscr{D}$ are equal and σ-finite on $\mathscr{D}$, then $\mu_1 = \mu_2$.

§18. Extension and Completion of Measures

In Section 16, we showed how a measure on a boolean algebra $\mathscr{F}$ can be extended (with the help of the outer measure) to the σ-algebra generated by it. In the process we extended it to the σ-algebra $\mathscr{B}^*$ of μ^*-measurable sets. It is natural to ask how big is $\mathscr{B}^*$ compared to $\mathscr{B}(\mathscr{F})$. We shall establish that $\mathscr{B}^*$ cannot be ' much ' larger than $\mathscr{B}(\mathscr{F})$. To this end we introduce a definition.

Definition. 18.1. Let $(X, \mathscr{B}, \mu)$ be a measure space. A subset E of X is said to be *μ-null* if there exists $A \in \mathscr{B}$ such that $E \subset A$, $A \in \mathscr{B}$ and $\mu(A) = 0$. The measure space is said to be *complete* if every μ-null set belongs to $\mathscr{B}$.

Exercise. 18.2. The class of all μ-null sets is closed under countable unions.

Proposition. 18.3. Let $(X, \mathscr{B}, \mu)$ be a measure space and let $\overline{\mathscr{B}} = \{E \Delta N, E \in \mathscr{B}, N \text{ is } \mu\text{-null}\}$. Then $\overline{\mathscr{B}}$ is a σ-algebra. If $\bar{\mu}$ is defined on $\overline{\mathscr{B}}$ by $\bar{\mu}(E \Delta N) = \mu(E)$ whenever $E \in \mathscr{B}$ and N is μ-null, then $\bar{\mu}$ is well-defined and $(X, \overline{\mathscr{B}}, \bar{\mu})$ is a complete measure space. (This is called the *completion of* $(X, \mathscr{B}, \mu)$).

Proof. If $E \in \mathscr{B}$, $N \subset A$, $A \in \mathscr{B}$ and $\mu(A) = 0$, then the identities

$$E \cup N = (E - A) \Delta [A \cap (E \cup N)],$$
$$E \Delta N = (E - A) \cup [A \cap (E \Delta N)],$$

show that

$$\overline{\mathscr{B}} = \{E \cup N, E \in \mathscr{B}, N \text{ is } \mu\text{-null}\}. \tag{18.1}$$

Hence $\overline{\mathscr{B}}$ is closed under countable union. Since $(E \Delta N)' = E' \Delta N$, it follows that $\overline{\mathscr{B}}$ is closed under complementation. Thus $\overline{\mathscr{B}}$ is a σ-algebra. To show that $\bar{\mu}$ is well defined, let $E_1 \Delta N_1 = E_2 \Delta N_2$, where $E_1, E_2 \in \mathscr{B}$ and N_1, N_2 are μ-null. Since Δ is an associative and commutative operation, we have

$$(E_1 \Delta E_2) \Delta (N_1 \Delta N_2) = (E_1 \Delta N_1) \Delta (E_2 \Delta N_2) = \varnothing.$$

Hence

$$(E_1 \Delta E_2) = (N_1 \Delta N_2).$$

Since $N_1 \Delta N_2 \subset N_1 \cup N_2$, it follows that $N_1 \Delta N_2$ is μ-null. Thus $\mu(E_1 \Delta E_2) = 0$. In other words $\mu(E_1) = \mu(E_2)$. Thus

$$\bar{\mu}(E_1 \Delta N_1) = \bar{\mu}(E_2 \Delta N_2).$$

Now Eq. (18.1) implies that $\bar{\mu}$ is countably additive on $\overline{\mathscr{B}}$. The last part is obvious and the proof is complete.

Proposition. 18.4. Let $(X, \mathscr{F}, \mu)$ be a boolean measure space and let $A \subset X$ be such that $\mu^*(A) < \infty$. Then there exists a set $B \in \mathscr{B}(\mathscr{F})$ such that (i) $A \subset B$; (ii) $\mu^*(A) = \mu^*(B)$; (iii) $\mu^*(C) = 0$, for every C such that $C \in \mathscr{B}(\mathscr{F})$ and $C \subset B - A$.

Proof. For any positive integer n, there exists a sequence $\{F_{nk}\}$, $k = 1, 2, \ldots$ of disjoint elements in $\mathscr{F}$ such that

$$B_n = \bigcup_k F_{nk} \supset A, \; \mu^*(A) + \frac{1}{n} \geqslant \sum_k \mu(F_{nk}) \geqslant \mu^*(A).$$

Since $\mu^* = \mu$ on $\mathscr{F}$ and μ^* is a measure on $\mathscr{B}(\mathscr{F})$, it follows that

$$\mu^*(A) \leqslant \mu^*(B_n) \leqslant \sum_k \mu(F_{nk}) \leqslant \mu^*(A) + \frac{1}{n}.$$

Let $B = \bigcap_{n=1}^{\infty} B_n$. Then $B \in \mathscr{B}(\mathscr{F})$ and $B \supset A$. Hence

$$\mu^*(A) \leqslant \mu^*(B) \leqslant \mu^*(B_n) \leqslant \mu^*(A) + \frac{1}{n}$$

for all n. Thus $\mu^*(A)=\mu^*(B)$. If now $C\subset B-A$ and $C\in\mathscr{B}(\mathscr{F})$, then $A\subset B-C$ and $\mu^*(A)\leqslant\mu^*(B-C)=\mu^*(B)-\mu^*(C)=\mu^*(A)-\mu^*(C)$. Hence $\mu^*(C)=0$. This completes the proof.

Definition. 18.5. For any set $A\subset X$, a set B satisfying properties (i), (ii) and (iii) of Proposition 18.4 is called a *measurable cover* of A.

Proposition. 18.6. Let $(X, \mathscr{F}, \mu)$ be a measure space and let $\mathscr{B}^*$ be the σ-algebra of μ^*-measurable sets. Then the measure space $(X, \mathscr{B}^*, \mu^*)$ is the completion of $(X, \mathscr{B}(\mathscr{F}), \mu^*)$, where $\mathscr{B}(\mathscr{F})$ is the σ-algebra generated by $\mathscr{F}$.

Proof. If $\overline{\mathscr{B}(\mathscr{F})}$ is the σ-algebra obtained by the completion of $(X, \mathscr{B}(\mathscr{F}), \mu^*)$, it follows from Exercise 18.2, that $\overline{\mathscr{B}(\mathscr{F})}\subset\mathscr{B}^*$. Let now $A\in\mathscr{B}^*$, $\mu^*(A)<\infty$. Let B be a measurable cover of A and let C be a measurable cover of $B-A$. Since μ^* is a measure on $\mathscr{B}^*$, we have

$$\mu^*(B-A)=\mu^*(B)-\mu^*(A)=0.$$

Hence $\mu^*(C)=0$. It is clear that $A=(B-C)\cup(AC)$. Since $B-C\in\mathscr{B}(\mathscr{F})$, $AC\subset C$, $C\in\mathscr{B}(\mathscr{F})$ and $\mu^*(C)=0$, it follows that $A\in\overline{\mathscr{B}(\mathscr{F})}$. The σ-finiteness of μ^* implies that $\overline{\mathscr{B}(\mathscr{F})}=\mathscr{B}^*$ and completes the proof.

Proposition. 18.7. Let $(X, \mathscr{F}, \mu)$ be a boolean measure space. For any μ^*-measurable set E such that $\mu^*(E)<\infty$ and any $\varepsilon>0$, there exists a set $F_\varepsilon\in\mathscr{F}$, such that $\mu^*(E\,\Delta\,F_\varepsilon)<\varepsilon$.

Proof. Let $\varepsilon>0$ be arbitrary. Then there exists a sequence $F_1, F_2,\ldots,$ of disjoint elements of $\mathscr{F}$ such that $\bigcup_n F_n\supset E$ and

$$\mu^*(E)+\frac{\varepsilon}{2}\geqslant\sum_{n=1}^{\infty}\mu^*(F_n)\geqslant\mu^*(E). \tag{18.2}$$

Since $\mu^*(E)<\infty$, the infinite series $\sum_n \mu^*(F_n)$ converges. Hence there exists an n_0 such that

$$\sum_{n>n_0}\mu^*(F_n)<\frac{\varepsilon}{2}. \tag{18.3}$$

Let $F_\varepsilon = \bigcup_{n=1}^{n_0} F_n$ and $E_\varepsilon = \bigcup_{n>n_0} F_n$. Since μ^* is countably additive, Eqs. (18.2) and (18.3) imply that

$$\mu^*(E_\varepsilon) < \frac{\varepsilon}{2},$$

$$\mu^*((E_\varepsilon \cup F_\varepsilon) - E) < \frac{\varepsilon}{2}.$$

Since $EF'_\varepsilon \subset E_\varepsilon$ and $F_\varepsilon E' \subset (E_\varepsilon \cup F_\varepsilon) - E$, it follows that $\mu^*(E \Delta F_\varepsilon) < \varepsilon$. Since $F_n \in \mathscr{F}$ for every n, $F_\varepsilon \in \mathscr{F}$. This completes the proof.

Remark. 18.8. Let $(X, \mathscr{B}, \mu)$ be a measure space. For any two sets $A, B \in \mathscr{B}$ such that $\mu(A) < \infty$, $\mu(B) < \infty$, let

$$d(A, B) = \mu(A \Delta B) = \int |\chi_A - \chi_B| \, d\mu.$$

(Note that the integrand is a simple function and the definition of the integral is as in section 3). Then

$$d(A, B) \leqslant d(A, C) + d(C, B).$$

If $d(A, B) = 0$, it is not necessary that $A = B$. In other words $d(.,.)$ is a 'pseudo metric' on $\mathscr{B}$.

Two sets $A, B \in \mathscr{B}$ are said to be *μ-equivalent* if $\mu(A \Delta B) = 0$. In such a case we write $A \sim B$. It is easy to verify that '$\sim$' defines an equivalence relation. For any A, let $\tilde{A}$ be the equivalence class of all sets equivalent to A. Let $\tilde{\mathscr{B}}$ denote the collection of all equivalence classes. Then the following holds:

(i) if $A_j \sim B_j$, $j = 1, 2, \ldots$, then $\bigcup_j A_j \sim \bigcup_j B_j$

(ii) if $A_j \sim B_j$, $j = 1, 2, \ldots$, then $\bigcap A_j \sim \bigcap_j B_j$

(iii) if $A \sim B$, then $A' \sim B'$.

Hence countable set operations in $\mathscr{B}$ filter down in a natural manner to operations in $\tilde{\mathscr{B}}$ and $\tilde{\mathscr{B}}$ becomes an algebra, where countable operations are possible. The algebra $\tilde{\mathscr{B}}$ is known as the *measure algebra* associated with μ.

Let $\mathscr{R} \subset \mathscr{B}$ be the class of all sets of finite μ-measure and $\tilde{\mathscr{R}} \subset \tilde{\mathscr{B}}$ be the associated collection of equivalence classes. In $\tilde{\mathscr{R}}$, define the metric $\tilde{d}$ by

$$\tilde{d}(\tilde{A}, \tilde{B}) = d(A, B) = \mu(A \Delta B).$$

Then $(\tilde{\mathscr{R}}, \tilde{d})$ is a metric space. (Later from a theorem of Riesz-Fischer, it would become clear that this is a complete metric space). The proofs of these assertions are left to the reader as exercises.

Exercise. 18.9. If $\mathscr{C}$ is a countable family of subsets of a set X, then the boolean algebra $\mathscr{F}$ generated by $\mathscr{C}$ is a countable family. If $\mathscr{B}$ is the σ-algebra generated by $\mathscr{F}$ and μ is a measure on $\mathscr{B}$, then the metric space $(\tilde{\mathscr{R}}, \tilde{d})$ constructed out of $(X, \mathscr{B}, \mu)$ in Remark 18.8 is separable.

§19. Measures on Metric Spaces

Let us now go back to Chapter 1 and look at the sample spaces occurring in Example 1.1 to Example 1.7. They vary through finite and countable sets, the real line, the Euclidean spaces and spaces of curves in an interval. Later in section 4 we met with products of such spaces. All these are examples of topological spaces. From the topological point of view open sets, closed sets, compact sets, etc., are very important. While doing probability theory on such sample spaces it would be desirable to include all the topologically important sets in the collection of all events. To this end we introduce the following definition.

Definition. 19.1. For any topological space X, the σ-algebra $\mathscr{B}_X$ generated by the class of all open subsets of X is called the *borel σ-algebra* of X. Any element of $\mathscr{B}_X$ is called a *borel set.*

Remark. 19.2. If X is a topological space and $Y \subset X$ is given the relative topology then $\mathscr{B}_Y = \mathscr{B}_X \cap Y$. (see Proposition 13.5). If X is a second countable topological space then $\mathscr{B}_X$ is generated by a countable base of open sets. If $X_1, X_2, \ldots$, is a sequence of second countable topological spaces then $X^\infty = X_1 \times X_2 \times \ldots$, is also a second countable topological space when it is equipped with the product topology. Further $\mathscr{B}_{X^\infty}$ is generated by the class of all open cylinders of the form

$$C = \{\mathbf{x} : \mathbf{x} = (x_1, x_2, \ldots), x_n \in X_n \text{ for all } n,$$
$$x_i \in G_i,\ i = 1, 2, \ldots, k\},$$

where G_i is an open subset of X_i, $i = 1, 2, \ldots k$ and k is a positive integer.

In the present section we shall see how a measure on the borel σ-algebra of any metric space is completely determined by its values on the class of all open sets or closed sets.

Example. 19.3. The space $X = R^k$ with its usual topology needs special mention, since most of the ordinary probability theory is done on this space. We denote any point of R^k by $\mathbf{x}=(x_1, x_2, \ldots, x_k)$, where x_i is the ith coordinate for every $i=1, 2, \ldots, k$. The class of all sets of the form

$$\{\mathbf{x} : a_i < x_i < b_i,\ i = 1, 2, \ldots, k\} \tag{19.1}$$

where a_i's and b_i's are rational numbers is a countable base for the topology of R^k. Thus $\mathscr{B}_{R^k}$ is generated by the class of rectangles of the form Eq. (19.1). Earlier in Exercise 4.5, we had introduced the class $\mathscr{I}^k$ of all rectangles of the form $I_1 \times I_2 \times \ldots \times I_k$, where every I_j is an interval of the type $(a, b]$, $(-\infty, a]$, (b, ∞) or $(-\infty, +\infty)$. We shall now show that

$$\mathscr{B}_{R^k} = \mathscr{B}(\mathscr{I}^k). \tag{19.2}$$

Indeed, this follows from the identities

$$\{\mathbf{x} : a_i < x_i < b_i,\ i = 1, 2, \ldots, k\}$$
$$= \bigcup_{n=1}^{\infty} \left\{\mathbf{x} : a_i < x_i \leqslant b_i - \frac{1}{n},\ i = 1, 2, \ldots, k\right\}$$

and

$$\{\mathbf{x} : a_i < x_i \leqslant b_i,\ i = 1, 2, \ldots, k\}$$
$$= \bigcap_{n=1}^{\infty} \left\{\mathbf{x} : a_i < x_i < b_i + \frac{1}{n},\ i = 1, 2, \ldots, k\right\}.$$

Equation (19.2) implies, in particular, that every measure on the boolean semi-algebra $\mathscr{I}^k$ extends uniquely to the borel σ-algebra. This means that one can define the measure of an arbitrary open or closed set.

Exercise. 19.4. Let F be a monotonic increasing right continuous function on R. Then there exists a unique σ-finite measure μ on $\mathscr{B}_R$ such that $\mu((a, b])=F(b)-F(a)$ for all $a, b \in R$.

Exercise. 19.5. If $F_1, F_2, \ldots, F_k$ are k monotonic increasing right continuous functions on R, then there exists a unique σ-finite measure μ on $\mathscr{B}_{R^k}$ such that

$$\mu(\{\mathbf{x} : a_i < x_i \leqslant b_i\}) = \prod_{i=1}^{k} [F_i(b_i) - F_i(a_i)].$$

Exercise. 19.6. Let $f(x_1, x_2, \ldots, x_k)$ be a continuous non-negative function on R^k such that the Riemann integral

$$\int_{-\infty}^{+\infty} \cdots \int_{-\infty}^{+\infty} f(x_1, x_2, \ldots, x_k)\, dx_1 dx_2 \ldots dx_k = 1.$$

Then there exists a unique probability measure μ on $\mathscr{B}_{R^k}$ such that

$$\mu(\{\mathbf{x} : a_i < x_i < b_i,\ i=1, 2,\ldots, k\})$$
$$= \int_{a_k}^{b_k} \cdots \int_{a_2}^{b_2} \int_{a_1}^{b_1} f(x_1, x_2,\ldots, x_k)\, dx_1 dx_2 \ldots dx_k.$$

Remark. 19.7. Let μ be a probability measure on $\mathscr{B}_{R^k}$. Let

$$F(\xi_1, \xi_2,\ldots, \xi_k) = \mu(\{\mathbf{x} : x_1 \leqslant \xi_1, x_2 \leqslant \xi_2,\ldots, x_k \leqslant \xi_k\}).$$

The function F of k variables $\xi_1, \xi_2,\ldots, \xi_k$ is called the *distribution function* of μ. It may be noted that F is monotonic increasing in each of the ξ_i's when the rest of the variables are fixed. Further,

$$\lim_{\xi_i \to -\infty} F(\xi_1, \xi_2,\ldots, \xi_i,\ldots, \xi_k) = 0 \text{ for every } i,$$

$$\lim_{\substack{\xi_1 \to \infty, \\ \xi_2 \to \infty, \\ \cdots\cdots \\ \xi_k \to \infty}} F(\xi_1, \xi_2,\ldots, \xi_k) = 1,$$

and F is a right continuous function in the variables ξ_j, i.e.,

$$\lim_{\substack{\xi_i \to a_i \\ \xi_i > a_i,\ i=1, 2,\ldots, k}} F(\xi_1, \xi_2,\ldots, \xi_k) = F(a_1, a_2,\ldots, a_k)$$

for all $a_1, a_2,\ldots, a_k$. However, a function F satisfying the above conditions need not be the distribution function of a probability measure.

For any function f of k variables $\xi_1, \xi_2,\ldots, \xi_k$ we shall write

$$(\Delta_h^{(i)} f)(\xi_1, \xi_2,\ldots, \xi_k)$$
$$= f(\xi_1, \xi_2,\ldots, \xi_{i-1}, \xi_i + h, \xi_{i+1},\ldots, \xi_k) - f(\xi_1, \xi_2,\ldots, \xi_k).$$

If now the function $F(\xi_1, \xi_2,\ldots, \xi_k)$ satisfies the additional condition

$$(\Delta_{h_k}^{(k)} \Delta_{h_{k-1}}^{(k-1)} \ldots \Delta_{h_1}^{(1)} F)(\xi_1, \xi_2,\ldots, \xi_k) \geqslant 0$$

for all $\xi_1, \xi_2,\ldots, \xi_k$ and all non-negative $h_1, h_2,\ldots, h_k$, then F is the distribution function of a probability measure μ. In such a case

$$\mu\{\mathbf{x} : \xi_i < x_i \leqslant \xi_i + h_i,\ i=1, 2,\ldots, k\}$$
$$= (\Delta_{h_k}^{(k)} \Delta_{h_{k-1}}^{(k-1)} \ldots \Delta_{h_1}^{(1)} F)(\xi_1, \xi_2,\ldots, \xi_k),$$

for all $\xi_1,\ldots, \xi_k$ and all non-negative $h_1, h_2,\ldots, h_k$.

Exercise. 19.8. Let F be defined on R^2 by

$$\begin{aligned} F(x, y) &= 0 \text{ if } x<1 \text{ or } y<1 \\ &= \alpha \text{ if } 1 \leqslant x < 2,\ 1 \leqslant y < 2, \\ &= \alpha + \beta \text{ if } 1 \leqslant x < 2,\ y \geqslant 2, \\ &= \alpha + \gamma \text{ if } x \geqslant 2,\ 1 \leqslant y < 2, \\ &= 1 \text{ if } x \geqslant 2,\ y \geqslant 2, \end{aligned}$$

where α, β, γ are positive and $1-\alpha>\beta>1-\alpha-\gamma>0$. Then F is monotonic increasing in x and y separately and right continuous. But F is not a probability distribution function.

Definition. 19.9. Let μ be a measure on the borel σ-algebra $\mathscr{B}_X$ of a topological space X. A borel subset A of X is said to be *μ-regular* if

$$\begin{aligned}\mu(A) &= \sup\{\mu(C), C\subseteq A, C \text{ closed}\}\\ &= \inf\{\mu(U), A\subset U, U \text{ open}\}.\end{aligned}$$

If every borel set is μ-regular we shall say that μ is *regular.*

We shall soon prove that if X is a metric space then every totally finite measure on $\mathscr{B}_X$ is regular. Hereafter by a *measure on a topological space* X, we shall mean a measure on $\mathscr{B}_X$. Now we state some definitions and prove an elementary proposition in topology.

Definition. 19.10. Let X be a topological space. A set $A\subset X$ is said to be a G_δ if there exists a sequence of open sets $\{U_n\}$ such that $U_1\supset U_2\supset\dots$ and $A=\bigcap_n U_n$. The complement of any G_δ set is called on F_σ set.

Proposition. 19.11. Let X be a metric space with metric d. For any $A\subset X$, let $d(x, A)=\inf_{y\in A} d(x,y)$. Then

$$|d(x, A)-d(x', A)| \leqslant d(x, x') \text{ for all } x, x'\in A. \tag{19.3}$$

In particular, $d(x, A)$ is a uniformly continuous function of x.

Proof. For any $y\in A$, we have

$$d(x, A) \leqslant d(x,y) \leqslant d(x, x')+d(x', y).$$

Hence

$$d(x, A) \leqslant d(x, x')+d(x', A).$$

Interchanging x, x' we get

$$d(x', A) \leqslant d(x, x')+d(x, A).$$

These two inequalities imply inequality (19.3) and complete the proof.

Corollary. 19.12. In a metric space X every closed subset is a G_δ. Every open set is an F_σ.

Proof. Let C be any closed set. Then

$$\begin{aligned}C &= \{x : d(x, C)=0\}\\ &= \bigcap_{n=1}^{\infty} \left\{x : d(x, C)< \frac{1}{n}\right\}.\end{aligned}$$

Since $d(x, C)$ is continuous, it follows that $\left\{x : d(x, C)< \frac{1}{n}\right\}$ is open. This completes the proof.

Proposition. 19.13. Every totally finite measure on a metric space X is regular.

Proof. If A is any μ-regular set it is clear that for any $\varepsilon>0$, we can find an open set $U_\varepsilon \supset A$ and a closed set $C_\varepsilon \subset A$ such that $\mu(U_\varepsilon - C_\varepsilon) < \varepsilon$. Conversely, if this is satisfied for every $\varepsilon > 0$, then A is μ-regular.

Now let $\mathscr{R}$ denote the class of all μ-regular borel sets. Since $\emptyset$ and X are open and closed it follows that they belong to $\mathscr{R}$. $\mathscr{R}$ is closed under complementation. Indeed, let $A \in \mathscr{R}$ and $\varepsilon > 0$. There exists an open set $U_\varepsilon \supset A$ and a closed set $C_\varepsilon \subseteq A$ such that $\mu(U_\varepsilon - C_\varepsilon) < \varepsilon$. Then $U'_\varepsilon \subset A' \subset C'_\varepsilon$, $C'_\varepsilon - U'_\varepsilon = U_\varepsilon - C_\varepsilon$, U'_ε is closed and C'_ε is open. This shows that $A' \in \mathscr{R}$.

$\mathscr{R}$ is closed under countable unions. Indeed, let $A_1, A_2, \ldots \in \mathscr{R}$ and $A = \bigcup_1^\infty A_i$. Let $\varepsilon>0$ be arbitrary. There exists an open set $U_{n,\varepsilon}$ and a closed set $C_{n,\varepsilon}$ such that $U_{n,\varepsilon} \supset A_n \supset C_{n,\varepsilon}$ and $\mu(U_{n,\varepsilon} - C_{n,\varepsilon}) < \frac{\varepsilon}{3^n}$. Let $U_\varepsilon = \bigcup_{n=1}^\infty U_{n,\varepsilon}$. Let $B = \bigcup_n C_{n,\varepsilon}$. Since μ is a measure we can choose a k so large that $\mu(B - \bigcup_{n=1}^k C_{n,\varepsilon}) < \frac{\varepsilon}{2}$ (see Proposition 15.10). Let $C_\varepsilon = \bigcup_{n=1}^k C_{n,\varepsilon}$. Then U_ε is open, C_ε is closed, $U_\varepsilon \supset A \supset C_\varepsilon$ and

$$\begin{aligned}\mu(U_\varepsilon - C_\varepsilon) &\leqslant \mu(U_\varepsilon - B) + \mu(B - C_\varepsilon)\\ &\leqslant \left[\sum_{n=1}^\infty \mu(U_{n,\varepsilon} - C_{n,\varepsilon})\right] + \frac{\varepsilon}{2}\\ &\leqslant \left(\sum_{n=1}^\infty \frac{\varepsilon}{3^n}\right) + \frac{\varepsilon}{2} = \varepsilon.\end{aligned}$$

Thus $\mathscr{R}$ is a σ-algebra. To complete the proof it is enough to prove that $\mathscr{R}$ contains all closed sets. If C is a closed set Corollary 19.12 implies that $C = \bigcap_n U_n$, where U_n is a decreasing sequence of open sets. By proposition 15.10, $\lim_{n\to\infty} \mu(U_n) = \mu(C)$. Hence, for any $\varepsilon>0$, we can find an n_0 such that $\mu(U_{n_0} - C) < \varepsilon$. If we put $U_\varepsilon = U_{n_0}$ and $C_\varepsilon = C$, then $U_\varepsilon \supset C \supset C_\varepsilon$ and $\mu(U_\varepsilon - C_\varepsilon) < \varepsilon$. Thus $C \in \mathscr{R}$ and the proof is complete.

Corollary. 19.14. If μ, ν are totally finite measures on a metric space X and $\mu(C)=\nu(C)$ for every closed set C, then $\mu=\nu$.

We can now ask the question whether the measure on a metric space is completely determined by its values for classes of sets smaller than the class of all closed subsets. As an example one may consider the class of all compact subsets. We shall investigate this situation in view of its importance in probability theory.

Definition. 19.15. A totally finite measure μ on a metric space X is said to be *tight* if, for each $\varepsilon>0$, there exists a compact set $K_\varepsilon \subset X$ such that $\mu(X-K_\varepsilon)<\varepsilon$.

Proposition. 19.16. Let X be a metric space and μ a tight measure on it. Then for any borel set E and any $\varepsilon>0$, there is a compact set $K_\varepsilon \subset E$ such that $\mu(E-K_\varepsilon)<\varepsilon$.

Proof. By Proposition 19.13 there exists a closed set $C_\varepsilon \subset E$ such that $\mu(E-C_\varepsilon)<\frac{\varepsilon}{2}$. Since μ is tight there is a compact set $K \subset X$ such that $\mu(X-K)<\frac{\varepsilon}{2}$. Let now $K_\varepsilon = K \cap C_\varepsilon$. Then $\mu(E-K_\varepsilon) \leqslant \mu(E-C_\varepsilon)+\mu(C_\varepsilon-K_\varepsilon) \leqslant \frac{\varepsilon}{2}+\mu(X-K)<\varepsilon$. Further K_ε is compact and the proof is complete.

Proposition. 19.17. Let X be a complete metric space and let $C \subseteq X$ be closed. Suppose, for each n, there exists an integer k_n such that $C \subset \bigcup_{j=1}^{k_n} \bar{S}_{nj}$, where each $\bar{S}_{nj}$ is a closed sphere of radius $\frac{1}{n}$ in X. Then C is compact.

Proof. Let $x_1, x_2, \ldots$ be any sequence of distinct points in C. We shall show that this has a limit point. Choose an integer $n_1(\leqslant k_1)$ such that $C \cap \bar{S}_{1n_1}=K_1$ contains infinitely many x_j's. Since $K_1 \subseteq \bigcup_1^{k_2} \bar{S}_{2j}$ we can find $n_2(\leqslant k_2)$ such that $K_1 \cap \bar{S}_{2n_2}=K_2$ contains infinitely many x_j's. We continue this procedure and construct $K_1 \supset K_2 \supset \ldots \supset K_n \supset \ldots$ such that the diameter of K_n is $\leqslant \frac{2}{n}$. The completeness of X implies that $\bigcap_n K_n$ consists of a single point x_0. Since $K_n \subset C$ it follows that $x_0 \in C$. Further every neighbourhood of x_0 contains infinitely many x_j's. This shows that C is compact.

Proposition. 19.18. Let X be a complete and separable metric space. Then every totally finite measure μ on X is tight.

Proof. Let d be the metric in X. Choose and fix $\varepsilon > 0$. For any integer n, the open spheres of radius $\frac{1}{n}$ around each point constitute a covering of X. Since X is separable, we can find countably many such spheres $S_{n1}, S_{n2}, \ldots$ such that $X = \bigcup_j S_{nj}$. Hence $X = \bigcup_j \bar{S}_{nj}$. By Proposition 15.10 we can find an integer k_n such that

$$\mu\left(X - \bigcup_{j=1}^{k_n} \bar{S}_{nj}\right) < \frac{\varepsilon}{2^n}.$$

Let

$$K_\varepsilon = \bigcap_{n=1}^{\infty} \bigcup_{j=1}^{k_n} \bar{S}_{nj}.$$

Then

$$\mu(X - K_\varepsilon) \leqslant \sum_{n=1}^{\infty} \frac{\varepsilon}{2^n} = \varepsilon.$$

By Proposition 19.17, the set K_ε is compact. This shows that μ is tight. The proof is complete.

Corollary. 19.19. Let μ be any totally finite measure on a complete and separable metric space X. Then for any borel set E, $\mu(E) = \sup \{\mu(K), K \subset E, K \text{ compact}\}$. In particular, if two totally finite measures μ, ν agree on all compact sets then $\mu = \nu$.

Exercise. 19.20. Consider the measure μ in $[0, 1]$ determined by

$$\mu(\{0\}) = 1,$$
$$\mu((a, b]) = \log b - \log a \text{ if } 0 < a < b \leqslant 1.$$

Then μ is a σ-finite measure in $[0, 1]$. For any open set $G \supset \{0\}$, $\mu(G) = \infty$. Thus μ is an irregular measure.

Exercise. 19.21. Let X be a complete and separable metric space and let $Y \subset X$ be a borel set. Consider Y as a metric space with the same metric as X. Then every totally finite measure on Y is tight. Further two totally finite measures on Y agree on all borel sets as soon as they agree on all compact sets.

Remark. 19.22. If two probability measures μ, ν on the borel σ-algebra $\mathscr{B}_X$ of a metric space have the property that

$\mu(S) = \nu(S)$ for every closed sphere $S \subset X$, is it true that $\mu = \nu$? (See Exercise 53.14).

Exercise. 19.23. Let X be a compact metric space and let $\{\mathscr{F}_n\}$ be an increasing sequence of boolean algebras of subsets of X such that $\mathscr{F}_n \subset \mathscr{B}_X$ for all $n=1, 2, \ldots$. Let μ_n be a probability measure on $\mathscr{F}_n$ such that

(i) $\mu_n(A) = \sup \{\mu_n(K), K \subset A, K \text{ closed}, K \in \mathscr{F}_n\}$

(ii) $\mu_m(A) = \mu_n(A)$ whenever $m < n$ and $A \in \mathscr{F}_m$.

Define the set function μ on $\bigcup_n \mathscr{F}_n = \mathscr{F}_\infty$ by the equation

$$\mu(A) = \mu_n(A) \text{ if } A \in \mathscr{F}_n,$$

for $n = 1, 2, \ldots$. Then μ is a probability measure on $\mathscr{F}_\infty$.

§20. Probability Contents

Till now we have studied the problem of extending additive functions on a boolean algebra to the σ-algebra generated by it. In the case of metric spaces we found that a totally finite measure on the σ-algebra is determined by its values on closed sets. It is only natural to ask whether an additive function on the class of closed sets can be extended to the boolean algebra and the σ-algebra generated by it. We shall investigate this problem in the present section.

Let X denote a metric space and let $\mathscr{C}_X$ and $\mathscr{G}_X$ denote respectively the class of all closed subsets and the class of all open subsets. We shall write C with or without indices to denote a general closed set. Similarly G with or without indices shall denote a general open set. For any $A \subset X$, $\bar{A}$ will be the closure of A.

Definition. 20.1. A map $\lambda : \mathscr{C}_X \to [0, 1]$ is called a *probability content* if

(i) $\lambda(\emptyset) = 0$, $\lambda(X) = 1$;

(ii) $\lambda(C_1) \leqslant \lambda(C_2)$ if $C_1 \subset C_2$;

(iii) $\lambda(C_1 \cup C_2) \leqslant \lambda(C_1) + \lambda(C_2)$ for all C_1, C_2 where equality holds whenever C_1, C_2 are disjoint.

A probability content is said to be *smooth* if for every $C \in \mathscr{C}_X$

$$\lambda(C) = \inf \{\lambda(\bar{G}), G \supset C, \text{G open}\}.$$

To every probability content λ we define a set function $\tilde{\lambda}$ on the class of *all* subsets in the following manner. Write

$$\tau_\lambda(G) = \sup \{\lambda(C) : C \subset G, C \in \mathscr{C}_X\}, G \in \mathscr{G}_X \tag{20.1}$$

$$\tilde{\lambda}(A) = \inf \{\tau_\lambda(G) : G \supset A, G \in \mathscr{G}_X\}, A \subset X. \tag{20.2}$$

We start with an elementary topological proposition.

Proposition. 20.2. Let $C_1, C_2 \in \mathscr{C}_X, C_1 \cap C_2 = \varnothing$. Then there exists a real continuous function f on X such that $0 \leqslant f(x) \leqslant 1$ for all x and $C_1 = \{x : f(x) = 0\}$, $C_2 = \{x : f(x) = 1\}$. In particular, there exist disjoint open sets G_1, G_2 such that $C_1 \subset G_1$, $C_2 \subset G_2$.

Proof. For any $x \in X$, $A \subset X$, we define the number $d(x, A)$ as in Proposition 19.11 in terms of the metric d of X. By the same proposition it follows that the function f defined by

$$f(x) = \frac{d(x, C_1)}{d(x, C_1) + d(x, C_2)}$$

is continuous, $0 \leqslant f(x) \leqslant 1$ and $\{x : f(x) = 0\} = C_1$, $\{x : f(x) = 1\} = C_2$. If we write $G_1 = \{x : f(x) < \frac{1}{2}\}$, $G_2 = \{x : f(x) > \frac{3}{4}\}$, the proof is complete.

Proposition. 20.3. Let λ be a probability content on $\mathscr{C}_X$. Then the function τ_λ defined by Eq. (20.1) satisfies the following properties:

(i) $\tau_\lambda(\varnothing) = 0$, $\tau_\lambda(X) = 1$;

(ii) $\tau_\lambda(G_1) \leqslant \tau_\lambda(G_2)$ if $G_1 \subset G_2$ and $G_1, G_2 \in \mathscr{G}_X$;

(iii) $\tau_\lambda(G_1 \cup G_2) \leqslant \tau_\lambda(G_1) + \tau_\lambda(G_2)$ for all $G_1, G_2 \in \mathscr{G}_X$.

Proof. Properties (i) and (ii) are obvious. To prove (iii) consider any closed set $C \subset G_1 \cup G_2$. Then CG'_1 and CG'_2 are disjoint closed sets. By Proposition 20.2 there exist two disjoint open sets G_α, G_β such that $CG'_1 \subset G_\alpha$, $CG'_2 \subset G_\beta$. Let $C_1 = CG'_\alpha$, $C_2 = CG'_\beta$. Then C_1, C_2 are closed sets, $C_1 \subset G_1$, $C_2 \subset G_2$ and $C_1 \cup C_2 = C \cap (G'_\alpha \cup G'_\beta) = C \cap (G_\alpha \cap G_\beta)' = C$. Hence,

$$\begin{aligned} \lambda(C) &\leqslant \lambda(C_1) + \lambda(C_2) \\ &\leqslant \tau_\lambda(G_1) + \tau_\lambda(G_2). \end{aligned}$$

Taking supremum over all $C \subset G_1 \cup G_2$ we get (iii). This completes the proof.

Proposition. 20.4. Let λ be a probability content on $\mathscr{C}_X$. Then the set function $\tilde{\lambda}$ defined by Eq. (20.2) satisfies the following properties:

(i) $\tilde{\lambda}(\varnothing) = 0$, $\tilde{\lambda}(X) = 1$;

(ii) $\tilde{\lambda}(G) = \tau_\lambda(G)$ for all $G \in \mathscr{G}_X$;

(iii) $\tilde{\lambda}(A) \leqslant \tilde{\lambda}(B)$ if $A \subset B$;

(iv) $\tilde{\lambda}(A \cup B) \leqslant \tilde{\lambda}(A) + \tilde{\lambda}(B)$ for all $A, B \subset X$;

(v) $\tilde{\lambda}(G) \leqslant \lambda(C)$ for all $G \in \mathcal{G}_X$, $C \in \mathcal{C}_X$ such that $G \subset C$;

(vi) $\tilde{\lambda}(A) = \tilde{\lambda}(A \cap G) + \tilde{\lambda}(A \cap G')$ for any $G \in \mathcal{G}_X$ and $A \subset X$.

Proof. As before, properties (i) to (iii) are straightforward consequences of the definition of $\tilde{\lambda}$. To prove (iv) we choose, for any $\varepsilon > 0$, open sets $G_1 \supset A$, $G_2 \supset B$ such that

$$\tau_\lambda(G_1) \leqslant \tilde{\lambda}(A) + \frac{\varepsilon}{2},$$

$$\tau_\lambda(G_2) \leqslant \tilde{\lambda}(B) + \frac{\varepsilon}{2}.$$

Since τ_λ is sub-additive we have

$$\begin{aligned}\tilde{\lambda}(A \cup B) &\leqslant \tau_\lambda(G_1 \cup G_2) \leqslant \tau_\lambda(G_1) + \tau_\lambda(G_2)\\ &\leqslant \tilde{\lambda}(A) + \tilde{\lambda}(B) + \varepsilon.\end{aligned}$$

The arbitrariness of ε implies (iv). To prove (v) choose any $C_1 \subset G$. Then $C_1 \subset C$ and $\lambda(C_1) \leqslant \lambda(C)$. Hence

$$\tilde{\lambda}(G) = \tau_\lambda(G) = \sup_{C_1 \subset G} \lambda(C_1) \leqslant \lambda(C).$$

This proves (v). To prove (vi) consider any open set $G_1 \supset A$. For any $\varepsilon > 0$ choose $C_1 \subset G \cap G_1$ such that

$$\tau_\lambda(G\, G_1) \leqslant \lambda(C_1) + \frac{\varepsilon}{2}.$$

Choose $C_2 \subset G_1 C'_1$ such that

$$\tau_\lambda(G_1\, C'_1) \leqslant \lambda(C_2) + \frac{\varepsilon}{2}.$$

Since C_1, C_2 are disjoint, λ is additive over disjoint closed sets and $G'G_1 \subset G_1 C'_1$, we have

$$\begin{aligned}\tilde{\lambda}(GG_1) + \tilde{\lambda}(G'G_1) &\leqslant \tau_\lambda(GG_1) + \tau_\lambda(C'_1 G_1)\\ &\leqslant \lambda(C_1) + \lambda(C_2) + \varepsilon\\ &= \lambda(C_1 \cup C_2) + \varepsilon.\end{aligned}$$

Since $C_1 \cup C_2 \subset G_1$, we have

$$\tilde{\lambda}(GG_1) + \tilde{\lambda}(G'G_1) \leqslant \tau_\lambda(G_1) + \varepsilon.$$

Since $G_1 \supset A$, (iii) implies

$$\tilde{\lambda}(GA) + \tilde{\lambda}(G'A) \leqslant \tau_\lambda(G_1) + \varepsilon.$$

Taking infimum over all such G_1, we have

$$\tilde{\lambda}(GA) + \tilde{\lambda}(G'A) \leqslant \tilde{\lambda}(A) + \varepsilon.$$

Since ε is arbitrary and $\tilde{\lambda}$ is sub-additive, the proof is complete.

Proposition. 20.5. Let X be a metric space and let $\mathscr{F}_X$ be the boolean algebra generated by the class $\mathscr{C}_X$ of all closed subsets of X. Suppose λ is a probability content on $\mathscr{C}_X$. Then the set function $\tilde{\lambda}$ defined by Eq. (20.2) is a probability distribution on $\mathscr{F}_X$. Further

$$\text{(i)}\quad \tilde{\lambda}(A) = \inf\{\tilde{\lambda}(G), G \supset A, G \text{ open}\} = \sup\{\tilde{\lambda}(C), C \subset A, C \text{ closed}\}, \quad \text{for all } A \in \mathscr{F}_X; \tag{20.3}$$

$$\text{(ii)}\quad \lambda(C) \leqslant \tilde{\lambda}(C) \text{ for all } C \in \mathscr{C}_X. \tag{20.4}$$

If λ is a smooth probability content then

$$\tilde{\lambda}(C) = \lambda(C) \text{ for all } C \in \mathscr{C}_X,$$

Proof. Let λ be a probability content and let $\tilde{\lambda}$ be defined by Eq. (20.2). Declare a set $E \subset X$ to be $\tilde{\lambda}$-measurable if, for any $A \subset X$,

$$\tilde{\lambda}(A) = \tilde{\lambda}(AE) + \tilde{\lambda}(AE').$$

Since $\tilde{\lambda}$ is subadditive it follows by the same arguments as in the proof of Proposition 16.7, that the class of all $\tilde{\lambda}$-measurable sets is a boolean algebra. By property (vi) of Proposition 20.4 every open set is $\tilde{\lambda}$-measurable. Further $\tilde{\lambda}$ is finitely additive on the boolean algebra of $\tilde{\lambda}$-measurable sets. This boolean algebra includes $\mathscr{F}_X$. This completes the proof of the first part. The first half of Eq. (20.3) follows from the definition of $\tilde{\lambda}$ and the fact that $\tilde{\lambda}(G) = \tau_\lambda(G)$ if G is open. Since $\tilde{\lambda}(C) = 1 - \tilde{\lambda}(C')$ and as C varies over $\mathscr{C}_X$, C' varies over $\mathscr{G}_X$ the second half follows form the first. To prove Eq. (20.4) we note that for any open set $G \supset C$, $\lambda(C) \leqslant \tau_\lambda(G)$. Taking infimum over all such G we obtain $\lambda(C) \leqslant \tilde{\lambda}(C)$.

To prove the last part, let λ be a smooth probability content. For any $\varepsilon > 0$, we can choose an open set $G \supset C$ such that

$$\lambda(\overline{G}) \leqslant \lambda(C) + \varepsilon.$$

Since $\overline{G}$ is closed we have from properties (iii) and (v) of Proposition 20.4,

$$\tilde{\lambda}(C) \leqslant \tilde{\lambda}(G) \leqslant \lambda(\overline{G}) \leqslant \lambda(C) + \varepsilon.$$

Since ε is arbitrary, we have $\tilde{\lambda}(C) \leqslant \lambda(C)$. This completes the proof of the proposition.

Remark. 20.6. We shall say that a probability distribution $\tilde{\lambda}$ on $\mathscr{F}_X$ is *regular* if it satisfies Eq. (20.3). Then the last part of the proposition implies that every smooth probability content on $\mathscr{C}_X$ possesses a unique extension to $\mathscr{F}_X$, which is a regular probability distribution.

Exercise. 20.7. Let λ be a probability content on $\mathscr{C}_X$. Suppose for any decreasing sequence $\{C_n\}$ of closed sets

$$\lambda(\bigcap_n C_n) = \lim_{n\to\infty} \lambda(C_n).$$

Then λ is smooth.

Proposition. 20.8. Let X be a compact metric space and let $\mathscr{F}_X$ be the boolean algebra generated by the class of all closed sets. Then any regular probability distribution on $\mathscr{F}_X$ is countably additive and hence can be extended uniquely to a probability measure on the borel σ-algebra $\mathscr{B}_X$.

Proof. Let μ be a regular probability distribution on $\mathscr{F}_X$. Suppose $A_n \in \mathscr{F}_X$, $A_1 \supset A_2 \supset \ldots \supset A_n \supset \ldots$, $\bigcap_n A_n = \emptyset$. Then we claim that $\lim_{n\to\infty} \mu(A_n) = 0$. Suppose this is not true. Then there exists a $\delta > 0$ such that $\mu(A_n) \geqslant \delta > 0$ for all n. By using the regularity of μ we choose a compact set $C_n \subset A_n$ such that

$$\mu(A_n - C_n) < \frac{\delta}{2^{n+1}},\ n = 1, 2, \ldots.$$

Then we have

$$\mu\left(\bigcap_{i=1}^n A_i - \bigcap_{i=1}^n C_i\right) \leqslant \sum_{i=1}^n \mu(A_i C_i')$$

$$\leqslant \sum_{i=1}^\infty \frac{\delta}{2^{i+1}} = \frac{\delta}{2}$$

Since $\mu(A_n) = \mu\left(\bigcap_{i=1}^n A_i\right) \geqslant \delta$, it follows that

$$\mu\left(\bigcap_{i=1}^n C_i\right) \geqslant \frac{\delta}{2}.$$

In other words $\bigcap_{i=1}^n C_i$ is non-empty for every n. Hence by th property of compactness $\bigcap_{i=1}^{\infty} C_i \neq \emptyset$. Since $C_i \subset A_i$ for all i, $\bigcap_i A_i \neq \emptyset$. This is a contradiction. Hence our claim is proved.

Now suppose $\{B_n\}$ is a sequence of disjoint sets in $\mathscr{F}_X$ such that $B = \bigcup_{n=1}^{\infty} B_n$ lies in $\mathscr{F}_X$. If we write $\widetilde{B}_n = B_1 \cup B_2 \cup \ldots \cup B_n$, then $\widetilde{B}_n$ increases to B or equivalently, $B - \widetilde{B}_n$ decreases to the empty set. Hence

$\lim_{n\to\infty} \mu(B - \widetilde{B}_n) = 0$. This implies that

$$\begin{aligned}\mu(B) &= \lim_{n\to\infty} \mu(\widetilde{B}_n) = \lim_{n\to\infty} \sum_{j=1}^{n} \mu(B_j)\\ &= \sum_{j=1}^{\infty} \mu(B_j).\end{aligned}$$

Thus μ is countably additive on $\mathscr{F}_X$. This completes the proof.

Corollary. 20.9. Every smooth probability content on the closed subsets of a compact metric space X can be extended uniquely to a probability measure on the borel σ-algebra $\mathscr{B}_X$.

Definition. 20.10. Let X be a metric space and let $\mathscr{K}_X$ be the class of all *compact* subsets of X. A map $\lambda : \mathscr{K}_X \to [0, \infty)$ is called a *compact content* if

(i) $\lambda(\emptyset) = 0$;

(ii) $\lambda(K_1) \leqslant \lambda(K_2)$ if $K_1 \subset K_2$;

(iii) $\lambda(K_1 \cup K_2) \leqslant \lambda(K_1) + \lambda(K_2)$ for all $K_1, K_2 \in \mathscr{K}_X$, where equality holds whenever K_1 and K_2 are disjoint.

We now observe that $\mathscr{K}_X \subset \mathscr{C}_X$. We shall now try to extend λ on $\mathscr{K}_X$ to the boolean algebra generated by $\mathscr{G}_X$. To this end we proceed exactly along the same lines as in the case of probability contents. Let

$$\eta_\lambda(G) = \sup \{\lambda(K) : K \subset G,\ K \in \mathscr{K}_X\},\ G \in \mathscr{G}_X \tag{20.5}$$

$$\hat{\lambda}(A) = \inf \{\eta_\lambda(G) : G \supset A,\ G \in \mathscr{G}_X\},\ A \subset X \tag{20.6}$$

Proposition. 20.11. Let X be a metric space and let $\mathscr{K}_X$ be the class of all compact subsets of X. Let λ be a compact content on $\mathscr{K}_X$. Then the function $\hat{\lambda}$ defined by Eq. (20.6) is finitely additive on the boolean algebra $\mathscr{F}_X$ generated by the class of all open subsets of X. For any $A \in \mathscr{F}_X$.

$$\hat{\lambda}(A) = \inf \{\hat{\lambda}(G),\ G \supset A,\ G \text{ open}\}. \tag{20.7}$$

For every compact set K,

$$\hat{\lambda}(K) \geqslant \lambda(K),\ K \in \mathscr{K}_X. \tag{20.8}$$

If $\lambda(X) = 1$, then $\hat{\lambda}(X) = 1$.

If X is locally compact then $\hat{\lambda}(K) < \infty$ and $\hat{\lambda}$ restricted to $\mathscr{K}_X \cap K$ is countably additive for every compact set $K \subset X$.

Proof. First of all, we observe that the function η_λ defined by Eq. (20.5) satisfies properties (ii) and (iii) of Proposition 20.3 (with τ_λ replaced by η_λ). It is clear that $\eta_\lambda(\varnothing) = 0$. Similarly $\hat{\lambda}$ satisfies properties (ii) to (vi) of Proposition 20.4, if $\tilde{\lambda}$ is replaced by $\hat{\lambda}$ and closed sets are replaced by compact sets. As in the proof of Proposition 20.5, we define a set E to be $\hat{\lambda}$-measurable if, for any $A \subset X$,

$$\hat{\lambda}(A) = \hat{\lambda}(AE) + \hat{\lambda}(AE').$$

Then $\hat{\lambda}$-measurable sets constitute a boolean algebra of subsets of X containing the class $\mathscr{G}_X$ of all open subsets of X. Equation (20.7) is an immediate consequence of the definitions. The inequality (20.8) is proved exactly in the same way as inequality (20.4).

Now let X be locally compact. If K is a compact set we can find an open set G such that $\bar{G}$ is compact and $\bar{G} \supset G \supset K$. Since $\hat{\lambda}(G) \leqslant \lambda(\bar{G}) < \infty$ it follows that $\hat{\lambda}(K) < \infty$. Now Eq. (20.7) implies that for any subset A of a compact set K such that $A \in \mathscr{F}_X$, we have

$$\hat{\lambda}(A) = \sup \{\hat{\lambda}(K_1),\ K_1 \subset A,\ K_1 \text{ compact}\}.$$

By proposition 20.8, $\hat{\lambda}$ restricted to $\mathscr{F}_X \cap K$ is countably additive. This completes the proof.

Corollary. 20.12. Let X be a locally compact second countable metric space and let $\mathscr{K}_X$ be the class of all compact subsets of X. Let λ be a compact content on $\mathscr{K}_X$. Then there exists a σ-finite measure μ on the borel σ-algebra $\mathscr{B}_X$ with the following properties:

(i) $\lambda(K) \leqslant \mu(K) < \infty$ for every $K \in \mathscr{K}_X$;

(ii) if $\mathscr{F}_X$ is the boolean algebra generated by the class of all open subsets of X and $A \in \mathscr{F}_X \cap K$, where K is a compact set then

$$\mu(A) = \inf \{\eta_\lambda(G),\ G \supset A,\ G \text{ open}\},$$

where

$$\eta_\lambda(G) = \sup \{\lambda(K_1),\ K_1 \subset G,\ K_1 \text{ compact}\}.$$

Proof. Construct the function $\hat{\lambda}$ on the class of all subsets of X as in Proposition 20.11. Choose an increasing sequence $\{G_n\}$ of open subsets such that $\overline{G}_n$ is compact for every n and $\bigcup_n G_n = X$. Let

$$H_1 = G_1,$$
$$H_n = G_n G'_{n-1} G'_{n-2} \dots G'_1, \; n=2, 3, \dots.$$

For any $A \in \mathscr{F}_X$, define

$$\mu(A) = \sum_{n=1}^{\infty} \hat{\lambda}(H_n \cap A).$$

Since $H_n \subset \overline{G}_n$ and $\overline{G}_n$ is compact it follows from Proposition 20.11 that $\hat{\lambda}(H_n \cap A)$ is countably additive in A for each fixed n. Hence μ is countably additive. If A is contained in a compact set K, then a finite number of the H_n's cover A. In such a case $\mu(A) = \lambda(A)$. This proves property (ii). Now μ can be extended uniquely to the borel σ-algebra $\mathscr{B}_X$. This completes the proof.

Remark. 20.13. In Proposition 20.11, suppose λ is smooth on *every* compact subset Y of X, considered as a metric space with the same metric as X. Then $\hat{\lambda}(K) = \lambda(K)$, for every compact set K. Under the same condition in Corollary 20.12 $\mu(K) = \lambda(K)$ for every compact set K.

§21. The Lebesgue Measure on the Real Line

Consider the monotonic increasing continuous function $F(x) \equiv x$ on the real line R. By Proposition 15.2 there exists a unique measure L on the boolean algebra $\mathscr{F}$ generated by the class of all intervals of the form $(a, b]$, $(a, \infty]$, $(-\infty, b]$, $(-\infty, +\infty)$, as a and b vary on R. Let $\mathscr{B}^*$ denote the class of all L^*-measurable sets. The set function L^* restricted to $\mathscr{B}^*$ is a σ-finite measure. We shall denote this measure by L itself. Then L is called the *Lebesgue measure on the real line*. Any element of $\mathscr{B}^*$ is called a *Lebesgue measurable set*. It may be recalled from Proposition 18.6 that $(R, \mathscr{B}^*, L)$ is also the completion of $(R, \mathscr{B}(\mathscr{F}), L)$, where $\mathscr{B}(\mathscr{F})$ is the σ-algebra generated by $\mathscr{F}$. Further the discussion in Example 19.3 implies that $\mathscr{B}(\mathscr{F})$ is also the borel σ-algebra of the real line with its usual topology.

By studying the Lebesgue measure carefully we shall establish the existence of sets which are not Lebesgue measurable. To this end we introduce some notation.

For any $E \subset R$, $a \in R$ we shall write

$$E+a = \{x+a,\ x \in E\},$$
$$aE = \{ax,\ x \in E\}.$$

If E and F are two subsets of R, we write

$$E+F = \{x+y,\ x \in E, y \in F\},$$
$$E-F = \{x-y,\ x \in E, y \in F\}.$$

Proposition. 21.1. If $E \subset R$ is Lebesgue measurable and $a \in R$, then $E+a$ is also Lebesgue measurable and $L(E)=L(E+a)$. Similarly aE is Lebesgue measurable and $L(aE) = |a|\, L(E)$.

Proof. If E is a finite disjoint union of intervals it is clear that $E+a$ is also such a set and

$$L(E+a) = L(E),\ L(aE) = |a|\, L(E).$$

By the definition of outer measure it follows that

$$\left.\begin{aligned} &L^*(A+a) = L^*(A), \\ &L^*(aA) = |a|\, L^*(A) \text{ for all } A \subset R,\ a \in R. \end{aligned}\right\} \tag{21.1}$$

If E is L^*-measurable then for any $A \subset R$, we have

$$L^*(A) = L^*(AE) + L^*(AE').$$

By Eq. (21.1)

$$\begin{aligned} L^*(A \cap (E+a)) &= L^*((A-a) \cap E), \\ L^*(A \cap (E+a)') &= L^*(A \cap (E'+a)) \\ &= L^*((A-a) \cap E'). \end{aligned}$$

Adding the two equations we have once again by Eq. (21.1)

$$\begin{aligned} &L^*(A \cap (E+a)) + L^*(A \cap (E+a)') \\ &\quad = L^*(A-a) = L^*(A). \end{aligned}$$

This shows that $E+a$ is L^*-measurable. To prove the second part we note that

$$\begin{aligned} L^*(A \cap (aE)) &= L^*(a[(a^{-1} A) \cap E]) \\ &= |a|\, L^*((a^{-1} A) \cap E), \\ L^*(A \cap (aE)') &= L^*(A \cap (aE')) \\ &= |a|\, L^*((a^{-1} A) \cap E'). \end{aligned}$$

Adding these two equations we obtain the L^*-measurability of aE. This completes the proof.

Proposition. 21.2. For any Lebesgue measurable set $E \subset R$ of finite Lebesgue measure, the function $L(E\Delta(E+x))$ is a uniformly continuous function of x in R.

Proof. Let

$$\begin{aligned} \mathscr{L} = \{E : &\ E \text{ is Lebesgue measurable, } L(E) < \infty, \\ &L(E\Delta(E+x)) \text{ is uniformly continuous in } x\}. \end{aligned}$$

Clearly every bounded interval belongs to $\mathscr{L}$. If $A, B \in \mathscr{L}$, $A \cap B = \emptyset$ then $A \cup B \in \mathscr{L}$. Indeed, we have

$$|\chi_{A\cup B}(y) - \chi_{(A\cup B)+x}(y)| \leqslant |\chi_A(y) - \chi_{A+x}(y)| + |\chi_B(y) - \chi_{B+x}(y)|.$$

By integration (see Section 3)

$$L((A \cup B) \,\Delta\, [(A \cup B)+x]) \leqslant L(A \,\Delta\, (A+x)) + L(B \,\Delta\, (B+x)). \tag{21.2}$$

Since $|L(A) - L(B)| \leqslant L(A \,\Delta\, B)$ for sets of finite measure, we have for any Lebesgue measurable set A of finite Lebesgue measure,

$$|L(A\Delta(A+x)) - L(A \,\Delta\, (A+y))| \leqslant L((A+x) \,\Delta\, (A+y)) = L(A \,\Delta\, [A+(y-x)]). \tag{21.3}$$

Inequalities (21.2) and (21.3) imply that $A \cup B \in \mathscr{L}$. Thus finite disjoint unions of bounded intervals belong to $\mathscr{L}$.

Let now E be any Lebesgue measurable set of finite Lebesgue measure. Then Proposition 18.7 implies that for any $\varepsilon > 0$, there exists a finite disjoint union F_ε of intervals such that

$$L(E \,\Delta\, F_\varepsilon) < \varepsilon.$$

Then

$$L(E\Delta(E+x)) \leqslant L(E\Delta F_\varepsilon) + L(F_\varepsilon \Delta(F_\varepsilon+x)) + L((F_\varepsilon+x)\,\Delta(E+x)) \leqslant 2\varepsilon + L(F_\varepsilon \,\Delta\, (F_\varepsilon+x)).$$

Hence $\lim_{x\to 0} L(E \,\Delta\, (E+x)) = 0$. Now inequality (21.3) implies that $L(E \,\Delta\, (E+x))$ is uniformly continuous in x.

Proposition. 21.3. For any set E of finite positive Lebesgue measure

$$0 \in \{x : L(E \,\Delta\, (E+x)) < L(E)\} \subset E-E.$$

Proof. Suppose $E \cap (E+a) = \emptyset$. Then

$$L(E \,\Delta\, (E+a)) = L(E) + L(E+a) = 2L(E).$$

This shows that, whenever $L(E \,\Delta\, (E+x)) < L(E)$, $E \cap (E+x) \neq \emptyset$. In other words $x \in E-E$. This completes the proof.

Proposition. 21.4. If $L(E) > 0$, the set $E-E$ contains a neighbourhood of 0.

Proof. This is an immediate consequence of Propositions 21.2 and 21.3.

Proposition. 21.5. There exists at least one set $A \subset R$ which is not Lebesgue measurable.

Proof. Let $D \subset R$ be any countable dense subgroup of the additive group R. (for example, D = the subgroup of all rational numbers). By the axiom of choice, pick exactly one point from each coset $D+x$ and thereby make a set A. Then

$$R=A+D=\bigcup_{d\in D}(A+d).$$

We claim that A cannot be Lebesgue measurable. Indeed, if A is Lebesgue measurable, at least one of the $A+d$'s must have positive Lebesgue measure and hence $L(A)=L(A+d)>0$. By Proposition 21.4, the set $A-A$ contains a neighbourhood of 0. Since D is dense the set $(A-A)\cap D$ contains a point $d_0\neq 0$. Then d_0 can be expressed as a_1-a_2 where $a_1, a_2\in A$. Thus a_1, a_2 are two distinct points of A belonging to the same coset. This contradicts the choice of A and hence our claim is proved. The proof is complete.

Exercise. 21.6. If $E\subset R$ is a borel set and $a\in R$, then $E+a$ and aE are borel sets. If μ is a measure on the borel space $(R, \mathscr{B}_R)$ such that $\mu((0, 1])=c<\infty$ and $\mu(A)=\mu(A+a)$ for all $A\in\mathscr{B}_R$ and $a\in R$, then $\mu=cL$, where L is the Lebesgue measure restricted to $\mathscr{B}_R$.

Exercise. 21.7. Let f be a real valued function on R such that $\{x:f(x)\leqslant a\}$ is a Lebesgue measurable set for some $a\neq 0$ and

$$f(x)+f(y)=f(x+y) \text{ for all } x, y\in R.$$

Then, (i) $f(rx)=rf(x)$ for all rational values of r and all $x\in R$; (ii) $\{x:f(x)\leqslant ra\}$ is a Lebesgue measurable set for every rational r; (iii) $\{x:|f(x)|<a\}$ is a Lebesgue measurable set of positive Lebesgue measure for every $a>0$; (iv) f is continuous; (v) $f(x)=cx$ for all x, where c is a constant. (Hint: Use Proposition 21.4).

CHAPTER THREE

Borel Maps

§22. Elementary Properties of Borel Maps

In Chapter 1 we saw many examples of simple random variables and their probability distributions. However, in the field of probability theory it is necessary to study random variables which take uncountably many values and that too in an abstract space. In view of this fact we introduce a more general definition.

Definition. 22.1. Let $(X_i, \mathscr{B}_i)$, $i=1, 2$ be borel spaces and let $f: X_1 \to X_2$ be a map. Then f is said to be a *borel* (or *measurable*) map if

$$f^{-1}(E) = \{x_1 : f(x_1) \in E\} \in \mathscr{B}_1 \text{ for every } E \in \mathscr{B}_2.$$

In the context of probability theory such a borel map f is also called an X_2-valued *random variable* on the sample space X_1. If further f is one-one, onto and the inverse of f is a borel map from $(X_2, \mathscr{B}_2)$ onto $(X_1, \mathscr{B}_1)$ then we say that f is a *borel isomorphism* between $(X_1, \mathscr{B}_1)$ and $(X_2, \mathscr{B}_2)$. If $X_1=X_2$ and $\mathscr{B}_1=\mathscr{B}_2$ such a borel isomorphism is called a *borel automorphism*.

Definition. 22.2. A borel space $(X, \mathscr{B})$ is said to be *standard* if there exists a complete and separable metric space Y and a borel subset $Z \subset Y$ such that $(X, \mathscr{B})$ is isomorphic to the borel space $(Z, \mathscr{B}_Y \cap Z)$.

Proposition. 22.3. If $(X_i, \mathscr{B}_i)$, $i=1, 2, 3$, are three borel spaces and $f: X_1 \to X_2$; $g: X_2 \to X_3$ are borel maps then the composed map $g \circ f: X_1 \to X_3$ defined by $(g \circ f)(x_1) = g(f(x_1))$, $x_1 \in X_1$ is a borel map. All borel automorphisms of a borel space $(X, \mathscr{B})$ constitute a group under the operation of composition.

Proof. The first part is an immediate consequence of the fact that

$$(g \circ f)^{-1}(E) = f^{-1}(g^{-1}(E)) \text{ for any } E \subset X_3.$$

The second part follows from the first and Definition 22.1.

Proposition. 22.4. Let $(X_i, \mathscr{B}_i)$, $i=1, 2$ be borel spaces such that $\mathscr{B}_2$ is generated by a class $\mathscr{E}$ of subsets of X_2. A map $f: X_1 \to X_2$ is borel if and only if $f^{-1}(E) \in \mathscr{B}_1$ for every $E \in \mathscr{E}$.

Proof. The 'only if' part is trivial. To prove the 'if' part consider a map $f: X_1 \to X_2$ such that $f^{-1}(E) \in \mathscr{B}_1$ for every $E \in \mathscr{E}$. Then the family

$$\mathscr{L} = \{A : f^{-1}(A) \in \mathscr{B}_1\}$$

is a σ-algebra which includes $\mathscr{E}$ and hence $\mathscr{B}_2$. This completes the proof.

Proposition. 22.5. Let X, Y be topological spaces and let $\mathscr{B}_X$, $\mathscr{B}_Y$ be their respective borel σ-algebras. Then any continuous map $f: X \to Y$ is a borel map.

Proof. Inverse images of open sets are open and open subsets of Y generate $\mathscr{B}_Y$. By the preceding proposition f is a borel map and the proof is complete.

Definition. 22.6. Let $(X, \mathscr{B})$ be a borel space and let R be the real line. A map $f: X \to R$ is called a real valued *borel* (or *measurable*) function if $f^{-1}(E) \in \mathscr{B}$ for every borel set $E \subset R$, i.e., $E \in \mathscr{B}_R$. Complex valued borel functions are defined similarly. An *extended real valued borel function* f on $(X, \mathscr{B})$ is a map $f: X \to R \cup \{+\infty\} \cup \{-\infty\}$ with the property: $f^{-1}(E) \in \mathscr{B}$ for every $E \in \mathscr{B}_R$; $f^{-1}(\{+\infty\}) \in \mathscr{B}$ and $f^{-1}(\{-\infty\}) \in \mathscr{B}$.

Definition. 22.7. Let $(X_i, \mathscr{B}_i)$, $i=1, 2$ be borel spaces. By a *borel rectangle* in the space $X_1 \times X_2$, we mean a set of the form $E_1 \times E_2$, where $E_i \in \mathscr{B}_i$, $i=1, 2$. All borel rectangles form a boolean semi-algebra. The smallest σ-algebra generated by the class of all borel rectangles is called the *product σ-algebra* and denoted by $\mathscr{B}_1 \times \mathscr{B}_2$. The borel space $(X_1 \times X_2, \mathscr{B}_1 \times \mathscr{B}_2)$ is called *the product borel space.*

Remark. 22.8. In Section 4 (See Exercise 4.6 and Remark 4.11) we saw how the product of an arbitrary family of boolean spaces is constructed. We have to modify the definition slightly in the case of borel spaces. Let $(X_\alpha, \mathscr{B}_\alpha)$, $\alpha \in \Gamma$ be a family of borel spaces. Let $\widetilde{X} = \prod_{\alpha \in \Gamma} X_\alpha$ be the cartesian product of the X_α's. As in Remark 4.11 we denote an arbitrary point in $\widetilde{X}$ by x and its 'αth coordinate' by $x(\alpha)$. Let

$$\pi_\alpha : \widetilde{X} \to X_\alpha, \; \pi_\alpha(x) = x(\alpha), \; \alpha \in \Gamma$$

be the projection map from $\widetilde{X}$ onto X_α. Since $\mathscr{B}_\alpha$'s are also boolean algebras we can form the product boolean algebra $\tilde{\mathscr{B}}$ as in Remark 4.11. The smallest σ-algebra generated by $\tilde{\mathscr{B}}$ is called the *product σ-algebra* and denoted by $\prod_{\alpha \in \Gamma} \mathscr{B}_\alpha$. The borel space $(\tilde{X}, \prod_{\alpha \in \Gamma} \mathscr{B}_\alpha)$ is called the *product borel space* and denoted by $\prod_{\alpha \in \Gamma} (X_\alpha, \mathscr{B}_\alpha)$. The product σ-algebra is the smallest σ-algebra which makes every projection map π_α a borel map.

Exercise. 22.9. Let (Ω, S), $(X_i, \mathscr{B}_i)$, $i=1, 2$ be borel spaces. Let $f_i : \Omega \to X_i$, $i=1, 2$ be borel maps. Then the map $f: \Omega \to X_1 \times X_2$ defined by $f(\omega)=(f_1(\omega), f_2(\omega))$, $\omega \in \Omega$ is a borel map from (Ω, S) into $(X_1 \times X_2, \mathscr{B}_1 \times \mathscr{B}_2)$.

Exercise. 22.10. Let X, Y be second countable topological spaces and let $X \times Y$ be the product topological space. Then $\mathscr{B}_{X \times Y} = \mathscr{B}_X \times \mathscr{B}_Y$. In particular, $\mathscr{B}_{R^2} = \mathscr{B}_R \times \mathscr{B}_R$; $\mathscr{B}_{R^n} = \mathscr{B}_R \times \mathscr{B}_R \times \ldots \times \mathscr{B}_R$ (n times). (The second countability assumption is essential!) If $\{X_n\}$, $n=1, 2, \ldots$ is a sequence of second countable topological spaces and $X^\infty = \prod_{n=1}^{\infty} X_n$ is the product topological space then $\mathscr{B}_{X^\infty} = \prod_{n=1}^{\infty} \mathscr{B}_{X_n}$. (Hint: if $\mathscr{B}_1$, $\mathscr{B}_2$ are σ-algebras generated by the families $\mathscr{E}_1$, $\mathscr{E}_2$ respectively then $\mathscr{B}_1 \times \mathscr{B}_2$ is generated by the family $\{E_1 \times E_2 : E_i \in B_i, i=1, 2\}$. See Proposition 13.11).

Exercise. 22.11. Let (Ω, S), $(X_i, \mathscr{B}_i)$, $i=1, 2, 3$ be borel spaces. Let $f_i : \Omega \to X_i$, $i=1, 2$ be borel maps and let $g : X_1 \times X_2 \to X_3$ be a borel map, where $X_1 \times X_2$ has the product σ-algebra $\mathscr{B}_1 \times \mathscr{B}_2$. Then the map $h : \Omega \to X_3$ defined by

$$h(\omega) = g(f_1(\omega), f_2(\omega)), \ \omega \in \Omega$$

is also a borel map.

Proposition. 22.12. If f, g are real (or complex) valued borel functions on a borel space $(X, \mathscr{B})$ then $f+g$ and fg are also borel functions.

Proof. Consider the maps $(\xi, \eta) \to \xi + \eta$ and $(\xi, \eta) \to \xi\eta$ from $R \times R$ onto R. These are continuous and hence by Proposition 22.5 borel maps. By Exercises 22.9 and 22.10 the map $x \to (f(x), g(x))$ is borel from X into $R \times R$. By Exercise 22.11, the maps $x \to f(x) + g(x)$ and $x \to f(x)\, g(x)$ are also borel. This completes the proof.

Exercise. 22.13. If f, g are extended real valued borel functions then $f+g$ and fg are also borel functions provided they are well defined.

Exercise. 22.14. Suppose the space of all $n \times n$ real matrices is given the topology of R^{n^2} and denoted by Υ. If $(X, \mathscr{B})$ is a borel space and $f: X \to \Upsilon$, $g: X \to \Upsilon$ are borel maps from $(X, \mathscr{B})$ into $(\Upsilon, \mathscr{B}_\Upsilon)$, then the map fg defined by $(fg)(x) = f(x)g(x)$ is borel. (In the language of probability theory we may say that the product of two matrix valued random variables is also a matrix valued random variable.)

§23. Borel Maps into Metric Spaces

In the preceding section (see Exercise 22.11) we saw how 'borel operations' on two borel maps lead once again to borel maps. In particular borel functions remain closed under the usual algebraic operations like addition, multiplication, subtraction and division. In the present section we shall prove that 'limits' of borel maps are also borel maps and show how a general borel map can be constructed as a limit of borel maps with finite or countable range. To this end we shall consider metric space valued borel maps.

As usual we denote by $\mathscr{B}_X$, the borel σ-algebra of any metric space X. Any X-valued borel map will be relative to $\mathscr{B}_X$.

Proposition. 23.1. Let X be a metric space. The borel σ-algebra $\mathscr{B}_X$ is the smallest σ-algebra with respect to which every real valued continuous function is borel.

Proof. Let $\mathscr{B}_0$ be the smallest σ-algebra with respect to which every real continuous function on X is borel. Then $\mathscr{B}_0$ is the σ-algebra generated by the family

$$\{f^{-1}(B),\ B \in \mathscr{B}_R,\ f \text{ any real continuous function on } X\}.$$

Let U be any open subset of R. Then $f^{-1}(U)$ is open in X and hence belongs to $\mathscr{B}_X$, whenever f is a real valued continuous function. By Proposition 22.4, $f^{-1}(B) \in \mathscr{B}_X$ for every $B \in \mathscr{B}_R$. Hence $\mathscr{B}_0 \subset \mathscr{B}_X$.

Let now V be any open set in X. Then the complement V' is closed. Consider the function $g(x) = d(x, V')$, which is defined as in Proposition 19.11. By the same proposition g is continuous and $V' = \{x : g(x) = 0\}$. Hence $V = g^{-1}(R - \{0\}) \in \mathscr{B}_0$. Hence $\mathscr{B}_X \subset \mathscr{B}_0$. This shows that $\mathscr{B}_0 = \mathscr{B}_X$ and the proof is complete.

Exercise. 23.2. Let X be a metric space. Then $\mathscr{B}_X$ is the smallest family of subsets of X which contains all the open (closed) subsets of X, and which is closed under countable intersections and countable unions. (Hint: use corollary 19.12).

Remark. 23.3. The following variation of the preceding exercise is sometimes useful. If X is a metric space then $\mathscr{B}_X$ is also the smallest family which contains all the open subsets of X, and which is closed under countable intersections and countable disjoint unions. To prove this claim we denote by $\mathscr{L}_0$ the smallest class containing all the open sets and closed under countable disjoint unions and countable intersections. The claim will be proved if $\mathscr{L}_0$ is closed under complementation.

By Corollary 19.12 every closed set is a G_δ and hence belongs to $\mathscr{L}_0$. Let

$$\mathscr{A} = \{E : E \subset X, E \text{ is either open or closed}\};$$

$$\mathscr{L}_1 = \{E : E \in \mathscr{L}_0, E' \in \mathscr{L}_0\}.$$

Then $\mathscr{A} \subset \mathscr{L}_1 \subset \mathscr{L}_0$. Now we shall prove that $\mathscr{L}_1$ is closed under countable unions and countable intersections. Indeed, let $A_i \in \mathscr{L}_1$, $i = 1, 2, \ldots$. Since $A_i \in \mathscr{L}_0$, $i = 1, 2, \ldots$, $\bigcap_i A_i \in \mathscr{L}_0$. Further $(\bigcap_i A_i)' = \bigcup_i A_i' = \bigcup_i (A_i' A_1 A_2 \ldots A_{i-1})$ is a countable disjoint union of sets in $\mathscr{L}_0$ and hence belongs to $\mathscr{L}_0$. Thus $\bigcap_i A_i \in \mathscr{L}_1$. Now $\bigcup_i A_i = \bigcup_i (A_i A_1' A_2' \ldots A_{i-1}')$ is a countable disjoint union of sets in $\mathscr{L}_0$ and hence belongs to $\mathscr{L}_0$. Further $(\bigcup_i A_i)' = \bigcap_i A_i' \in \mathscr{L}_0$. Thus $\bigcup_i A_i \in \mathscr{L}_1$. Hence $\mathscr{L}_1 = \mathscr{L}_0$ and the claim is proved.

Proposition. 23.4. Let Ω, S) be a borel space and let X be a metric space. A map $f : \Omega \to X$ is borel if and only if, for every real continuous function g on X, $g \circ f$ is a borel function.

Proof. If f is borel and g is continuous then $g \circ f$ is borel. Conversely, if $g \circ f$ is borel for every continuous g then $(g \circ f)^{-1} (B) = f^{-1} (g^{-1} (B)) \in$ S for every $B \in \mathscr{B}_R$. Since, by Proposition 23.1, sets of the form $g^{-1} (B)$ generate $\mathscr{B}_X$ as g varies over all continuous functions and B over $\mathscr{B}_R$, it follows that f is borel and the proof is complete.

Definition. 23.5. A sequence $\{f_n\}$ of maps from a set Ω into a metric space X is said to *converge pointwise* to a map f if $\lim_{n\to\infty} d(f_n(\omega), f(\omega)) = 0$ for every $\omega \in \Omega$, where d is the metric in X. It is said to *converge uniformly* to f if $\lim_{n\to\infty} \sup_{\omega \in \Omega} d(f_n(\omega), f(\omega)) = 0.$

Proposition. 23.6. Let $\{f_n\}$ be a sequence of extended real valued borel functions on a borel space (Ω, S). Then the functions $\sup_n f_n(\omega)$, $\inf_n f_n(\omega)$, $\overline{\lim_{n\to\infty}} f_n(\omega)$, $\underline{\lim_{n\to\infty}} f_n(\omega)$, $\omega \in \Omega$ are all borel.

Proof. Let $\tilde{f}(\omega) = \sup_n f_n(\omega)$ for all $\omega \in \Omega$. Then $\tilde{f}$ is an extended real valued function. Further,

$$\{\omega : \tilde{f}(\omega) \leqslant x\} = \bigcap_n \{\omega : f_n(\omega) \leqslant x\}, \; x \in R,$$

$$\{\omega : \tilde{f}(\omega) = +\infty\} = \bigcap_{N=1}^{\infty} \bigcup_n \{\omega : f_n(\omega) > N\},$$

$$\{\omega : \tilde{f}(\omega) = -\infty\} = \bigcap_{N=1}^{\infty} \bigcap_n \{\omega : f_n(\omega) \leqslant -N\}.$$

Since every entrant on the right side of the above equations is an element of S, the sets on the left hand side also belong to S. Since intervals of the form $(-\infty, x]$ generate $\mathscr{B}_R$, it follows that $\tilde{f}$ is borel. Since $\inf_n f_n(\omega) = -\sup_n (-f_n(\omega))$, it is clear that $\inf_n f_n$ is borel. If we now note that

$$\overline{\lim_{n\to\infty}} f_n(\omega) = \inf_n \sup_{k\geqslant n} f_k(\omega),$$

$$\underline{\lim_{n\to\infty}} f_n(\omega) = -\overline{\lim_{n\to\infty}} (-f_n(\omega)),$$

the proof is complete.

Corollary. 23.7. If $\{f_n\}$ is a sequence of extended real valued borel functions on a borel space (Ω, S) and f_n converges to a limit f pointwise, then f is borel.

Proof. It is only necessary to note that the limit function f is also the same as $\overline{\lim_{n\to\infty}} f_n$.

Exercise. 23.8. Let $\{f_n\}$ be a sequence of extended real valued borel functions on a borel space (Ω, S). Then $\{\omega : f_n(\omega)$ converges as $n\to\infty\}$ is an element of S.

Proposition. 23.9. Let $\{f_n\}$ be a sequence of borel maps from the borel space (Ω, S) into a metric space X. If f_n converges pointwise to a limit f, then f is a borel map.

Proof. Let g be any real valued continuous function X. Since $\lim_{n\to\infty} f_n(\omega)=f(\omega)$, $\lim_{n\to\infty} g(f_n(\omega))=g(f(\omega))$. By Corollary 23.7, $g \circ f$ is borel. Since this holds for every continuous g it follows from Proposition 23.4 that f is borel. This completes the proof.

Definition. 23.10. Let (Ω, S) be a borel space and let X be any metric space. A map $s:\Omega\to X$ is said to be *simple* if it is borel and takes only a finite number of values.

Remark. 23.11. Every simple map s is of the following form: there exists a finite partition of Ω into sets $A_1, A_2, \ldots, A_k$, $A_i\in S$, $i=1, 2, \ldots, k$ and a set of points $x_1, x_2, \ldots, x_k$ in X such that

$$s(\omega) = x_i \text{ if } \omega\in A_i,\ i=1, 2, \ldots, k.$$

Proposition. 23.12. Let (Ω, S) be a borel space and let X be a compact metric space. If $f: \Omega\to X$ is a borel map then there exists a sequence of simple maps $\{s_n\}$ such that s_n converges to f uniformly.

Proof. Let $\varepsilon > 0$ be arbitrary. Then the space X can be written as a finite union of open spheres of radius $\frac{1}{2}\varepsilon$. Hence X can be written as a finite disjoint union of borel sets of diameter $\leqslant\varepsilon$. Let $X= \bigcup_{i=1}^{N} A_i$, where A_i are disjoint borel sets of diameter $\leqslant\varepsilon$. Then $\Omega = \bigcup_{i=1}^{N} f^{-1}(A_i)$. Let $x_i\in A_i$, $i = 1, 2, \ldots, N$ be arbitrary. Define a simple function s_ε as follows:

$$s_\varepsilon(\omega) = x_i \text{ if } \omega \in f^{-1}(A_i),\ i=1, 2, \ldots, N.$$

Since f is a borel map, $f^{-1}(A_i)$ are in S. If $\omega\in f^{-1}(A_i)$, then $f(\omega)\in A_i$ and $d(f(\omega), x_i) \leqslant \varepsilon$, where d is the metric in X. Thus

$$\sup_\omega d(f(\omega), s_\varepsilon(\omega)) \leqslant \varepsilon.$$

It is clear that the sequence $s_{1/n}$ converges to f uniformly as $n\to\infty$. This completes the proof.

Proposition. 23.13. Let f be an extended real valued function on the borel space (Ω, S). Then there is a sequence $\{s_n\}$ of simple functions on Ω converging pointwise to f. If f is non-negative then s_n can be chosen to be non-negative and monotonically increasing.

Proof. To prove the first part we define for any $N > 0$,

$$\begin{aligned} f_N(\omega) &= f(\omega) \text{ if } |f(\omega)| < N,\\ &= -N \text{ if } f(\omega) \leqslant -N,\\ &= N \text{ if } f(\omega) \geqslant N. \end{aligned}$$

Then $\lim_{N\to\infty} f_N(\omega) = f(\omega)$ for every $\omega \in \Omega$. Since f_N takes values in the compact interval $[-N, N]$ we apply Proposition 23.12 and construct a simple function s_N such that

$$\sup_{\omega} |f_N(\omega) - s_N(\omega)| \leqslant \frac{1}{N}.$$

It is then clear that s_N converges pointwise to f as $N \to \infty$.

Now suppose $f(\omega) \geqslant 0$ for all ω. Then the sequence $\{f_N\}$ defined above monotonically increases to f. Define

$$t_N(\omega) = \max\left\{0, s_1(\omega)-1, s_2(\omega)-\tfrac{1}{2}, \ldots, s_N(\omega)-\frac{1}{N}\right\}.$$

Then t_N is also simple and $s_N(\omega) - \frac{1}{N} \leqslant t_N(\omega) \leqslant f_N(\omega)$ for all ω.

Hence $t_N(\omega)$ increases monotonically to $f(\omega)$ as $n \to \infty$ for every ω. This completes the proof.

Exercise. 23.14. Let (Ω, S) be a borel space and let f be a borel map from Ω into a separable metric space X. Then there exists a sequence $\{f_n\}$ of borel maps from Ω into X such that each f_n takes only a countable number of values and f_n converges uniformly to f as $n \to \infty$.

§24. Borel Maps on Measure Spaces

We shall now study properties of borel maps in relation to a measure. Let (Ω, S, μ) be a σ-finite measure space and let $(X, \mathscr{B})$ be a borel space. For any borel map $f: \Omega \to X$, define the function μf^{-1} on $\mathscr{B}$ by the equation

$$(\mu f^{-1})(B) = \mu[f^{-1}(B)], B \in \mathscr{B}.$$

If $B_1, B_2, \ldots$ is a sequence of disjoint sets in $\mathscr{B}$, then $f^{-1}(B_1), f^{-1}(B_2), \ldots$ is a disjoint sequence in S and $f^{-1}(\bigcup_i B_i) = \bigcup_i f^{-1}(B_i)$. Hence

$$\mu f^{-1}(\bigcup_i B_i) = \sum_i \mu f^{-1}(B_i).$$

In other words μf^{-1} is countably additive. Since $f^{-1}(X) = \Omega$, it follows that μf^{-1} is totally finite whenever μ is. Further μf^{-1} is a probability measure whenever μ is.

Whenever μf^{-1} is a σ-finite measure on $\mathscr{B}$ we say that μf^{-1} is the *measure induced by* the borel map f. If μ is a probability measure we say that μf^{-1} is the *probability distribution of the* X-*valued random variable* f.

Remark. 24.1. In probability theory any statistical experiment is described by a probability space (Ω, S, μ). The performance of the experiment leads to an observation $\omega \in \Omega$. The probability that the event $\omega \in E$ (where $E \in S$) takes place is equal to $\mu(E)$. Then we evaluate a statistical characteristic f at the point ω. This characteristic f may take values in any abstract space X. The probability that this characteristic takes a value lying in $F \subset X$ is the number $\mu\{\omega: f(\omega) \in F\} = \mu(f^{-1}(F)) = (\mu f^{-1})(F)$. In order that these statements may be meaningful it is necessary to assume that X has a borel structure and f is borel map on Ω.

Definition. 24.2. Let (Ω, S, μ) be a σ-finite measure space and let $(X, \mathscr{B})$ be a borel space. Two borel maps f, g from Ω into X are said to be μ-equivalent if there exists a set $N \in S$ such that $\mu(N) = 0$ and $f(\omega) = g(\omega)$ for all $\omega \notin N$. Whenever there is no confusion we simply say that f and g are equivalent and write $f \sim g$. The relation '$\sim$' is indeed an equivalence.

Proposition. 24.3. Let (Ω, S) be a borel space and let X be any separable metric space. For any two borel maps f, g from Ω into X, the set $\{\omega : f(\omega) \neq g(\omega)\} \in S$.

Proof. If d is the metric in X, then

$$\{\omega : f(\omega) \neq g(\omega)\} = \{\omega : d(f(\omega), g(\omega)) \neq 0\}.$$

Since d is a continuous function on the product topological space $X \times X$, d is a borel function on $(X \times X, \mathscr{B}_{X \times X})$. By Exercise 22.10, $\mathscr{B}_{X \times X} = \mathscr{B}_X \times \mathscr{B}_X$. Hence by Exercise 22.11 the map $\omega \to d(f(\omega), g(\omega))$ is borel on (Ω, S). This shows that $\{\omega : f(\omega) \neq g(\omega)\} \in S$.

Exercise. 24.4. Let (Ω, S, μ) be a σ-finite measure space and let f, g be equivalent borel maps on Ω into a borel space $(X, \mathscr{B})$. If μf^{-1} is σ-finite so is μg^{-1} and $\mu f^{-1} = \mu g^{-1}$.

Exercise. 24.5. If $\{f_n\}$, $\{g_n\}$ are two sequences of borel maps from (Ω, S, μ) into a metric space X, $f_n \sim g_n$ for every n and $f_n \to f$, $g_n \to g$ pointwise as $n \to \infty$, then $f \sim g$.

Exercise. 24.6. Let f_i, g_i, $i=1, 2$, be extended real valued borel functions on (Ω, S, μ). Suppose $f_1 \sim f_2$, $g_1 \sim g_2$. Then (i) $f_1 + g_1 \sim f_2 + g_2$; (ii) $f_1 g_1 \sim f_2 g_2$ if these functions are well defined. If $\{f_n\}$ and $\{g_n\}$ are sequences of extended real valued borel functions and $f_n \sim g_n$ for every n, then (i) $\overline{\lim}\, f_n \sim \overline{\lim}\, g_n$; (ii) $\underline{\lim}\, f_n \sim \underline{\lim}\, g_n$.

Remark. 24.7. Hereafter by a borel map on a measure space $(\Omega, \mathsf{S}, \mu)$, we shall actually mean the equivalence class to which it belongs. Thus, in order to specify a borel map in a measure space $(\Omega, \mathsf{S}, \mu)$, it is enough to define it outside a set $\mathcal{N} \in \mathsf{S}$ with $\mu(\mathcal{N})=0$.

By the preceding exercise it follows in particular that extended real valued borel functions on $(\Omega, \mathsf{S}, \mu)$ are closed under $\overline{\lim}$, $\underline{\lim}$, addition, multiplication, etc., provided they are defined.

Definition. 24.8. Let $(\Omega, \mathsf{S}, \mu)$ be a σ-finite measure space and let $\{f_n\}$ be a sequence of borel maps from Ω into a separable metric space X with metric d. f_n is said to *converge in measure* to a borel map f if, for every $E \subset \Omega$ such that $E \in \mathsf{S}$ and $\mu(E) < \infty$,

$$\lim_{n\to\infty} \mu(E \cap \{\omega : d(f_n(\omega), f(\omega)) > \varepsilon\}) = 0$$

for all $\varepsilon > 0$. f_n is said to *converge almost everywhere* to a borel map f if

$$\mu\{\omega : \lim_{n\to\infty} d(f_n(\omega), f(\omega)) \neq 0\} = 0.$$

In such a case we write

$$f_n \to f \qquad \text{a.e. } (\mu).$$

If μ is a probability measure, convergence in measure is called *convergence in probability* and convergence almost every where is called *almost sure convergence* or *convergence with probability one.* If convergence in measure or almost everywhere takes place in the space $(\Upsilon, \Upsilon \cap \mathsf{S}, \mu)$ where $\Upsilon \subset \Omega$ and $\Upsilon \in \mathsf{S}$, we say that f_n converges in measure or almost everywhere accordingly on the set Υ.

Definition. 24.9. A sequence $\{f_n\}$ of borel maps from $(\Omega, \mathsf{S}, \mu)$ into a separable metric space X with metric d is said to be *fundamental in measure* if, for every $E \in \mathsf{S}$ with $\mu(E) < \infty$ and every $\varepsilon > 0$, we have

$$\lim_{m,\, n\to\infty} \mu(E \cap \{\omega : d(f_m(\omega), f_n(\omega)) > \varepsilon\}) = 0.$$

Remark. 24.10. It may be noted that we have assumed X to be a separable metric space in Definitions 24.8 and 24.9 to ensure that $d(f(\omega), g(\omega))$ is a borel function in ω whenever f and g are borel maps from Ω into X.

Proposition. 24.11. (Egorov's theorem). Let $(\Omega, \mathsf{S}, \mu)$ be a totally finite measure space. Let $\{f_n\}$ be a sequence of borel maps from (Ω, S) into a separable metric space X with metric d such that f_n converges almost everywhere to a borel map f. Then, for any $\varepsilon > 0$, there exists a set $\mathcal{N}_\varepsilon \in \mathsf{S}$ such that $\mu(\mathcal{N}_\varepsilon) < \varepsilon$ and

$$\lim_{n\to\infty} \sup_{\omega \in \mathcal{N}\varepsilon} d(f_n(\omega), f(\omega)) = 0 \tag{24.1}$$

Proof. By neglecting a set of measure zero (if necessary) we may assume without loss of generality that f_n converges to f pointwise on Ω. Let

$$F_n^{\,m} = \bigcap_{i=n}^{\infty} \{\omega : d(f_i(\omega), f(\omega)) < 1/m\}.$$

Then

$$F_1^{\,m} \subset F_2^{\,m} \subset \ldots.$$

Since f_n converges to f

$$\bigcup_n F_n^{\,m} = \Omega, \text{ for } m = 1, 2, \ldots.$$

Since μ is totally finite, there exists $n_0(m)$ such that

$$\mu\left(F^{m'}_{n_0(m)}\right) < \frac{\varepsilon}{2^m}.$$

Let

$$\mathcal{N}_\varepsilon = \bigcup_{m=1}^{\infty} F^{m'}_{n_0(m)}.$$

Then $\mu(\mathcal{N}_\varepsilon) < \varepsilon$. Suppose $\omega \notin \mathcal{N}_\varepsilon$. Then $\omega \in F^m_{n_0(m)}$ for all $m = 1, 2, \ldots$. Hence

$$d(f_i(\omega), f(\omega)) < \frac{1}{m} \text{ for every } i \geqslant n_0(m).$$

In other words Eq. (24.1) holds. This completes the proof.

Proposition. 24.12. Let (Ω, S, μ) be a totally finite measure space and let $\{f_n\}$ be a sequence of borel maps from Ω into a separable metric space X. If f is a borel map from Ω into X then $f_n \to f$ a.e. (μ) if and only if, for every $\varepsilon > 0$,

$$\lim_{n\to\infty} \mu\left(\bigcup_{m=n}^{\infty} E_m(\varepsilon)\right) = 0 \tag{24.2}$$

where

$$E_n(\varepsilon) = \{\omega : d(f_n(\omega), f(\omega)) > \varepsilon\},\ n = 1, 2, \ldots. \tag{24.3}$$

In particular, convergence almost everywhere implies convergence in measure.

Proof. It is clear that $f_n(\omega) \not\to f(\omega)$ as $n \to \infty$ for a particular ω if and only if, for some $\varepsilon > 0$, $\omega \in E_n(\varepsilon)$ for infinitely many n. If $D = \{\omega : f_n(\omega) \not\to f(\omega)\}$ then

$$\begin{aligned} D &= \bigcup_{\varepsilon>0} \bigcap_{n=1}^{\infty} \bigcup_{m=n}^{\infty} E_m(\varepsilon) \\ &= \bigcup_{k=1}^{\infty} \bigcap_{n=1}^{\infty} \bigcup_{m=n}^{\infty} E_m\left(\frac{1}{k}\right). \end{aligned}$$

Thus $f_n \to f$ a.e. (μ) if and only if $\mu(D)=0$, i.e.,

$$\mu\left(\bigcap_{n=1}^{\infty} \bigcup_{m=n}^{\infty} E_m\left(\frac{1}{k}\right)\right)=0 \text{ for every } k=1, 2,\ldots.$$

By Proposition 15.10 this holds, if and only if,

$$\lim_{n\to\infty} \mu\left(\bigcup_{m=n}^{\infty} E_m\left(\frac{1}{k}\right)\right)=0 \text{ for every } k=1, 2,\ldots.$$

This holds if and only if, for every $\varepsilon>0$, Eq. (24.2) holds. The last part is an immediate consequence of Eqs. (24.2) and (24.3). This completes the proof.

Proposition. 24.13. (Borel-Cantelli lemma). Let (Ω, S, P) be a probability space and let $\{A_n\}$ be a sequence of events such that

$$\sum_{n=1}^{\infty} P(A_n) < \infty.$$

Then

$$P\left(\bigcap_{m=1}^{\infty} \bigcup_{n=m}^{\infty} A_n\right)=0,$$

i.e., with probability one A_n occurs only for finitely many n.

If $\{A_n\}$ is a sequence of mutually independent events and

$$\sum_{n=1}^{\infty} P(A_n) = \infty, \text{ then}$$

$$P\left(\bigcap_{m=1}^{\infty} \bigcup_{n=m}^{\infty} A_n\right)=1,$$

i.e., with probability one A_n occurs for infinitely many n.

If f_n, f are random variables on Ω with values in a separable metric space X with metric d and

$$\sum_{n=1}^{\infty} P\{\omega: d(f_n(\omega), f(\omega)) > \varepsilon\} < \infty$$

for every $\varepsilon > 0$, then $f_n \to f$ a.e. (P).

Proof. Suppose $\sum_{n=1}^{\infty} P(A_n) < \infty$. Since

$$P\left(\bigcap_{m=1}^{\infty} \bigcup_{n=m}^{\infty} A_n\right) \leqslant \sum_{n=k}^{\infty} P(A_n),$$

for every k we get the first part of the proposition by letting $k \to \infty$ in the above inequality.

Suppose A_n's are mutually independent and

$$\sum_{n=1}^{\infty} P(A_n) = \infty.$$

Then $$\prod_{n=k}^{\infty} (1 - P(A_n)) = 0$$

for every k. Since the complements of A_n are also mutually independent we have

$$P(\bigcap_{n=k}^{\infty} A'_n) = \prod_{n=k}^{\infty} P(A'_n) = 0 \text{ for all } k.$$

Thus

$$P(\bigcup_{k=1}^{\infty} \bigcap_{n=k}^{\infty} A'_n) = 0.$$

Or equivalently

$$P(\bigcap_{k=1}^{\infty} \bigcup_{n=k}^{\infty} A_n) = 1.$$

This proves the second part.

The last part is an immediate consequence of the first part and the definition of almost everywhere convergence.

Remark. 24.14. As an application of Proposition 24.12, we shall take a second look at Corollary 11.5 of Chapter 1. Suppose $s_1, s_2, \ldots, s_n, \ldots$ are independent simple random variables on a probability space (Ω, S, P), where P is now a probability measure on the σ-algebra S. Since S is, in particular, a boolean algebra the discussion of Corollary 11.5 holds. If $\mathbf{E}s_i = m_i$, $\mathbf{V}(s_i) = \sigma_i^2$, $i = 1, 2, \ldots$ and $\sum_{k=1}^{\infty} k^{-2} \sigma_k^2 < \infty$, then for any $\varepsilon > 0$, $\delta > 0$ we deduce that

$$P\left\{\frac{|S_n - M_n|}{n} \leqslant \varepsilon \text{ for every } n \geqslant N\right\} \geqslant 1 - \delta$$

for some N depending on ε, δ, where S_n denotes $s_1 + s_2 + \ldots + s_n$. In other words

$$\lim_{N\to\infty} P\left\{\frac{|S_n - M_n|}{n} > \varepsilon \text{ for some } n \geqslant N\right\} = 0.$$

Now Proposition 24.12 implies that the sequence of random variables $\frac{S_n - M_n}{n}$ converges to zero with probability one. We summarise in the form of a proposition.

Proposition. 24.15. (Strong Law of Large Numbers) If $s_1, s_2, \ldots$ is a sequence of independent simple random variables on a probability space (Ω, S, P), $\mathbf{E}s_i = m_i$, $i = 1, 2, \ldots$,

$$V(s_i) = \sigma_i^2 \text{ and } \sum_{k=1}^{\infty} k^{-2}\sigma_k^2 < \infty, \text{ then}$$

$$\lim_{n \to \infty} \frac{(s_1+s_2+\ldots+s_n)-(m_1+m_2+\ldots+m_n)}{n} = 0$$

with probability one.

Corollary. 24.16. If $s_1, s_2, \ldots$ is a sequence of independent simple random variables on a probability space (Ω, S, P) with $\mathbf{E}s_i = m$, $\mathbf{V}(s_i) = \sigma^2$ for all $i = 1, 2, \ldots$ then $\frac{1}{n}(s_1+s_2+\ldots+s_n)$ converges to m with probability one.

Proposition. 24.17. Let (Ω, S, μ) be a totally finite measure space. If $\{f_n\}$ is a sequence of borel maps on Ω into a separable metric space X with metric d and f_n converges to a borel map f in measure then $\{f_n\}$ is fundamental in measure. If f_n converges in measure to another borel map g, then $f \sim g$.

Proof. We have for any $\varepsilon > 0$,

$$\{\omega : d(f_n(\omega), f_m(\omega)) > \varepsilon\} \subset$$
$$\{\omega : d(f_n(\omega), f(\omega)) + d(f(\omega), f_m(\omega)) > \varepsilon\} \subset$$
$$\left\{\omega : d(f_n(\omega), f(\omega)) > \frac{\varepsilon}{2}\right\} \cup \left\{\omega : d(f_m(\omega), f(\omega)) > \frac{\varepsilon}{2}\right\}.$$

Hence the first part follows immediately. To prove the second part we note that

$$\{\omega : d(f(\omega), g(\omega)) > \varepsilon\} \subset$$
$$\left\{\omega : d(f(\omega), f_n(\omega)) > \frac{\varepsilon}{2}\right\} \cup \left\{\omega : d(f_n(\omega), g(\omega)) > \frac{\varepsilon}{2}\right\}.$$

Since the measure of the set on the right hand side can be made arbitrarily small by making n large, we have

$$\mu\{\omega : d(f(\omega), g(\omega)) > \varepsilon\} = 0.$$

Since ε is arbitrary the proof is complete.

Proposition. 24.18. Let (Ω, S, μ) be a totally finite measure space and let X be a complete and separable metric space. Suppose $\{f_n\}$ is a sequence of borel maps from Ω into X, which is fundamental in measure. Then $\{f_n\}$ has a subsequence which converges almost everywhere.

Proof. For any integer k let $\bar{n}(k)$ be an integer such that

$$\mu \left\{ \omega : d(f_n(\omega), f_m(\omega)) \geqslant \frac{1}{2^k} \right\} < \frac{1}{2^k},$$

whenever $n \geqslant \bar{n}(k)$, $m \geqslant \bar{n}(k)$. The existence of such an $\bar{n}(k)$ follows from the fact that $\{f_n\}$ is fundamental in measure. Let

$$n_1 = \bar{n}(1),\ n_2 = \max\ (n_1+1,\ \bar{n}(2)),\ n_3 = \max\ (n_2+1,\ \bar{n}(3)), \ldots.$$

Then $n_1 < n_2 < n_3 < \ldots$ and $\{f_{n_k}\}$ is a subsequence of $\{f_n\}$. Let

$$g_k = f_{n_k},\ k = 1, 2, \ldots,$$

$$E_k = \left\{ \omega : d(g_k(\omega), g_{k+1}(\omega)) \geqslant \frac{1}{2^k} \right\}.$$

If $j \geqslant i \geqslant k$ and $\omega \notin \bigcup_{j=k}^{\infty} E_j$ then

$$\begin{aligned} d(g_i(\omega), g_j(\omega)) &\leqslant \sum_{r=i}^{j-1} d(g_r(\omega), g_{r+1}(\omega)) \\ &\leqslant \frac{1}{2^i} + \frac{1}{2^{i+1}} + \ldots \\ &= \frac{1}{2^{i-1}}. \end{aligned}$$

This shows that outside $E_k \cup E_{k+1} \cup \ldots$, the sequence $\{g_n(\omega)\}$ is a Cauchy sequence in X. The completeness of X implies that there exists a $g(\omega) \in X$ such that $g_n(\omega) \to g(\omega)$ as $n \to \infty$. Since k is arbitrary it follows that $g_n(\omega)$ converges to $g(\omega)$ for every $\omega \notin \bigcap_{k=1}^{\infty} (E_k \cup E_{k+1} \cup \ldots) = E$, say. We have

$$\begin{aligned} \mu(E) &= \lim_{k\to\infty} \mu(E_k \cup E_{k+1} \cup \ldots) \\ &\leqslant \lim_{k\to\infty} \mu(E_k) + \mu(E_{k+1}) + \ldots \\ &\leqslant \lim_{k\to\infty} \frac{1}{2^k} + \frac{1}{2^{k+1}} + \ldots \\ &= 0. \end{aligned}$$

For $\omega \in E$, define $g(\omega)$ to be some fixed point in X. Then $g_n \to g$ a.e. This completes the proof.

Proposition. 24.19. Let (Ω, S, μ) be a totally finite measure space and let X be a complete and separable metric space. Suppose $\{f_n\}$ is a sequence of borel maps from Ω into X which is fundamental in measure. Then there exists a borel map f such that f_n converges to f in measure.

Proof. By the preceding proposition there exists a subsequence $\{f_{n_k}\}$ which converges almost everywhere to a borel map f. We have, for any $\varepsilon>0$,

$$\left\{\omega : d(f_n(\omega), f(\omega)) \geqslant \varepsilon\right\} \subset \left\{\omega : d(f_n(\omega), f_{n_k}(\omega)) \geqslant \frac{\varepsilon}{2}\right\} \cup$$

$$\left\{\omega : d(f_{n_k}(\omega), f(\omega)) \geqslant \frac{\varepsilon}{2}\right\}.$$

The measure of the first set on the right hand side can be made arbitrarily small by choosing n and n_k large. By Proposition 24.12 the measure of the second set tends to zero as $k\to\infty$. This completes the proof.

Exercise. 24.20. Let f_{rn} be the characteristic function of the interval $\left[\frac{r-1}{n}, \frac{r}{n}\right]$, $r=1, 2,\ldots, n$ in the space $[0, 1]$ with the Lebesgue measure (restricted to $[0, 1]$). Consider the sequence $f_{11}, f_{12}, f_{22}, f_{13}, f_{23}, f_{33},\ldots, f_{1n}, f_{2n},\ldots, f_{nn},\ldots$. This sequence converges in measure to 0 but not almost everywhere. Produce a subsequence which converges almost everywhere.

Proposition. 24.21. Let X be a complete and separable metric space and let Y be a compact metric space. Suppose μ is a probability measure on the borel σ-algebra $\mathscr{B}_X$ of X and f is a borel map from X to Y. Then, for any $\varepsilon>0$, there exists a compact set $K_\varepsilon \subset X$ such that (i) $\mu(K_\varepsilon)>1-\varepsilon$; (ii) f restricted to K_ε is continuous.

Proof. First, let us prove the proposition when $f=s$ is a simple map. Suppose $X = \bigcup_{i=1}^{n} A_i$, where $A_1, A_2,\ldots, A_n$ are disjoint borel sets and

$$s(x) = y_i \text{ if } x \in A_i,\ i = 1, 2,\ldots, n.$$

By Corollary 19.19 we can find a compact set $K_i \subset A_i$, $i=1, 2, \ldots, n$ such that $\mu(A_i \cap K'_i) < \frac{\varepsilon}{n}$, $i=1, 2,\ldots, n$. Put $K_\varepsilon = \bigcup_i K_i$. Since K_i are disjoint compact sets and $s(x)=y_i$ for all $x\in K_i$, it follows that s restricted to K_ε is continuous. Further $\mu(K_\varepsilon)>1-\varepsilon$. Thus the proposition is proved for simple maps.

If f is an arbitrary borel map we can, by Exercise 23.14, construct a sequence $\{s_n\}$ of simple maps such that s_n converges uniformly to f as $n \to \infty$. By the discussion in the preceding paragraph there exists a compact set $K_n \subset X$ such that (i) $\mu(K_n) > 1 - \frac{\varepsilon}{2^n}$, $n=1, 2, \ldots$ (ii) s_n restricted to K_n is continuous. Let $K_\varepsilon = \bigcap_{n=1}^{\infty} K_n$. Then K_ε is compact and s_n restricted to K_ε is continuous for every n. Further

$$\mu(K_\varepsilon) = 1 - \mu(\bigcup_n K_n') \geqslant 1 - \sum_n \mu(K_n') \geqslant 1 - \sum_{n=1}^{\infty} \frac{\varepsilon}{2^n} = 1 - \varepsilon.$$

Since s_n converges uniformly to f on K_ε it follows that f restricted to K_ε is also continuous. This completes the proof.

Corollary. 24.22. (Lusin's theorem). Let X and Y be complete and separable metric spaces and let μ be a probability measure on $\mathscr{B}_X$. If f is a borel map from X to Y, then, for any $\varepsilon > 0$, there exists a compact set $K_\varepsilon \subset X$ such that (i) $\mu(K_\varepsilon) > 1 - \varepsilon$; (ii) f restricted to K_ε is continuous.

Proof. By a theorem of Urysohn and Alexandroff in general topology ([12], pp. 215-16) there exists a compact metric space Z such that $Y \subset Z$ and Y is a G_δ set in Z. In particular Y is a borel set in Z. Thus f can be considered as a borel map from X into the compact metric space Z. Now the required result follows from Proposition 24.21.

Corollary. 24.23. Let X, Y be complete and separable metric spaces and let $f: X \to Y$ be a borel map. Let μ be any probability measure on X and let $f(X)$ denote the range $\{f(x), x \in X\}$ of the map f. Then there exists a borel set $Y_1 \subset f(x)$ such that $(\mu f^{-1})(Y_1) = 1$. Indeed, Y_1 may be chosen to be a countable union of compact sets.

Proof. By Corollary 24.22 we choose a compact set $K_\varepsilon \subset X$ for every $\varepsilon > 0$ such that $\mu(K_\varepsilon) > 1 - \varepsilon$ and f restricted to K_ε is continuous. Let $X_1 = \bigcup_{n=1}^{\infty} K_{1/n}$ and $Y_1 = \bigcup_{n=1}^{\infty} f(K_{1/n})$. Since f is continuous on $K_{1/n}$ it follows that $f(K_{1/n})$ is compact. Thus Y_1 is an F_σ set. Further

$$f^{-1}(Y_1) = \bigcup_{n=1}^{\infty} f^{-1}(f(K_{1/n})) \supset \bigcup_{n=1}^{\infty} K_{1/n} = X_1.$$

It is clear that $\mu(X_1) = 1$ and the proof is complete.

Remark. 24.24. It may be noted that $f(X)$ may not be a borel set. Such sets are known as analytic sets. In this connection the reader may refer to Chapter 1 of [17].

Corollary. 24.25. Let X, Y be complete and separable metric spaces and let f be a one-one borel map from X into Y. If μ is a probability measure on X, then there exists a borel set $X_1 \subset X$ such that $\mu(X_1)=1$ and f restricted to X_1 is a borel isomorphism from X_1 onto the image $f(X_1)=\{f(x), x \in X_1\}$ and $f(X_1)$ is a borel subset of Y.

Proof. We choose X_1, Y_1 as in the proof of Corollary 24.24. Since f is continuous on each $K_{1/n}$ it follows that f restricted to $K_{1/n}$ is a homeomorphism between $K_{1/n}$ and $f(K_{1/n})$. Since this holds for every n it follows that f^{-1} restricted to Y_1 is a borel map onto X_1. This completes the proof.

Exercise. 24.26. Corollary 24.25 holds whenever X and Y are borel subsets of complete and separable metric spaces $\bar{X}$ and $\bar{Y}$ respectively.

Remark. 24.27. It is a theorem of Kuratowski that if X and Y are uncountable borel subsets of complete and separable metric spaces $\bar{X}$ and $\bar{Y}$ respectively then there exists a borel isomorphism between X and Y. (This is trivial if X and Y are both countably infinite or of same finite cardinality.) A proof of this result is quite arduous and the interested reader may refer to Chapter 1 of [17].

§25. Construction of Lebesgue Measure and Other Measures in the Unit Interval through Binary, Decimal and other k-ary Expansions

Let X be the finite set $\{0, 1, 2, \ldots, k-1\}$ and let $\Omega = X^\infty$, be the product of countable copies of X. Any point of X^∞ can be written as $\mathbf{x}=(x_1, x_2, \ldots)$, where $x_n \in X$ for every n. We give the discrete topology to X and the product topology to X^∞. If we define

$$\begin{aligned}\rho(x_1, x_2) &= 1 \text{ if } x_1 \neq x_2,\ x_1, x_2 \in X,\\ &= 0 \text{ otherwise,}\end{aligned}$$

and

$$d(\mathbf{x}, \mathbf{y}) = \sum_{j=1}^{\infty} \rho(x_j, y_j) 2^{-j},$$

then X^∞ becomes a compact metric space with metric d. The borel σ-algebra in X^∞ is the smallest σ-algebra containing all the cylinders. Let π be the map from X^∞ into the unit interval [0, 1] defined by

$$\pi(\mathbf{x}) = \sum_{j=1}^{\infty} x_j k^{-j}.$$

Since the projections $\mathbf{x} \to x_j$ are continuous and the above infinite series is uniformly convergent it follows that π is a continuous map from X^∞ into the unit interval. It may be noted that every $t \in [0, 1]$ can be expanded as

$$t = \sum_{j=1}^{\infty} x_j(t) k^{-j},$$

where $x_j(t) \in X$ for all j. However, each t may have either one or two such expansions. If in the above expansion all but a finite number of the $x_j(t)$ vanish we shall call it a terminating expansion. Each $t \in (0, 1]$ has a *unique* non-terminating expansion. This is called the *k-ary expansion* of the number t. (If $k=2$ or 10, it is called *binary* or *decimal* expansion accordingly.) Let $\mathcal{N} \subset X^\infty$ be defined by

$$\mathcal{N} = \bigcup_{n=1}^{\infty} \{\mathbf{x} : x_j = 0 \text{ for all } j \geqslant n\}. \tag{25.1}$$

Consider the image under π of the set

$$\{\mathbf{x} : x_1 = a_1, x_2 = a_2, \ldots, x_n = a_n\}, \; a_i \in X, \; i = 1, 2, \ldots, n.$$

It consists of all points of the form

$$\frac{a_1}{k} + \frac{a_2}{k^2} + \ldots + \frac{a_n}{k^n} + \frac{1}{k^n}\left(\frac{x_{n+1}}{k} + \frac{x_{n+2}}{k^2} + \ldots\right)$$

where $x_{n+1}, x_{n+2}, \ldots$, vary in X. Let

$$\frac{m}{k^n} = \frac{a_1}{k} + \frac{a_2}{k^2} + \ldots + \frac{a_n}{k^n}.$$

Then

$$\pi\{\mathbf{x} : x_1 = a_1, x_2 = a_2, \ldots, x_n = a_n\} = \left[\frac{m}{k^n}, \frac{m+1}{k^n}\right]. \tag{25.2}$$

Conversely, if m is a non-negative integer strictly less than k^n, we can express $\frac{m}{k^n}$ as $\frac{a_1}{k} + \frac{a_2}{k^2} + \ldots + \frac{a_n}{k^n}$, $a_i \in X$. In such a case

$$\pi^{-1}\left(\left[\frac{m}{k^n}, \frac{m+1}{k^n}\right]\right) = \{\mathbf{x} : x_1 = a_1, x_2 = a_2, \ldots, x_n = a_n\}. \tag{25.3}$$

Points of the form $\frac{m}{k^n}$, where m, n vary over positive integers are dense in [0, 1]. Hence intervals of the form $\left[\frac{m}{k^n}, \frac{m+1}{k^n}\right]$ generate the borel σ-algebra in [0, 1]. Equations (25.2) and (25.3) show that the map π is a one-one borel map from $X^\infty - \mathcal{N}$ onto (0, 1] such that π^{-1} is also a borel map, where $\mathcal{N}$ is defined by Eq. (25.1).

We shall now construct measures on X^∞ by adopting the procedure outlined in Examples 15.5 and 15.6. Let $p_i \geqslant 0$, $i=0, 1, 2,\ldots, k-1$; $\Sigma p_i = 1$ be a probability distribution on X. Let us denote this distribution by $\mathbf{p}$. Define the measure $\mu_\mathbf{p}$ on X^∞ by

$$\mu_\mathbf{p}\{\mathbf{x} : x_1=a_1,\ x_2=a_2,\ldots,\ x_n=a_n\} = p_{a_1}\, p_{a_2} \ldots p_{a_n}, \text{ for all } n \text{ and } a_1, a_2,\ldots, a_n \in X.$$

There exists such a measure by Proposition 15.4 and Corollary 16.9. If we define $s_i(\mathbf{x}) = x_i$, then $s_1, s_2,\ldots$ are mutually independent simple random variables with the same probability distribution $\mathbf{p}$. Let $\nu_\mathbf{p}$ be the probability measure $\mu_\mathbf{p}\pi^{-1}$ induced by the map π. Consider the special case when $p_0 = p_1 = \ldots = p_{k-1} = \frac{1}{k}$. Let us denote the corresponding distribution by $\mathbf{e}$ (to indicate equidistribution). Then

$$\mu_\mathbf{e}\{\mathbf{x} : x_1=a_1,\ x_2=a_2,\ldots,\ x_n=a_n\}) = \frac{1}{k^n}.$$

By Eq. (25.3)

$$\nu_\mathbf{e}\left(\left[\frac{m}{k^n}, \frac{m+1}{k^n}\right]\right) = \mu_\mathbf{e}\left(\pi^{-1}\left[\frac{m}{k^n}, \frac{m+1}{k^n}\right]\right) = \frac{1}{k^n}.$$

Thus $\nu_\mathbf{e}$ agrees with the Lebesgue measure for all intervals of the type $\left[\frac{m}{k^n}, \frac{m+1}{k^n}\right]$. Since $\nu_\mathbf{e}$ of any single point set is zero it follows that $\nu_\mathbf{e}$ and Lebesgue measure agree for all intervals of the form $\left(\frac{m}{k^n}, \frac{m+1}{k^n}\right]$. By the uniqueness part of the extension theorem it follows that $\nu_\mathbf{e}$ is, indeed, the restriction of Lebesgue measure in [0, 1].

Remark. 25.1. The above mentioned result has an important statistical interpretation. The Lebesgue measure in [0, 1] is a probability measure. It is called the *uniform distribution* in the unit interval. To 'generate' a random variable ζ with uniform distribution we can adopt the following procedure. Generate a squence of independent random variables $\zeta_1, \zeta_2,\ldots$ such that each ζ_n takes the values $0, 1, 2,\ldots, k-1$ with the same probability $\frac{1}{k}$ and write

$\zeta = \sum_{1}^{\infty} \frac{\zeta_n}{k^n}$. (In particular one may choose k to be 2.) Each ζ_n is called a *random number* between 0 and $k-1$. ζ is called a random number in the interval [0, 1].

Remark. 25.2. In many practical problems it is necessary to generate a random variable with a given probability distribution μ on the real line. Let $F(t)=\mu((-\infty, t])$ be the probability distribution function corresponding to the measure μ. Suppose $F(t)$ is a strictly increasing continuous function of the variable t. That is, $F(t)<F(s)$ whenever $t<s$. Then F maps the extended real line $[-\infty, +\infty]$ in a one-one manner onto the unit interval [0, 1] if we define $F(-\infty)=0, F(+\infty)=1$. In such a case the inverse function $\xi \to F^{-1}(\xi)$ from [0, 1] to the extended real line $[-\infty, +\infty]$ is defined by the equation $F(F^{-1}(\xi))=\xi$. Further F^{-1} is also a continuous map. If we have a random variable ζ with uniform distribution in [0, 1], then the random variable $F^{-1}(\zeta)$ has the probability distribution function $F(t)$, $-\infty< t<\infty$. Indeed, $P\{F^{-1}(\zeta) \leqslant t\}=P\{\zeta \leqslant F(t)\}=F(t)$.

The disadvantage of this method lies in the fact that one may have to use a statistical table of the distribution function F to read the value of $F^{-1}(\zeta)$ against the value of ζ. There are many situations where it is necessary to simulate a large sample of random variables with a given distribution F. If there is a 'limit theorem' which yields the distribution F from simpler random variables one may avoid the use of tables. We shall illustrate, for example, the case of normal distribution function $\Phi(x)$. We can now use the limit theorem of Section 6. Choose a sequence of independent and identically distributed random variables $\zeta_1, \zeta_2, \ldots$ with the bionomial distribution $P(\zeta_n=0)=P(\zeta_n=1)=\frac{1}{2}$ for all n. Choose a 'large' positive integer N. Write

$$\tilde{\zeta}_1 = \frac{\zeta_1+\zeta_2+\ldots+\zeta_N - \dfrac{N}{2}}{\frac{1}{2}\sqrt{N}}$$

$$\tilde{\zeta}_2 = \frac{\zeta_{N+1}+\zeta_{N+2}+\ldots+\zeta_{2N} - \dfrac{N}{2}}{\frac{1}{2}\sqrt{N}}$$

....

$$\tilde{\zeta}_j = \frac{\zeta_{(j-1)N+1}+\zeta_{(j-1)N+2}+\dots+\zeta_{jN}-\dfrac{N}{2}}{\frac{1}{2}\sqrt{N}}$$

....

By Proposition 6.1, $\tilde{\zeta}_1$, $\tilde{\zeta}_2$... are approximately normally distributed with mean zero and variance unity, provided N is 'reasonably' large.

Exercise. 25.3. Suppose $\zeta_1, \zeta_2, \dots$ is a sequence of independent random variables with uniform distribution in [0, 1]. Determine a procedure to generate a sequence $\{\tilde{\zeta}_1, \tilde{\zeta}_2, \dots\}$ of independent random variables with the same probability distribution which is 'approximately Poisson' with parameter λ.

Remark. 25.4. Let us now go back to the measures $\mu_{\mathbf{p}}$ and $\nu_{\mathbf{p}}$ constructed in the discussion before Remark 25.1. We shall now partition the space X^∞ and [0, 1] into an uncountable number of sets $\{A_{\mathbf{p}}\}$ and $\{B_{\mathbf{p}}\}$ respectively such that $\mu_{\mathbf{p}}(A_{\mathbf{p}})=1$ and $\nu_{\mathbf{p}}(B_{\mathbf{p}})=1$ for all (non-degenerate) probability distributions $\mathbf{p}$ on 0, 1, 2,..., $k-1$. To this end we write

$$\delta_{ij} = 1 \text{ if } i=j,$$
$$= 0 \text{ if } i \neq j.$$

For any $i \in X$ and any $q > 0$, let

$$A_i(q) = \left\{\mathbf{x}: \lim_{n\to\infty} \frac{\delta_{ix_1}+\delta_{ix_2}+\dots+\delta_{ix_n}}{n} = q\right\}$$

$$A_{\mathbf{p}} = \bigcap_{i=0}^{k-1} A_i(p_i).$$

Let us consider only those $\mathbf{p}$ where no p_i equals unity. (These are indeed the non-degenerate distributions mentioned in the beginning of the paragraph). It is clear that whenever the distributions $\mathbf{p}$ and $\mathbf{q}$ are distinct, $A_{\mathbf{p}} \cap A_{\mathbf{q}} = \emptyset$ for $\mathbf{p} \neq \mathbf{q}$. By the strong law of large numbers $\mu_{\mathbf{p}}(A_{\mathbf{p}})=1$. Thus the 'masses' of the different distributions $\mu_{\mathbf{p}}$ are concentrated in disjoint borel sets. Further, for any single point $\mathbf{x}$, $\mu_{\mathbf{p}}(\{\mathbf{x}\})=0$. This shows that $\nu_{\mathbf{p}}=\mu_{\mathbf{p}}\,\pi^{-1}$ has its mass concentrated in $B_{\mathbf{p}}=\pi(A_{\mathbf{p}} \cap (X^\infty - N))$, where N is defined by Eq. (25.1). Thus the measures $\nu_{\mathbf{p}}$ have their mass concentrated in disjoint sets $B_{\mathbf{p}}$ and $\nu_{\mathbf{p}}(\{t\})=0$ for every $t \in [0, 1]$. In particular, $\nu_{\mathbf{p}}$ and the Lebesgue measure $\nu_{\mathbf{e}}$ have their mass concentrated in disjoint sets whenever $\mathbf{p} \neq \mathbf{e}$. By this procedure we have constructed a wide variety of measures in the unit interval [0, 1]. This construction shows that the uncountable set $B'_{\mathbf{e}}$ has Lebesgue measure zero.

Exercise. 25.5. Let $E \subset [0, 1]$ be defined by

$$E = \{ \sum_{j=1}^{\infty} 3^{-j} x_j, \text{ where each } x_j \text{ is either 0 or 1} \}.$$

Then $L(E) = 0$ but

$$E + E = \{\xi + \eta,\ \xi \in E,\ \eta \in E\} = [0, 1].$$

(E is known as the Cantor set in honour of the German mathematician G. Cantor, who was the founder of modern set theory.)

§26. Isomorphism of Measure Spaces

We begin with a definition.

Definition. 26.1. Two probability spaces $(X_i, \mathscr{B}_i, \mu_i)$, $i=1, 2$ are said to be *isomorphic* if there is a map $T : X_1 \to X_2$ and a pair of subsets N_i, $i=1, 2$, such that

(i) $N_i \subset X_i$, $N_i \in \mathscr{B}_i$ and $\mu_i(N_i) = 0$, $i=1, 2$;

(ii) T is a borel isomorphism from $(X_1 - N_1, \mathscr{B}_1 \cap (X_1 - N_1))$ onto $(X_2 - N_2, \mathscr{B}_2 \cap (X_2 - N_2))$;

(iii) $\mu_1 T^{-1} = \mu_2$.

The aim of the present section is to establish that any probability space $(X, \mathscr{B}, \mu)$, where X is a complete and separable metric space, $\mathscr{B}$ is its borel σ-algebra and μ is a probability measure such that $\mu(\{x\}) = 0$ for all $x \in X$, is isomorphic to the unit interval with its borel σ-algebra and Lebesgue measure.

Proposition. 26.2. Let μ be a probability measure on $(R, \mathscr{B}_R)$ such that $\mu(\{x\}) = 0$ for every single point set $\{x\}$, $x \in R$. Let I denote the unit interval [0, 1] and let L be the Lebesgue measure on it. Then $(R, \mathscr{B}_R, \mu)$ and $(I, \mathscr{B}_I, L)$ are isomorphic.

Proof. Let $F(x) = \mu((-\infty, x])$ be the distribution function of $\boldsymbol{\mu}$. Since $\mu(\{x\}) = 0$ it follows that F is a continuous and monotonically increasing function of x. Let

$$x_\alpha^- = \inf \{x : F(x) = \alpha\},$$
$$x_\alpha^+ = \sup \{x : F(x) = \alpha\}, \quad \alpha \in [0, 1].$$

If $0<\alpha<\beta<1$, then $x_\alpha^- \leqslant x_\alpha^+ < x_\beta^- \leqslant x_\beta^+$. We claim that the set $\{\alpha : x_\alpha^- < x_\alpha^+\}$ is countable. Indeed,

$$\{\alpha : x_\alpha^- < x_\alpha^+\} = \bigcup_{n=1}^{\infty} \bigcup_{k=1}^{\infty} \left\{\alpha : -n \leqslant x_\alpha^- < x_\alpha^+ \leqslant n,\ x_\alpha^+ - x_\alpha^- \geqslant \frac{1}{k}\right\}$$

and each set occurring within the union sign on the right hand side is finite because more than a finite number of disjoint intervals of length $\geqslant 1/k$ connot be found in a finite interval $[-n, n]$. We write

$$J = \{\alpha : x_\alpha^- < x_\alpha^+\},$$
$$f(\alpha) = x_\alpha^-, \text{ for all } \alpha \in [0, 1].$$

Then f is a map from $[0, 1]$ into $[-\infty, +\infty]$. Clearly f is one to one. For $\xi \in R$,

$$\{\alpha : x_\alpha^- \leqslant \xi\} = [0, F(\xi)]. \tag{26.1}$$

Further

$$f([\alpha, \beta]) = [x_\alpha^-, x_\beta^-] - \bigcup_{\gamma \in J \cap [\alpha, \beta]} (x_\gamma^-, x_\gamma^+] \tag{26.2}$$

$$f([0, 1]) = R - \bigcup_{\gamma \in J} (x_\gamma^-, x_\gamma^+]. \tag{26.3}$$

We write

$$\mathcal{N}_1 = \{0\} \cup \{1\},\ \mathcal{N}_2 = \bigcup_{\gamma \in J} (x_\gamma^-, x_\gamma^+]. \tag{26.4}$$

Since $\mu((x_\alpha^-, x_\alpha^+])=0$ for every α, it follows that $\mu(\mathcal{N}_2)=0$. Clearly $L(\mathcal{N}_1)=0$. Eqs. (26.1) and (26.2) imply that f is a borel isomorphism between $(0, 1)$ and $R-\mathcal{N}_2$. Further Eq. (26.1) implies that

$$Lf^{-1}((-\infty, \xi]) = \mu((-\infty, \xi]),\ \xi \in R.$$

Hence $Lf^{-1}=\mu$. This shows that f is an isomorphism between $(I, \mathcal{B}_I, L)$ and $(R, \mathcal{B}_R, \mu)$.

Definition. 26.3. Let X be a metric space. A measure μ on $\mathcal{B}_X$ is said to be *nonatomic* if $\mu(\{x\})=0$ for every single point set $\{x\}$, $x \in X$.

Proposition. 26.4. Let X be a complete and separable metric space. Suppose there exists a one to one borel map f from X into the unit interval $I=[0, 1]$. If μ is a nonatomic probability measure on $\mathcal{B}_X$, then the probability spaces $(X, \mathcal{B}_X, \mu)$ and $(I, \mathcal{B}_I, L)$ are isomorphic.

Proof. Since f is $1-1$, we have

$$(\mu f^{-1})(\{a\}) = \mu(\{f^{-1}(a)\})=0,\ a \in [0, 1].$$

Thus μf^{-1} is a nonatomic measure in [0, 1]. Let $\varepsilon > 0$ be arbitrary. By Lusin's theorem we can find a compact set $K_\varepsilon \subset X$ such that $\mu(K_\varepsilon) > 1 - \varepsilon$ and f restricted to K_ε is continuous. Since f is also $1-1$ it follows that f is a homeomorphism on K_ε. Hence f is a borel isomorphism on the F_σ subset $X_1 = \bigcup_{n=1}^{\infty} K_{1/n}$. If we write $N_1 = X - X_1$, $N_2 = I - f(X_1)$ it follows that $\mu(N_1) = 0$, $\mu f^{-1}(N_2) = 0$ and f is a borel isomorphism between $X - N_1$ and $I - N_2$. Thus $(X, \mathscr{B}_X, \mu)$ and $(I, \mathscr{B}_I, \mu f^{-1})$ are isomorphic. Now an application of Proposition 26.2 shows that $(I, \mathscr{B}_I, \mu f^{-1})$ and $(I, \mathscr{B}_I, L)$ are isomorphic. This completes the proof.

Proposition. 26.5. Let $I = [0, 1]$ be the unit interval and let I^∞ be the compact metric space which is a countable infinite product of I. Then there exists a one to one borel map f from $I\infty$ into I.

Proof. Let D denote the set consisting of two points 0 and 1. Let D_∞ denote the countable product of D and let $B \subset D^\infty$ denote the subset of all those dyadic sequences in which 1 occurs infinitely often. The discussion at the beginning of Section 25 shows that there is $1-1$ borel isomorphism between B and I. (In fact apply that discussoin for the case $k=2$). This shows that I^∞ and B^∞ are borel isomorphic. The proof will be complete if we show that there exists a one-to-one map from B^∞ into B. To this end let $(\mathbf{d}^{(1)}, \mathbf{d}^{(2)}, \ldots) \in B^\infty$, where

$$\mathbf{d}^{(j)} = (d_{j1}, d_{j2}, \ldots, d_{jn}, \ldots),$$

$\mathbf{d}_{jn} = 1$ for infinitely many n and every $j = 1, 2, \ldots$. Consider the dyadic sequence obtained by going along the route shown below:

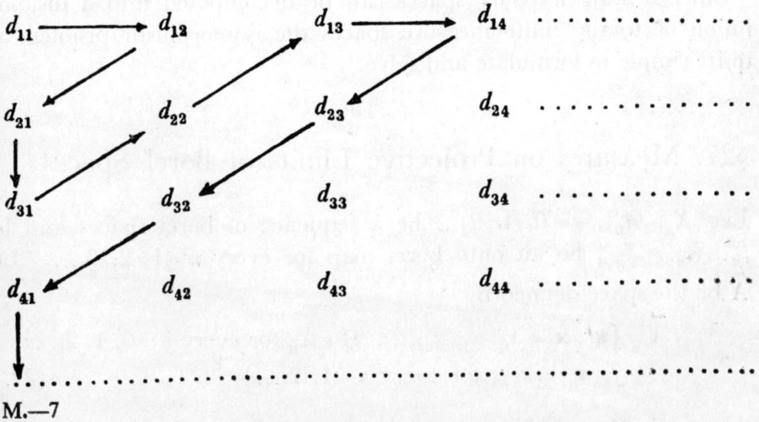

In this dyadic sequence the element 1 occurs infinitely often. We denote this sequence by $f(\mathbf{d}^{(1)}, \mathbf{d}^{(2)},\ldots)$. Then f is a $1-1$ borel map from B^∞ into B. This completes the proof.

Proposition. 26.6. (Isomorphism theorem). Let X be a complete and separable metric space and let μ be a nonatomic probability measure on $\mathscr{B}_X$. Then $(X, \mathscr{B}_X, \mu)$ is isomorphic to $(I, \mathscr{B}_I, L)$ where I is the unit interval and L is the Lebesgue measure.

Proof. By Proposition 26.4 it is enough to construct a $1-1$ borel map from X into the unit interval. Proposition 26.5 shows that it is enough to construct a $1-1$ borel map from X into I^∞. By the theorem of Urysohn and Alexandroff in general topology there exists a homeomorphism between X and a G_δ subset of I^∞. This completes the proof of the Theorem.

Exercise. 26.7. Let X be a complete and separable metric space and let Y be a borel subset of X. If μ is a nonatomic probability measure on $\mathscr{B}_Y$, then $(Y, \mathscr{B}_Y, \mu)$ is isomorphic to the space $(I, \mathscr{B}_I, L)$, where I is the unit interval and L is the Lebesgue measure.

Remark. 26.8. Let X be any metric space and let μ be a probability measure on $\mathscr{B}_X$. A point $x \in X$ is called an *atom* of μ if $\mu(\{x\}) > 0$. It is an exercise for the reader to show that there cannot be more than a countable number of atoms for μ. Further $\mu = p\lambda + q\nu$, where $0 \leqslant p, q \leqslant 1$, $p+q=1$, λ is a nonatomic probability measure, and ν is a 'purely atomic' probability measure, i.e., $\nu(A)=1$ where A is a countable subset of X. Together with Proposition 26.6 and Exercise 26.7 this remark reveals the structure of an arbitrary probability space when the sample space is a borel subset of a complete and separable metric space.

Since σ-finite measure spaces can be decomposed into a disjoint union of totally finite measure spaces the isomorphism problem is quite simple to formulate and solve.

§27. Measures on Projective Limits of Borel Spaces

Let $(X_n, \mathscr{B}_n)$, $n=0, 1, 2,\ldots$ be a sequence of borel spaces and let $f_n : X_n \to X_{n-1}$ be an onto borel map for every $n=1, 2, 3,\ldots$. Let $\tilde{X}$ be the space defined by

$$\tilde{X}=\{\mathbf{x} : \mathbf{x}=(x_0, x_1, x_2,\ldots),\ x_n \in X_n \text{ for every } n=0, 1, 2,\ldots;\ f_n(\mathbf{x}_n)=x_{n-1} \text{ for every } n=1, 2,\ldots\}. \quad (27.1$$

We shall now define a borel structure on $\widetilde{X}$. Define the map

$$\begin{aligned} \pi_n &: \widetilde{X} \to X_n,\ n=0, 1, 2, \ldots \\ \pi_n(\mathbf{x}) &= x_n. \end{aligned} \tag{27.2}$$

Let

$$\pi_n^{-1}(\mathscr{B}_n) = \{\pi_n^{-1}(E), E \in \mathscr{B}_n\}$$

Proposition. 27.1. $\pi_n^{-1}(\mathscr{B}_n)$ is an increasing sequence of σ-algebras in the space $\widetilde{X}$.

Proof. Let $A \in \pi_n^{-1}(\mathscr{B}_n)$, i.e., $A = \{\mathbf{x} : x_n \in E\}$, where $E \in \mathscr{B}_n$. Since $f_{n+1}(x_{n+1}) = x_n$ it follows that $A = \{\mathbf{x} : f_{n+1}(x_{n+1}) \in E\} = \{\mathbf{x} : x_{n+1} \in f_{n+1}^{-1}(E)\}$. Since f_{n+1} is a borel map, $A = \pi_{n+1}^{-1}(f_{n+1}^{-1}(E)) \in \pi_{n+1}^{-1}(\mathscr{B}_{n+1})$. Thus $\pi_n^{-1}(\mathscr{B}_n) \subset \pi_{n+1}^{-1}(\mathscr{B}_{n+1})$. $\pi_n^{-1}(\mathscr{B}_n)$ is a σ-algebra in $\widetilde{X}$ since it is the inverse image of a σ-algebra. This completes the proof.

Definition. 27.2. Let $\widetilde{\mathscr{B}}$ be the σ-algebra generated by the class $\widetilde{\mathscr{F}} = \bigcup_{n=0}^{\infty} \pi_n^{-1}(\mathscr{B}_n)$. The borel space $(\widetilde{X}, \widetilde{\mathscr{B}})$ is called the *projective limit* of the sequence of borel spaces $(X_n, \mathscr{B}_n)$, $n=0, 1, 2, \ldots$ through the maps $f_1, f_2, \ldots$, ($\widetilde{\mathscr{F}}$ is a boolean algebra of subsets of $\widetilde{X}$).

Definition. 27.3. Let $(X_n, \mathscr{B}_n, \mu_n)$, $n=0, 1, 2, \ldots$ be a sequence of probability spaces and let $f_n : X_n \to X_{n-1}$, $n=1, 2, \ldots$ be a sequence of onto borel maps. The sequence $\{\mu_n\}$ is said to be *consistent* with respect to $\{f_n\}$ if $\mu_n f_n^{-1} = \mu_{n-1}$, $n=1, 2, \ldots$.

Proposition. 27.4. (Daniel-Kolmogorov consistency theorem). Let X_n be a sequence of complete and separable metric spaces for $n=0, 1, 2, \ldots$ and let $f_n : (X_n, \mathscr{B}X_n) \to (X_{n-1}, \mathscr{B}X_{n-1})$ be a sequence of onto borel maps. Suppose μ_n is a probability measure on $\mathscr{B}X_n$ for every n such that $\{\mu_n\}$ is consistent under f_n. Then there exists a probability measure $\widetilde{\mu}$ on the projective limit $(\widetilde{X}, \widetilde{\mathscr{B}})$ of the borel spaces $(X_n, \mathscr{B}X_n)$ such that $\widetilde{\mu}\, \pi_n^{-1} = \mu_n$ for every n.

Proof. Let us denote by $\mathscr{B}_n$ the σ-algebra $\mathscr{B}X_n$. Let $\widetilde{\mathscr{F}}$ and $\widetilde{\mathscr{B}}$ be as in Definition 27.2. We now define $\widetilde{\mu}$ on $\widetilde{\mathscr{F}}$ by

$$\widetilde{\mu}(A) = \mu_n(E) \text{ if } A = \pi_n^{-1}(E),\ E \in \mathscr{B}_n, \tag{27.3}$$

where π_n is defined by Eq. (27.2). We claim that $\widetilde{\mu}$ is well-defined. Indeed, let

$$\begin{aligned} A &= \pi_m^{-1}(E) = \pi_n^{-1}(F),\ m<n, \\ E &\in \mathscr{B}_m, F \in \mathscr{B}_n. \end{aligned}$$

For any point $\mathbf{x} \in A$, $\mathbf{x}=(x_0, x_1,\ldots)$, $x_m \in E$, $x_n \in F$. Then $(f_{m+1} \circ f_{m+2} \circ \ldots \circ f_n)(x_n)=x_m$. Hence $E=f_n^{-1}(f_{n-1}^{-1}(\ldots(f_{m+1}^{-1}(F))$. Now the consistency of $\{\mu_n\}$ implies that $\mu_n(E)=\mu_m(F)$. Thus $\tilde{\mu}$ is well-defined. Now we claim that $\tilde{\mu}$ is finitely additive on $\tilde{\mathscr{F}}$. Indeed, let $A, B \in \tilde{\mathscr{F}}$ and $A \cap B = \emptyset$. If $A \in \pi_m^{-1}(\mathscr{B}_m)$, $B \in \pi_n^{-1}(\mathscr{B}_n)$, and $m < n$ then Proposition 27.1 implies that A and $B \in \pi_n^{-1}(\mathscr{B}_n)$. Let

$$A = \pi_n^{-1}(E),\ B = \pi_n^{-1}(F);\ E, F \in \mathscr{B}_n.$$

Since $A \cap B = \emptyset$ it follows that $\pi_n^{-1}(E \cap F) = \emptyset$ and hence $E \cap F = \emptyset$. Thus

$$\begin{aligned}\tilde{\mu}(A \cup B) &= \tilde{\mu}(\pi_n^{-1}(E \cup F))\\ &= \mu_n(E \cup F) = \mu_n(E) + \mu_n(F)\\ &= \tilde{\mu}(A) + \tilde{\mu}(B).\end{aligned}$$

Thus $\tilde{\mu}$ is finitely additive.

To prove the countable additivity of $\tilde{\mu}$ on $\tilde{\mathscr{F}}$ it is enough to prove that for any sequence $A_0 \supset A_1 \supset A_2 \supset \ldots$ such that $\bigcap_{n=0}^{\infty} A_n = \emptyset$, $A_n \in \pi_n^{-1}(\mathscr{B}_n)$, $n = 0, 1, 2, \ldots,$

$$\lim_{n \to \infty} \tilde{\mu}(A_n) = 0. \tag{27.4}$$

(In this context see the proof of Proposition 20.8). Let $A_n=\pi_n^{-1}(E_n)$, $E_n \in \mathscr{B}_n$. If Eq. (27.4) does not hold there exists a $\delta > 0$ such that

$$\tilde{\mu}(A_n) \geqslant \delta,\ n = 0, 1, 2, \ldots. \tag{27.5}$$

Then

$$\mu_n(E_n) \geqslant \delta,\ n = 0, 1, 2, \ldots. \tag{27.6}$$

Now Corollary 19.19 and Corollary 24.22 imply that there exists a compact set $K_n \subset E_n$ such that

(i) $\mu_n(E_n - K_n) < \dfrac{\delta}{2^{n+1}}$,

(ii) f_n restricted to K_n is continuous,

for every $n = 1, 2, 3, \ldots.$ Let

$$B_n = \bigcap_{j=1}^{n} \pi_j^{-1}(K_j),\ n = 1, 2, \ldots. \tag{27.7}$$

Then $A_n B_n' \subset \bigcup_{j=1}^{n} (A_j \cap \pi_j^{-1}(K_j')) = \bigcup_{j=1}^{n} \pi_j^{-1}(E_j K_j')$.

Hence

$$\tilde{\mu}(A_n B_n') \leqslant \sum_{j=1}^{n} \tilde{\mu}(\pi_j^{-1}(E_j K_j'))$$
$$= \sum_{j=1}^{n} \mu_j (E_j K_j')$$
$$\leqslant \sum_{j=1}^{n} \frac{\delta}{2^{j+1}} \leqslant \frac{\delta}{2}.$$

Since $B_n \subset A_n$, Eq. (27.5) implies that

$$\tilde{\mu}(B_n) \geqslant \frac{\delta}{2}, \; n=1, 2, 3, \ldots$$

This shows in particular that $B_n \neq \varnothing$ for every $n \geqslant 1$. Choose a point $\mathbf{x}^{(n)} = (x_{n0}, x_{n1}, x_{n2}, \ldots) \in B_n$ for each $n=1, 2, \ldots$. By the definition of B_n in Eq. (27.7), it follows that

$$x_{nj} \in K_j, j=1, 2, \ldots, n; \; n=1, 2, \ldots$$
$$f_j(x_{nj}) = x_{n,j-1}, j=1, 2, \ldots \tag{27.8}$$

Since each K_j is compact we can select (by the diagonal procedure) $n_1 < n_2 < n_3, \ldots$ such that

$$\lim_{k\to\infty} x_{n_k j} = x_j, \; j=1, 2, \ldots$$

exists for every j, where the limit $x_j \in K_j, j=1, 2, \ldots$ Since f_j restricted to K_j is continuous, (27.8) implies that

$$f_j(x_j) = x_{j-1}, \; j=2, 3, \ldots$$

Define $x_0 = f_1(x_1)$. Then $\mathbf{x} = (x_0, x_1, x_2, \ldots) \in \tilde{X}$ and $x_j \in K_j$ for $j=1$, $2, \ldots$ Thus $\mathbf{x} \in \bigcap_{n=1}^{\infty} B_n \subset \bigcap_{n=1}^{\infty} A_n$, which contradicts the assumption that $\bigcap_{n=1}^{\infty} A_n = \varnothing$. Thus Eq. (27.4) holds. Now that $\tilde{\mu}$ is countably additive on $\tilde{\mathscr{F}}$, it follows from Corollary 16.9 that it extends to a probability measure on $\tilde{\mathscr{B}}$. Equation (27.3) implies that $\tilde{\mu}\pi_n^{-1} = \mu_n$ for all n. This completes the proof.

Exercise. 27.5. Proposition 27.4 holds when the sequence $(X_n, \mathscr{B}_{X_n})$ is replaced by a sequence $(X_n, \mathscr{B}_n)$ of standard borel spaces.

Example. 27.6. Let $\{Y_n\}$ be a sequence of complete and separable metric spaces and let

$$X_n = Y_0 \times Y_1 \times Y_2 \times \ldots \times Y_n, \; n=0, 1, 2, \ldots.$$

Let $f_n : X_n \to X_{n-1}$ be the projection map defined by

$$f_n(y_0, y_1, \ldots, y_n) = (y_0, y_1, \ldots, y_{n-1}).$$

Then the projective limit $\tilde{X}$ can be identified with the cartesian product $Y_0 \times Y_1 \times Y_2 \times \ldots$ as follows. If $x_n = (y_0, y_1, \ldots, y_n)$ and $f_n(x_n) = x_{n-1}$, then x_{n-1} is given by $(y_0, y_1, \ldots, y_{n-1})$. Thus we define

$$f(x_0, x_1, x_2, \ldots) = (y_0, y_1, y_2, \ldots)$$

where $x_n = (y_0, y_1, \ldots, y_n)$. A sequence $\{\mu_n\}$, where μ_n is a probability measure on X_n is consistent if $\mu_n f_n^{-1} = f_{n-1}$, $n=1, 2, \ldots$. (In this case we say that μ_{n-1} is the *marginal distribution* of $(y_0, y_1, \ldots, y_{n-1})$ when μ_n is the distribution of $(y_0, y_1, \ldots, y_n)$. By Proposition 27.4 it then follows that there exists a probability measure $\mu (= \tilde{\mu} f^{-1})$ on $Y_0 \times Y_1 \times \ldots \times Y_n \times \ldots$ with the property $\mu \tau_n^{-1} = \mu_n$ for all $n=0, 1, 2, \ldots$ where τ_n is the projection map defined by

$$\tau_n(y_0, y_1, y_2, \ldots) = (y_0, y_1, y_2, \ldots, y_n)$$

where $(y_0, y_1, y_2, \ldots) \in Y_0 \times Y_1 \times Y_2 \times \ldots \times Y_n \times \ldots$. Obviously the measure μ is defined on the boral σ-field of the product metric space $Y_0 \times Y_1 \times Y_2 \times \ldots$.

Remark. 27.7. In the study of continuous time stochastic processes it is necessary to construct measures on product spaces of the form $\prod_{t \in I} Y_t$, where I is an interval of the real line. However, the data that could be observed refers to only a finite number of time points $t_1 < t_2 < \ldots < t_k$, $t_j \in I$ for all $j=1, 2, \ldots, k$. Suppose we have a knowledge of the probability distribution $\mu_{t_1, t_2, \ldots, t_k}$ (for every finite set $F = \{t_1, t_2, \ldots, t_k\}$) on the spaces $Y_{t_1} \times Y_{t_2} \times \ldots \times Y_{t_k}$. The family of distributions $\{\mu_{t_1, t_2, \ldots, t_k}\}$ is said to be *consistent* if the projection map

$$\Pi^{t_1 t_2 \ldots t_{k+1}}_{t_1 t_2 \ldots t_k} : Y_{t_1} \times Y_{t_2} \times \ldots \times Y_{t_{k+1}} \to Y_{t_1} \times Y_{t_2} \times \ldots \times Y_{t_k}$$

defined by

$$\Pi^{t_1 t_2 \ldots t_{k+1}}_{t_1 t_2 \ldots t_k} (y_{t_1}, y_{t_2}, \ldots, y_{t_{k+1}}) = (y_{t_1}, y_{t_2}, \ldots, y_{t_k})$$

has the property

$$\mu_{t_1 t_2 \ldots t_{k+1}} \left(\Pi^{t_1 t_2 \ldots t_{k+1}}_{t_1 t_2 \ldots t_k} \right)^{-1} = \mu_{t_1 t_2 \ldots t_k}$$

for all $t_1 < t_2 < \ldots < t_{k+1}$, $t_1, t_2, \ldots, t_{k+1} \in I$ and all k. Now we can ask the following question: is there a unique probability measure μ on $\prod_{t \in I} Y_t$ such that

$$\mu \Pi^{-1}_{t_1 t_2 \ldots t_k} = \mu_{t_1 t_2 \ldots t_k} \text{ for all } t_1 < t_2 \ldots < t_k$$

where $\Pi_{t_1 t_2 \ldots t_k}$ is the projection map defined by

$$\Pi_{t_1 t_2 \ldots t_k} (y(\cdot)) = (y(t_1), y(t_2), \ldots, y(t_k))$$

from $\prod_{t \in I} Y_t$ onto $Y_{t_1} \times Y_{t_2} \times \ldots \times Y_{t_k}$, where $y(.)$ is the element whose t th coordinate is $y(t) \in Y_t$? The σ-field in $\prod_{t \in I} Y_t$ under consideration is the smallest σ-field which makes all the projections $\pi_{t_1 t_2 \ldots t_k}$ borel maps. One can deduce the existence of such a probability measure μ by appealing to Proposition 27.4. We can proceed as follows: consider all finite subsets of the interval I. Let us denote this family by Γ. Γ is what is called a directed family. If $F, G \in \Gamma$ we write $F \leqslant G$ if $F \subset G$. To each $F \in \Gamma$, write

$$X(F) = Y_{t_1} \times Y_{t_2} \times \ldots Y_{t_k}, F = \{t_1, t_2, \ldots, t_k\}.$$

Suppose all the Y_t's are complete and separable metric spaces. Then $X(F)$ is also a complete and separable metric space. Further when $F \leqslant G$ we have the projection map

$$\pi_F^G : X(G) \to X(F)$$

defined in a natural manner. Indeed if $t_1 < t_2 < \ldots < t_k < t_{k+1} < \ldots t_l$ and $F = \{t_1, t_2, \ldots, t_k\}$, $G = \{t_1, \ldots, t_l\}$, then

$$\pi_F^G (y_{t_1}, y_{t_2}, \ldots, y_{t_l}) = (y_{t_1}, y_{t_2}, \ldots, y_{t_k}).$$

If $F \leqslant G \leqslant H$, then

$$\pi_F^G \pi_G^H = \pi_F^H.$$

If we write μ_F for the measure $\mu_{t_1 t_2 \ldots t_k}$ (where $F = \{t_1, t_2, \ldots, t_k\}$) then the consistency condition implies that

$$\mu_G \left(\pi_F^G\right)^{-1} = \mu_F \text{ for all } F \leqslant G, F, G \in \Gamma.$$

In view of these properties we can introduce the following definitions:

Suppose Γ is any directed set with the partial ordering $\leqslant$ and $(X_\gamma, \mathscr{B}_\gamma, \mu_\gamma)$, $\gamma \in \Gamma$ is a family of probability spaces. For any pair $\gamma_1, \gamma_2 \in \Gamma$ such that $\gamma_1 \leqslant \gamma_2$, let $f_{\gamma_1}^{\gamma_2}$ be a borel map from X_{γ_2} onto X_{γ_1} such that

$$f_{\gamma_1}^{\gamma_2} f_{\gamma_2}^{\gamma_3} = f_{\gamma_1}^{\gamma_3} \text{ for all } \gamma_1 \leqslant \gamma_2 \leqslant \gamma_3.$$

We say that the measures $\{\mu_\gamma\}$ are *consistent* if

$$\mu_{\gamma_2} \left(f_{\gamma_1}^{\gamma_2}\right)^{-1} = \mu_{\gamma_1} \text{ whenever } \gamma_1 \leqslant \gamma_2.$$

We define the set X_Γ as the set of all maps

$$\mathbf{x} : \Gamma \to \bigcup_{\gamma \in \Gamma} X_\gamma$$

satisfying the properties:

$$\mathbf{x}(\gamma) \in X_\gamma \text{ for every } \gamma \in \Gamma;$$

$$f_{\gamma_1}^{\gamma_2}(\mathbf{x}(\gamma_2)) = \mathbf{x}(\gamma_1) \text{ for all } \gamma_1 \leqslant \gamma_2, \gamma_1, \gamma_2 \in \Gamma.$$

Let π_γ be the map from X_Γ onto X_γ defined by

$$\pi_\gamma(\mathbf{x}) = \mathbf{x}(\gamma).$$

Let $\mathscr{B}_\Gamma$ be the smallest σ-algebra in X_Γ with respect to which every map π_γ is a borel map from X_Γ onto X_γ. Then $(X_\Gamma, \mathscr{B}_\Gamma)$ is called the *projective limit* of the borel spaces $(X_\gamma, \mathscr{B}_\gamma)$, $\gamma \in \Gamma$. If we write

$$\mathscr{F}_\Gamma = \bigcup_{\gamma \in \Gamma} \overset{-1}{\pi_\gamma}(\mathscr{B}_\gamma),$$

then $\mathscr{F}_\Gamma$ is a boolean algebra which generates $\mathscr{B}_\Gamma$. For every $A \in \mathscr{F}_\Gamma$, $A = \overset{-1}{\pi_\gamma}(E)$ for some $E \in \mathscr{B}_\Gamma$. Define

$$\mu_\Gamma(A) = \mu_\Gamma(E).$$

Then μ_Γ is a well defined finitely additive function on $\mathscr{F}_\Gamma$. This is proved exactly as in Proposition 27.4. To show that μ_Γ is countably additive on $\mathscr{F}_\Gamma$ it is enough to prove that $\lim_{n\to\infty} \mu_\Gamma(A_n) = 0$ whenever $A_1 \supset A_2 \ldots$, $\bigcap_1^\infty A_n = \varnothing$ and $A_n \in \mathscr{F}_\Gamma$ for all n. In such a case we can find $\gamma_1 \leqslant \gamma_2 \leqslant \ldots \gamma_n \leqslant \ldots$ such that $A_n \in \overset{-1}{\pi_{\gamma_n}}(\mathscr{B}_{\gamma_n})$. Now suppose that all the X_γ's are complete and separable metric spaces and $\mathscr{B}_\gamma$'s are their respective borel σ-algebras. If we write $X_n = X_{\gamma_n}$, $\mathscr{B}_n = \mathscr{B}_{\gamma_n}$, $f_n = f_{\gamma_{n-1}}^{\gamma_n}$, $\mu_n = \mu_{\gamma_n}$, then the proof of Proposition 27.4 applies to this sequence and $\mu_\Gamma(A_n) \to 0$ as $n \to \infty$. Thus μ_Γ becomes countably additive and extends to a unique probability measure on $(X_\Gamma, \mathscr{B}_\Gamma)$ such that $\mu_\Gamma \pi_\gamma^{-1} = \mu_\gamma$ for all γ. We summarise this in the form of a proposition.

Proposition. 27.8. Let Γ be a directed set and let $(X_\gamma, \mathscr{B}_\gamma, \mu_\gamma)$, $\gamma \in \Gamma$ be a family of probability spaces where X_γ is a complete and separable metric space and $\mathscr{B}_\gamma$ its borel σ-algebra. Let $f_{\gamma_1}^{\gamma_2} : X_{\gamma_2} \to X_{\gamma_1}$ be a borel map from X_{γ_2} onto X_{γ_1} for every pair $\gamma_1 \leqslant \gamma_2$, $\gamma_1, \gamma_2 \in \Gamma$ such that $f_{\gamma_1}^{\gamma_2} f_{\gamma_2}^{\gamma_3} = f_{\gamma_1}^{\gamma_3}$ for all $\gamma_1 \leqslant \gamma_2 \leqslant \gamma_3$, $\gamma_1, \gamma_2, \gamma_3 \in \Gamma$, and $\mu_{\gamma_2}\left(f_{\gamma_1}^{\gamma_2}\right)^{-1} = \mu_{\gamma_1}$. Then there exists a unique probability measure μ_Γ on the projective limit $(X_\Gamma, \mathscr{B}_\Gamma)$ such that $\mu_\Gamma \pi_\gamma^{-1} = \mu_\gamma$, where π_γ is the natural projection from X_Γ onto X_γ for every γ.

Exercise. 27.9. Proposition 27.8 holds under the assumption that $(X_\gamma, \mathscr{B}_\gamma)$ are just standard borel spaces.

Example. 27.10. Let $\{\Upsilon_\alpha, \alpha \in I\}$ be a family of complete and separable metric spaces. Let Γ be the class of all finite subsets of I, partially ordered by inclusion. Then Γ is a directed set. Let

$$X_F = \prod_{\alpha \in F} \Upsilon_\alpha$$

and for $F \subset G$, let f_F^G be the natural projection map from X_G onto X_F. If μ_F is a probability measure on X_F for every $F \in \Gamma$ and $\{\mu_F\}$ is consistent, then there exists a probability measure μ on $\left(\prod_{\alpha \in \Gamma} \Upsilon_\alpha, \prod_{\alpha \in \Gamma} \mathscr{B}_\alpha\right)$ which induces μ_F on X_F through every natural projection from $\prod_{\alpha \in \Gamma} \Upsilon_\alpha$ on X_F.

Example. 27.11. Let any point of the n-dimensional real Euclidean space R^n be denoted by $\mathbf{x}$ in the form of a column vector

$$\mathbf{x} = \begin{pmatrix} x_1 \\ x_2 \\ \vdots \\ x_n \end{pmatrix}, \; x_i \in R, \; i = 1, 2, \ldots, n, \tag{27.9}$$

where x_i is called the ith coordinate of $\mathbf{x}$. For any $\mathbf{m} \in R^n$ and $n \times n$ real positive definite matrix $(\sigma_{ij}) = \Sigma$, $1 \leqslant i, j \leqslant n$, let

$$\phi_n(\mathbf{x}; \mathbf{m}, \Sigma) = \frac{1}{(2\pi)^{n/2} (\det \Sigma)^{\frac{1}{2}}} \exp\left[-\tfrac{1}{2} (\mathbf{x}-\mathbf{m})' \Sigma^{-1} (\mathbf{x}-\mathbf{m})\right] \tag{27.10}$$

where $(\mathbf{x}-\mathbf{m})'$ denotes the transpose of $\mathbf{x}-\mathbf{m}$. It is known from the theory of gamma integrals, (see the proof of Proposition 53.8) that

$$\int_{-\infty}^{+\infty} \cdots \int_{-\infty}^{+\infty} \phi_n(\mathbf{x}; \mathbf{m}, \Sigma)\, dx_1\, dx_2 \ldots dx_n = 1, \tag{27.11}$$

$$\int_{-\infty}^{+\infty} \phi_n(\mathbf{x}; \mathbf{m}, \Sigma)\, dx_n = \phi_{n-1}(\mathbf{x}^0; \mathbf{m}^0, \Sigma^0) \tag{27.12}$$

where

$$\mathbf{x}^0 = \begin{pmatrix} x_1 \\ x_2 \\ \cdot \\ \cdot \\ \cdot \\ x_{n-1} \end{pmatrix}, \quad \mathbf{m}^0 = \begin{pmatrix} m_1 \\ m_2 \\ \cdot \\ \cdot \\ \cdot \\ m_{n-1} \end{pmatrix}; \tag{27.13}$$

$$\Sigma^0 = ((\sigma_{ij})), \; 1 \leqslant i, j \leqslant n-1. \tag{27.14}$$

By Exercise 19.6, it is clear that there exists a probability measure $\mu_n(.; \mathbf{m}, \Sigma)$ on the borel σ-algebra of R^n such that

$$\mu_n(\{\mathbf{x}: a_i < x_i \leqslant b_i \text{ for all } i = 1, 2, \ldots, n\}; \mathbf{m}, \Sigma) = \int_{a_n}^{b_n} \cdots \int_{a_2}^{b_2} \int_{a_1}^{b_1} \phi_n(\mathbf{x}; \mathbf{m}, \Sigma)\, dx_1, dx_2 \ldots dx_n, \text{ for}$$

$$\text{all } a_1, a_2, \ldots, a_n, b_1, b_2, \ldots, b_n \text{ such that } a_i < b_i,\ i=1, 2, \ldots, n. \tag{27.15}$$

The measure μ_n thus defined is called the *multivariate normal distribution* with mean $\mathbf{m}$ and covariance matrix Σ. In all these definitions we use only the Riemann integrals. Once again from the theory of gamma integrals it is known that

$$\int_{-\infty}^{\infty} \cdots \int_{-\infty}^{+\infty} x_i\, \phi_n(\mathbf{x}; \mathbf{m}, \Sigma)\, dx_1\, dx_2 \ldots dx_n = m_i, \tag{27.16}$$

$$\int_{-\infty}^{+\infty} \cdots \int_{-\infty}^{+\infty} (x_i - m_i)(x_j - m_j)\, \phi_n(\mathbf{x}; \mathbf{m}, \Sigma)\, dx_1\, dx_2 \ldots dx_n = \sigma_{ij}. \tag{27.17}$$

Now consider the projection map

$$\pi : \begin{pmatrix} x_1 \\ x_2 \\ \cdot \\ \cdot \\ \cdot \\ x_n \end{pmatrix} \rightarrow \begin{pmatrix} x_1 \\ x_2 \\ \cdot \\ \cdot \\ \cdot \\ x_{n-1} \end{pmatrix}$$

from R^n to R^{n-1}. Then Eq. (27.12) implies that

$$\mu_n(\pi^{-1}(E); \mathbf{m}, \Sigma) = \mu_{n-1}(E; \mathbf{m}^0, \Sigma^0), \quad \text{for all } E \in \mathscr{B}_{R^{n-1}}, \tag{27.18}$$

where $\mathbf{m}^0$ and Σ^0 are defined by Eq. (27.13) and (27.14) respectively. Thus the multivariate normal distribution (in R^n) with mean $\mathbf{m}$ and covariance matrix Σ has 'marginal distribution' which is again multivariate normal with mean $\mathbf{m}^0$ and covariance matrix Σ^0 in R^{n-1}.

Using Eq. (27.18) we shall construct a probability measure on infinite products of the real line. Let $K(s, t)$, $t, s \in I$ be a real valued function on an interval I with the property:

$$\sum_{1 \leq i,j \leq n} a_i\, a_j\, K(t_i, t_j) \geqslant 0 \tag{27.19}$$

for all positive integers n and all $t_1, t_2 \dots t_n \in I$ and all real numbers $a_1, a_2, \dots, a_n$. Such a function is called a *positive definite kernel.* If equality is attained only when the a_i's are all zero we say that $K(\dots)$ is a *strictly positive definite kernel.* As examples one may consider the functions:

$$K_1(s, t) = \sigma^2 \min(s, t),\ s, t \in (0, \infty),$$
$$K_2(s, t) = \exp\{-\tfrac{1}{2}(s-t)^2\},\ s, t \in (-\infty, +\infty).$$

Let now $K(s, t)$ be a strictly positive definite kernel on I. Let $m(t)$ be a real valued function on I. For any $t_1, t_2, \dots, t_k$, consider the normal probability distribution $\mu_{t_1 t_2 \cdots t_k}$ which has mean vector and variance covariance matrix

$$\begin{pmatrix} m(t_1) \\ m(t_2) \\ \cdot \\ \cdot \\ \cdot \\ m(t_k) \end{pmatrix}, \qquad ((K(t_i, t_j)), 1 \leqslant i, j \leqslant k,$$

respectively. Then the family $\{\mu_{t_1 t_2 \cdots t_k}\}$, where $\{t_1, t_2, \dots, t_k\}$ varies over all finite subsets of I, is consistent. Then the product borel space

$$\left(\prod_{t \in I} R_t, \prod_{t \in I} \mathscr{B}_t\right)$$

where each $R_t = R$, $\mathscr{B}_t = \mathscr{B}_R$ admits a probability measure μ such that

$$\mu \pi^{-1}_{t_1 t_2 \dots t_k} = \mu_{t_1 t_2 \cdots t_k},$$

where $\pi_{t_1 t_2 \cdots t_k}$ is the projection

$$x(.) \to (x(t_1), x(t_2), \dots, x(t_k))$$

from $\prod_{t \in I} R_t$ onto $R \times R \times \dots \times R$, the product being k-fold. Here an arbitrary element $x(.)$ of $\prod_{t \in I} R_t$ is a real valued function on I. Thus we have constructed a measure on the space of all real valued functions in an interval I, where any finite number of coordinates have a multivariate normal distribution. Multivariate normal distributions are also called *gaussian distributions*, in honour of the German mathematician Gauss. The measure space $\left(\prod_{t \in I} R_t, \prod_{t \in I} \mathscr{B}_t, \mu\right)$ is said to describe a *gaussian stochastic process* with *mean function* $m(t)$, $t \in I$ and *covariance kernel* $K(s, t)$, $s, t \in I$.

CHAPTER FOUR

Integration

§28. Integration of Non-negative Functions

In the very first chapter we have seen the usefulness of integration of simple functions on a boolean space. In many problems of probability and statistics random variables which are not necessarily simple, do arise and it is necessary to define the ' average value ' or ' expectation ' of such quantities. This can be achieved by extending the notion of integral further. It is also worth noting that mechanical concepts like centre of mass, moment of inertia, work, etc., can be formulated precisely in terms of integrals. However, in the initial stages of its development, the theory of integrals received its first push from the hands of the French mathematician H. Lebesgue on account of many new problems that arose in analysing the convergence properties of Fourier series. In the present chapter we shall introduce the idea of integral with respect to a measure on any borel space and investigate its basic properties.

Throughout this chapter we shall denote by $(X, \mathscr{B}, \mu)$ a fixed σ-finite measure space. By a borel function on this measure space we shall refer to the equivalence class of an extended real valued borel function. It may be recalled that any such function is specified as soon as it is defined almost everywhere, i.e., outside a subset $\mathcal{N}$ of μ-measure zero. For any $E \in \mathscr{B}$ and any property under consideration we shall say ' property holds almost everywhere on E with respect to μ' and write

$$\text{property a.e. on } E(\mu),$$

if there exists a μ-null subset $\mathcal{N}$ of E such that the property holds in the set E-$\mathcal{N}$. If E is the whole space we omit the phrase ' on E'.

Let s be any non-negative simple function on $(X, \mathscr{B}, \mu)$. Then there exists a partition of X into disjoint sets $A_1, A_2, \ldots, A_k$ which belong to $\mathscr{B}$ and k numbers $a_1, a_2, \ldots, a_k$ in $[0, \infty]$ such that $s = \sum_{i=1}^{k} a_i \chi_{A_i}$.

We define the integral of s over any set $E \in \mathscr{B}$ as the number $\sum_{i=1}^{k} a_i \, \mu\,(A_i\, E)$ and write

$$\int_E s\, d\, \mu = \sum_{i=1}^{k} a_i\, \mu(A_i\, E). \tag{28.1}$$

When the integration is over the whole space X we write $\int s\, d\mu$ instead of $\int_X s\, d\mu$. In particular, we have the identities:

$$\int \chi_E\, d\mu = \mu(E); \quad \int_F \chi_E\, d\mu = \mu(E \cap F), E, F \in \mathscr{B}. \tag{28.2}$$

Formula (28.1) shows that $\int_E s\, d\mu$ is a non-negative countably additive function in E. In particular, it is also countably sub-additive in E.

Definition. 28.1. Let f be a non-negative borel function on $(X, \mathscr{B}, \mu)$. The *integral* of f over E is defined as the number

$$\int_E f\, d\, \mu = \sup \{ \int_E s\, d\mu,\ s \geqslant 0,\ s \text{ simple},\ s \leqslant f \text{ on } E\},\ E \in \mathscr{B}. \tag{28.3}$$

Sometimes we write $\int_E f(x)\, d\mu(x)$ or $\int_E f(x)\, \mu\,(dx)$ instead of $\int_E f\, d\mu$ in order to indicate the variable of integration.

Remark. 28.2. It is clear from the above definition that for any two non-negative borel functions f and g on $(X, \mathscr{B}, \mu)$ such that $f = g$ a.e. on E,

$$\int_E f\, d\, \mu = \int_E g\, d\, \mu\,.$$

Proposition. 28.3. For any non-negative borel function f on $(X, \mathscr{B}, \mu)$, the map $E \to \int_E f\, d\, \mu$ is non-negative, monotonic, increasing and countably additive on $\mathscr{B}$.

Proof. Only the last part needs a proof. To this end, let $E_1, E_2, \ldots,$ be a sequence of sets from $\mathscr{B}$ and let $E = \bigcup_i E_i$. For any non-negative simple function s such that $s \leqslant f$ on E, we have from Eqs. (28.1) and (28.3)

$$\int_E s\, d\, \mu \leqslant \sum_i \int_{E_i} s\, d\, \mu \leqslant \sum_i \int_{E_i} f\, d\, \mu.$$

Now taking supremum over all such $s \leqslant f$, we get

$$\int_E f\, d\, \mu \leqslant \sum_i \int_{E_i} f\, d\, \mu\,.$$

In other words the left hand side of the above inequality is countably sub-additive as a function of E. By Proposition 15.9 it is enough to prove finite additivity in order to complete the proof. To this end, let E_1, E_2 be two disjoint sets in $\mathscr{B}$. Let s_1, s_2 be two non-negative simple functions such that $s_i \leqslant f$ on E_i, $i = 1, 2$, and

$$\int_{E_i} f\,d\mu \leqslant \int_{E_i} s_i\,d\mu + \frac{\varepsilon}{2}, \tag{28.4}$$

where ε is a fixed arbitrary positive number. Let

$$\begin{aligned} S &= s_1 \text{ on } E_1 \\ &= s_2 \text{ on } E_2 \\ &= 0 \text{ on } (E_1 \cup E_2)'. \end{aligned}$$

Then $S \leqslant f$ on $E_1 \cup E_2$ and $S \geqslant 0$. Adding inequalities (28.4) over $i=1, 2$, we have from Eqs. (28.1) and (28.3)

$$\begin{aligned} \int_{E_1} f\,d\mu + \int_{E_2} f\,d\mu &\leqslant \varepsilon + \int_{E_1} S\,d\mu + \int_{E_2} S\,d\mu \\ &= \varepsilon + \int_{E_1 \cup E_2} S\,d\mu \\ &\leqslant \varepsilon + \int_{E_1 \cup E_2} f\,d\mu. \end{aligned}$$

Since ε is arbitrary and $\int_E f\,d\mu$ is sub-additive we have

$$\int_{E_1 \cup E_2} f\,d\mu = \int_{E_1} f\,d\mu + \int_{E_2} f\,d\mu.$$

This completes the proof.

Proposition. 28.4. (Lebesgue's monotone convergence theorem) Let $0 \leqslant f_1 \leqslant f_2 \leqslant \dots \leqslant f_n \leqslant \dots$ be borel functions on $(X, \mathscr{B}, \mu)$ and let $\lim_{n \to \infty} f_n = f$. Then

$$\lim_{n \to \infty} \int_E f_n\,d\mu = \int_E f\,d\mu \text{ for all } E \in \mathscr{B}. \tag{28.5}$$

Proof. Let a denote the left hand side of Eq. (28.5). Since $f_n \leqslant f$ for all n, we have

$$a \leqslant \int_E f\,d\mu. \tag{28.6}$$

Let s be a non-negative simple function such that $s \leqslant f$ on E and let $0 < c < 1$ be a constant. We write

$$E_n = \{x : x \in E, 0 \leqslant c\,s(x) \leqslant f_n(x)\}.$$

Then E_n increases to E. Hence

$$\int_E f_n\,d\mu \geqslant \int_{E_n} f_n\,d\mu \geqslant c \int_{E_n} s\,d\mu.$$

Taking limits and using the preceding proposition we have

$$a \geqslant c \int_E s\, d\mu.$$

Now taking supremum over all non-negative simple functions $s \leqslant f$ and then letting $c \to 1$, we have

$$a \geqslant \int_E f\, d\mu.$$

This together with inequality (28.6) completes the proof.

Proposition. 28.5. If f_1, f_2 are non-negative borel functions on $(X, \mathscr{B}, \mu)$ then

$$\int_E (f_1+f_2)\, d\mu = \int_E f_1 d\mu + \int_E f_2 d\mu,\ E \in \mathscr{B}.$$

Proof. If f_1, f_2 are simple this is just property (i) of Proposition 3.6 when P is replaced by μ. If they are not simple we can construct two sequences $\{s_n\}$, $\{s'_n\}$ of non-negative simple functions monotonically increasing respectively to f_1 and f_2 as $n \to \infty$. (See Proposition 23.13). Then $s_n + s'_n$ increases to $f_1 + f_2$. By the preceding proposition we have

$$\begin{aligned}\int_E (f_1+f_2) d\mu &= \lim_{n\to\infty} \int_E (s_n+s'_n)\, d\mu \\ &= \lim_{n\to\infty} \int_E s_n d\mu + \lim_{n\to\infty} \int_E s'_n d\mu \\ &= \int_E f_1 d\mu + \int_E f_2 d\mu.\end{aligned}$$

This completes the proof.

Proposition. 28.6. For any non-negative borel function f on $(X, \mathscr{B}, \mu)$ and any constant $c \geqslant 0$,

$$\int_E c f\, d\mu = c \int_E f\, d\mu,\ E \in \mathscr{B}.$$

Proof. It is left to the reader.

Proposition. 28.7. For any non-negative borel function f on $(X, \mathscr{B}, \mu)$, $\int_E f\, d\mu = 0$ if and only if $f(x) = 0$. a.e. x in E.

Proof. Suppose f does not vanish a.e. in E. Then there exists a constant $c > 0$ and a set $F \subset E$ such that $\mu(F) > 0$ and $f(x) \geqslant c$ for all $x \in F$. Then

$$\int_E f\, d\mu \geqslant \int_F f\, d\mu \geqslant c \int_F 1\, d\mu = c\mu(F) \neq 0.$$

This proves the 'only if' part. The 'if' part is left to the reader.

Definition. 28.8. Let f be any borel function on $(X, \mathscr{B}, \mu)$. For any $E \in \mathscr{B}$, the *essential supremum of f over E* is defined as the number

$$\operatorname*{ess\,sup}_E f = \inf \{ \sup_{x \in F} f(x), F \in \mathscr{B}, F \subset E, \mu(EF') = 0 \}.$$

The *essential infimum of f over E* is defined as the number

$$\operatorname*{ess\,inf}_E f = \sup \{ \inf_{x \in F} f(x), F \in \mathscr{B}, F \subset E, \mu(EF') = 0 \}.$$

It is clear that

$$\operatorname*{ess\,inf}_E f = -\operatorname*{ess\,sup}_E (-f).$$

Exercise. 28.9. For any borel function f on $(X, \mathscr{B}, \mu)$ and for any $E \in \mathscr{B}$ there exists a set $F \subset E$ such that $\mu(EF') = 0$ and

$$\operatorname*{ess\,sup}_E f = \sup_F f.$$

Exercise. 28.10. For any non-negative borel function f on $(X, \mathscr{B}, \mu)$ and $E \in \mathscr{B}$,

$$\mu(E) \operatorname*{ess\,inf}_E f \leqslant \int_E f\, d\mu \leqslant \mu(E) \operatorname*{ess\,sup}_E f.$$

Exercise. 28.11. If μ_1, μ_2 are two σ-finite measures on $(X, \mathscr{B})$ and p, q are non-negative numbers

$$\int_E f\, d(p\mu_1 + q\mu_2) = p \int_E f\, d\mu_1 + q \int_E f\, d\mu_2.$$

Exercise. 28.12. If f is a non-negative borel function on $(X, \mathscr{B}, \mu)$ and $\int_E f\, d\mu < \infty$ for some $E \in \mathscr{B}$, then

$$\mu(\{x : x \in E, f(x) = \infty\}) = 0,$$

i.e., $f(x) < \infty$ a.e. x on E.

Proposition. 28.13. Let $(X, \mathscr{B})$, $(Y, \mathscr{C})$ be two borel spaces and let $T: X \to Y$ be a borel map. Suppose μ is a σ-finite measure on $\mathscr{B}$ such that μT^{-1} is a σ-finite measure on $\mathscr{C}$. Then for any non-negative borel function f on Y and any $F \in \mathscr{C}$,

$$\int_{T^{-1}(F)} (f \circ T)\, d\mu = \int_F f\, d\mu T^{-1}$$

where $f \circ T$ is the composition of f and T.

Proof. First, let $f = \chi_B$ where $B \in \mathscr{C}$. Then

$$\begin{aligned}
\int_F \chi_B\, d\mu T^{-1} &= \mu T^{-1} (B \cap F) \\
&= \mu((T^{-1} B) \cap (T^{-1} F)) \\
&= \int_{T^{-1}F} \chi_{T^{-1}B}\, d\mu \\
&= \int_{T^{-1}F} (\chi_B \circ T)\, d\mu.
\end{aligned}$$

Thus the proposition holds for characteristic functions. Since non-negative simple functions are non-negative linear combinations of characteristic functions the proposition holds for all non-negative simple functions. If now f is any non-negative borel function on Y there exists a sequence $\{s_n\}$ of simple functions on Y monotonically increasing to f. Then $\{s_n \circ T\}$ is a sequence of non-negative simple functions on X monotonically increasing to $f \circ T$. Hence an application of Proposition 28.4 completes the proof.

Remark. 28.14. If in Proposition 28.13, $X=Y$, $\mathscr{B}=\mathscr{C}$ and $\mu T^{-1}=\mu$, then T is called a *μ-measure preserving transformation.* If T preserves μ we have

$$\int (f \circ T)\, d\mu = \int f\, d\mu.$$

If this equation holds for every non-negative borel function f, then T preserves μ.

§29. Integration of Borel Functions

Now we shall try to carry over all the results of the preceding section to borel functions which are not necessarily non-negative.

Proposition. 29.1. Let f be any borel function on the borel space $(X, \mathscr{B})$ and let

$$\begin{aligned} f^+(x) &= f(x) && \text{if } f(x) > 0, \\ &= 0 && \text{if } f(x) \leqslant 0, \\ f^-(x) &= -f(x) && \text{if } f(x) < 0, \\ &= 0 && \text{if } f(x) \geqslant 0. \end{aligned}$$

Then f^+ and f^- are non-negative borel functions and

$$\begin{aligned} f(x) &= f^+(x) - f^-(x), \\ |f(x)| &= f^+(x) + f^-(x). \end{aligned}$$

(The functions f^+ and f^- are called the *positive* and *negative parts* of f respectively.)

Proof. This is a straightforward consequence of definitions and hence left to the reader.

Definition. 29.2. A borel function f on $(X, \mathscr{B}, \mu)$ is said to be *integrable over a set* E in $\mathscr{B}$ if

$$\int_E |f|\, d\mu < \infty.$$

In such a case, the *integral of* f *over* E *with respect to* μ is defined as the number

$$\int_E f\,d\mu = \int_E f^+\,d\mu - \int_E f^-\,d\mu. \tag{29.1}$$

If $E = X$, the whole space, we write $\int f\,d\mu$ instead of $\int_X f\,d\mu$.

Remark. 29.3. If f is integrable over E, it follows from the inequalities $f^+ \leqslant |f|$ and $f^- \leqslant |f|$ that f^+ and f^- are integrable and hence the right hand side of Eq. (29.1) is well defined and

$$\begin{aligned} \left| \int_E f\,d\mu \right| &\leqslant \int_E (f^+ + f^-)\,d\mu \\ &= \int_E |f|\,d\mu. \end{aligned}$$

Remark. 29.4. If f_1, f_2 are two borel functions on $(X, \mathscr{B}, \mu)$ and $f_1 = f_2$ a.e. on E, then f_1 is integrable over E if and only if f_2 is and $\int_E f_1 d\mu = \int_E f_2 d\mu$. Thus the integral does not depend on the values of a function on a set of measure zero. To define the integral over E it is enough to define the function almost everywhere on E.

Proposition. 29.5. If f, g are two borel functions on $(X, \mathscr{B}, \mu)$, which are integrable over E, then for any real constants a, b the function $af+bg$ is defined a.e. on E, integrable over E and

$$\int_E (af+bg)\,d\mu = a \int_E f\,d\mu + b \int_E g\,d\mu. \tag{29.2}$$

Proof. Since f, g are integrable over E, it follows from Exercise 28.12 that $|f(x)| < \infty$, $|g(x)| < \infty$ a.e. on E. Hence $af+bg$ is defined a.e. on E. Further

$$|af+bg| \leqslant |a|\,|f| + |b|\,|g| \text{ a.e. on } E.$$

Since

$$\int_E (|a|\,|f| + |b|\,|g|)\,d\mu = |a| \int_E |f|\,d\mu + |b| \int_E |g|\,d\mu < \infty,$$

the integrability of $af+bg$ over E follows. This proves the first part. Proposition 29.1 and Equation (29.1) imply that, whenever f is integrable over E and c is a constant, cf is also integrable over E and

$$\int_E cf\,d\mu = c \int_E f\,d\mu.$$

Thus it is enough to prove the last part of the proposition when $a=b=1$. We have

$$\begin{aligned} f+g &= (f+g)^+ - (f+g)^- \\ &= f^+ - f^- + g^+ - g^-. \end{aligned}$$

Hence

$$(f+g)^+ + f^- + g^- = (f+g)^- + f^+ + g^+.$$

Now both sides are sums of non-negative functions which are integrable over E. By Proposition 28.5.

$$\int_E (f+g)^+ \, d\mu + \int_E f^- \, d\mu + \int_E g^- \, d\mu$$
$$= \int_E (f+g)^- \, d\mu + \int_E f^+ \, d\mu + \int_E g^+ \, d\mu,$$

and all the terms above are finite. Now Eq. (29.1) implies (29.2). The proof is complete.

Exercise. 29.6. If $\int |f| \, d\mu < \infty$, $E_1, E_2, \ldots$ are disjoint sets in $\mathscr{B}$, then

$$\int_{\bigcup_i E_i} f \, d\mu = \sum_{i=1}^{\infty} \int_{E_i} f \, d\mu.$$

Exercise. 29.7. If f is a non-negative borel function on $(X, \mathscr{B}, \mu)$ and the set function ν on $\mathscr{B}$, defined by

$$\nu(E) = \int_E f \, d\mu$$

is σ-finite, then a borel function g on $(X, \mathscr{B}, \mu)$ is integrable with respect to ν over E if and only if gf is integrable with respect to μ over E and in such a case

$$\int_E g \, d\nu = \int_E gf \, d\mu.$$

Exercise. 29.8. If f is a borel function on $(X, \mathscr{B}, \mu)$ such that $\int_E f \, d\mu = 0$ for every $E \in \mathscr{B}$, then $f = 0$ a.e. (μ).

Proposition. 29.9. Let $(X, \mathscr{B})$, $(Y, \mathscr{C})$ be two borel spaces and let $T : X \to Y$ be a borel map. Suppose μ is a σ-finite measure on $\mathscr{B}$ such that μT^{-1} is also a σ-finite measure. Let f be a borel function on Y. Then f is integrable with respect to μT^{-1} over a set $F \in \mathscr{C}$ if and only if $f \circ T$ is integrable with respect to μ over $T^{-1}(F)$ and in such a case

$$\int_{T^{-1}(F)} f \circ T \, d\mu = \int_F f \, d\mu T^{-1}. \tag{29.3}$$

Proof. This follows immediately from the equation $|f \circ T| = |f| \circ T$, Propositions 28.13 and 29.1.

Remark. 29.10. If $X = Y$, $\mathscr{B} = \mathscr{C}$, $\mu T^{-1} = \mu$ then for any borel function f on X, $f \circ T$ is integrable if and only if f is integrable and

$$\int f \circ T \, d\mu = \int f \, d\mu.$$

Till now we have studied the 'algebraic' properties of $\int_E f\,d\mu$ with respect to f. Now we shall have a look at the 'continuity' properties of integrals with respect to f.

We know that for any two sequences $\{a_n\}$ and $\{b_n\}$ of non-negative numbers,

$$\varliminf_{n\to\infty} (a_n+b_n) \geqslant \varliminf_{n\to\infty} a_n + \varliminf_{n\to\infty} b_n.$$

Our first result is a generalisation of the above property when addition is replaced by integration.

Proposition. 29.11. (Fatou's lemma). Let $\{f_n\}$ be a sequence of non-negative borel functions on $(X, \mathscr{B}, \mu)$. Then

$$\varliminf_{n\to\infty} \int_E f_n\,d\mu \geqslant \int_E (\varliminf_{n\to\infty} f_n)\,d\mu \tag{29.4}$$

for all $E \in \mathscr{B}$.

Proof. Let $g_n(x) = \inf\{f_i(x), i \geqslant n\}$, $x \in X$. Then $g_n(x)$ increases monotonically to the limit $\varliminf_{n\to\infty} f_n(x)$. Hence by the Lebesgue monotone convergence theorem (Proposition 28.4) we have

$$\lim_{n\to\infty} \int_E g_n d\mu = \int_E (\varliminf_{n\to\infty} f_n)\,d\mu. \tag{29.5}$$

Since $g_n \leqslant f_n$ for every n, $\int_E g_n\,d\mu \leqslant \int_E f_n\,d\mu$ for every n. This together with Eq. (29.5) implies Eq. (29.4) and completes the proof.

Proposition. 29.12. (Lebesgue dominated convergence theorem.) Let $\{f_n\}$ be a sequence of borel functions on $(X, \mathscr{B}, \mu)$ converging in measure to f. Suppose there exists a non-negative borel function g such that $|f_n| \leqslant g$ a.e. on E and g is integrable over E. Then f is integrable over E and

$$\lim_{n\to\infty} \int_E f_n\,d\mu = \int_E f\,d\mu.$$

Proof. By going to a subsequence, if necessary, we may by Propositions 24.17 and 24.18 assume that f_n converges to f everywhere. We may also assume that $|f_n(x)| \leqslant g(x)$ for all x. Then

$$g(x) - f_n(x) \geqslant 0;\ g(x) + f_n(x) \geqslant 0 \text{ for all } x.$$

By the preceding proposition

$$\varliminf_{n\to\infty} \int_E (g-f_n)\,d\mu \geqslant \int_E \varliminf_{n\to\infty} (g-f_n)\,d\mu$$
$$= \int_E (g-f)\,d\mu.$$

But the left hand side of the above inequality is the same as $\int_E g\,d\mu - \overline{\lim_{n\to\infty}} \int_E f_n\,d\mu$. Since g is integrable over E and $|f| \leqslant g$, $|f_n| \leqslant g$, it follows that f_n and f are also integrable over E and hence

$$\overline{\lim_{n\to\infty}} \int_E f_n\,d\mu \leqslant \int_E f\,d\mu.$$

Similarly, applying Fatou's lemma to the non-negative functions $g+f_n$ we obtain

$$\underline{\lim_{n\to\infty}} \int_E f_n\,d\mu \geqslant \int_E f\,d\mu.$$

Combining the last two inequalities we obtain the required result.

Example. 29.13. Let $(R, \mathscr{B}_R, L)$ be the measure space where L is the Lebesgue measure on the real line. In this case we write $\int_E f(x)\,dx$ or $\int_E f\,dx$ for $\int_E f\,dL$. Let

$$\begin{aligned} f_n(x) &= n \text{ if } n \leqslant x \leqslant n+\frac{1}{n} \\ &= 0 \text{ otherwise.} \end{aligned}$$

Then

$$\int f_n(x)\,dx = 1 \text{ for all } n.$$

But $f_n \to 0$ a.e. (L) as $n\to\infty$. Thus

$$\lim_{n\to\infty} \int f_n\,dx \neq \int (\lim_{n\to\infty} f_n)\,dx.$$

This example shows that the condition of domination by an integrable function g cannot be dropped in Proposition 29.12. It also shows that strict inequality can hold in Fatou's lemma.

Proposition. 29.14. Let $\{f_n\}$ be a sequence of non-negative borel functions on $(X, \mathscr{B}, \mu)$ and let f_n converge to f in measure. If f_n, f are integrable over E and $\int_E f_n\,d\mu \to \int_E f\,d\mu$, then $\lim_{n\to\infty} \int_E |f_n - f|\,d\mu = 0$.

Proof. We note that $(f-f_n)^+ \leqslant f$, f is integrable over E and $(f-f_n)^+ \to 0$ in measure as $n\to\infty$. Hence by Proposition 29.12,

$$\lim_{n\to\infty} \int_E (f-f_n)^+\,d\mu = 0. \tag{29.6}$$

Since $(f-f_n) = (f-f_n)^+ - (f-f_n)^-$ and $\int_E (f-f_n)\,d\mu \to 0$, it follows that

$$\lim_{n\to\infty} \int_E (f-f_n)^-\,d\mu = 0. \tag{29.7}$$

Now adding Eqs. (29.6) and (29.7), we get

$$\lim_{n\to\infty} \int_E |f_n - f|\,d\mu = 0.$$

This completes the proof.

Remark. 29.15. Let $(X, \mathscr{B}, \mu)$ be a σ-finite measure space and let f_n, f be non-negative borel functions on X which are integrable over X. Suppose $\int f_n d\mu = \int f\, d\mu = 1$ for all n. Define

$$\nu_n(E) = \int_E f_n d\mu, \quad \nu(E) = \int_E f\, d\mu.$$

Then ν_n and ν are probability measures on $\mathscr{B}$, f_n and f are called their *densities* with respect to μ. If $f_n \to f$ in μ-measure, the preceding proposition implies that $\int |f_n - f|\, d\mu \to 0$ as $n \to \infty$. In particular,

$$\begin{aligned} |\nu_n(E) - \nu(E)| &\leqslant \left| \int_E (f_n - f) d\mu \right| \\ &\leqslant \int |f_n - f|\, d\mu \quad \text{for all } E \in \mathscr{B}, \end{aligned}$$

and hence

$$\lim_{n\to\infty} \sup_{E \in \mathscr{B}} |\nu_n(E) - \nu(E)| = 0.$$

More precisely $\sup_{E \in \mathscr{B}} |\nu_n(E) - \nu(E)| = \frac{1}{2} \int |f_n - f|\, d\mu$ (see Billingsley [1], p. 224). In other words, convergence in measure of probability densities (with respect to μ) implies the uniform convergence of the probability measures over $\mathscr{B}$.

Exercise. 29.16 Let $\{f_n\}$ be a sequence of integrable borel functions on a totally finite measure space $(X, \mathscr{B}, \mu)$ such that $f_n \to f$ a.e. (μ) and for any $\varepsilon > 0$, there corresponds a $\delta > 0$ such that

$$\sup_n \left| \int_E f_n\, d\mu \right| < \varepsilon, \text{ whenever } \mu(E) < \delta, \tag{29.8}$$

for all $E \in \mathscr{B}$. Then f is integrable with respect to μ and

$$\lim_{n\to\infty} \int f_n\, d\mu = \int f\, d\mu.$$

(Condition (29.8) is usually called *uniform absolute continuity*.)

§30. Integration of Complex Valued Functions

Let $\mathfrak{C}$ denote the complex plane and let $\mathfrak{C} \cup \{\infty\}$ be the extended complex plane. $\mathfrak{C}$ has the metric topology arising from the metric d defined by $d(z_1, z_2) = |z_1 - z_2|$ for all $z_1, z_2 \in \mathfrak{C}$.

As usual $\mathscr{B}\mathfrak{C}$ will denote the borel σ-algebra on $\mathfrak{C}$. The extended complex plane will be equipped with the smallest σ-algebra containing the class $\mathscr{B}\mathfrak{C}$ and the single point set $\{\infty\}$. Let now f be any borel map from a measure space $(X, \mathscr{B}, \mu)$ into the extended complex plane. We call any such map a *complex valued borel function.* We write

$|f(x)| = +\infty$ if $f(x) = \infty$. Otherwise $|f(x)|$ is the usual modulus of the complex number.

A complex valued borel function f on $(X, \mathscr{B}, \mu)$ is said to be *integrable* over E with respect to μ if $\int_E |f|\, d\mu < +\infty$. In such a case $|f(x)| < \infty$ a.e. on E. Then we can write $f(x) = f_1(x) + i f_2(x)$, where $f_1(x)$ and $f_2(x)$ are the real and imaginary parts of $f(x)$ a.e. on E. The integral of f over E is then defined by

$$\int_E f\, d\mu = \int_E f_1 d\mu + i \int_E f_2 d\mu.$$

Note that the integrability of f_1 and f_2 follows from the fact that $|f_1| \leqslant |f|$, $|f_2| \leqslant |f|$ and $|f|$ is integrable over E.

Exercise. 30.1. If f is a complex valued borel function which is integrable over E, then

$$\left| \int_E f\, d\mu \right| \leqslant \int_E |f|\, d\mu.$$

(Hint: Prove it for simple functions and then approximate.)

Remark. 30.2. Remark 29.4, Propositions 29.5, 29.9 and 29.12 and Exercises 29.6, 29.7 and 29.8 carry over to the complex case easily.

§31. Integration with Respect to a Probability Measure

Let $(X, \mathscr{B}, \mu)$ be a probability space. Real valued borel functions on X are called *random variables*. If f is a random variable on X, μf^{-1} is a probability measure on the real line. It is called the *distribution* of f. If $f_1, f_2, \ldots, f_n$ are n random variables, then the map $\mathbf{f} : x \to (f_1(x), f_2(x), \ldots, f_n(x))$ is borel from X into R^n. The probability measure $\mu \mathbf{f}^{-1}$ is called the *joint distribution* of the random variables $f_1, f_2, \ldots, f_n$.

If f is a random variable on X and f is integrable over X (with respect to μ) we say that the *expectation* of f (with respect to μ) exists and write

$$\mathbf{E}f = \int f\, d\mu$$

and call $\mathbf{E}f$ the *expectation* of f. By Proposition 29.9

$$\mathbf{E}f = \int_R t\, d\mu f^{-1}(t).$$

If ϕ is a borel function on R and $\mathbf{E}\phi(f)$ exists then

$$\mathbf{E}\phi(f) = \int_R \phi(t)\, d\mu f^{-1}(t).$$

These definitions extend in an obvious manner to complex valued random variables.

Proposition. 31.1. Let $(X, \mathscr{B}, \mu)$ be a probability space. Then the following properties hold good:

(i) if f is a non-negative random variable then $\mathbf{E}f \geqslant 0$;

(ii) if $f_1 = f_2$ a.e. then $\mathbf{E}f_1$ exists whenever $\mathbf{E}f_2$ exists and $\mathbf{E}f_1 = \mathbf{E}f_2$;

(iii) if $\mathbf{E}f_1$ and $\mathbf{E}f_2$ exist and a and b are constants, then $\mathbf{E}(af_1 + bf_2)$ exists and equals $a\mathbf{E}f_1 + b\mathbf{E}f_2$;

(iv) if $\{f_n\}$ is a sequence of random variables, $|f_n| \leqslant g$ for all n, $\mathbf{E}g$ exists and f_n converges in probability to f, then $\mathbf{E}f$ exists and

$$\lim_{n \to \infty} \mathbf{E}\,|f_n - f| = 0,$$

$$\lim_{n \to \infty} \mathbf{E}f_n = \mathbf{E}f.$$

Proof. Immediate.

§32. Riemann and Lebesgue Integrals

We shall now establish that Riemann integrability implies integrability with respect to Lebesgue measure and then the two integrals are the same.

Let $[a, b]$ be a bounded interval and let f be a bounded real valued function on it. Let $a = t_0 < t_1 < t_2 \ldots < t_n = b$ be a partition of $[a, b]$ into n intervals. Let

$$M_1 = \sup_{x \in [t_0, t_1]} f(x);$$

$$M_j = \sup_{x \in (t_{j-1}, t_j]} f(x), \ j = 2, 3, \ldots, n;$$

$$m_1 = \inf_{x \in [t_0, t_1]} f(x);$$

$$m_j = \sup_{x \in (t_{j-1}, t_j]} f(x), \ j = 2, 3, \ldots, n.$$

Let us denote by γ the partition of $[a, b]$ into intervals $[t_0, t_1]$, $(t_2, t_3]$, ..., $(t_{n-1}, t_n]$ and write

$$\begin{aligned} S^{\gamma}(x) &= M_j \text{ if } x \in (t_{j-1}, t_j], \ j = 2, 3, \ldots, n; \\ &= M_1 \text{ if } x \in [t_0, t_1]; \\ s^{\gamma}(x) &= m_j \text{ if } x \in (t_{j-1}, t_j], \ j = 2, 3, \ldots, n \\ &= m_1 \text{ if } x \in [t_0, t_1], \end{aligned}$$

$$S^{\overline{\gamma}} = \sum_{j=1}^{n} M_j\,(t_j - t_{j-1}),$$

$$s^{\overline{\gamma}} = \sum_{j=1}^{n} m_j\,(t_j - t_{j-1}).$$

$S^{\overline{\gamma}}$ and $s^{\overline{\gamma}}$ are called the upper and lower sums of the partition γ. If f is Riemann integrable over $[a, b]$, then there exists a sequence of partitions $\{\gamma_n\}$, $n = 1, 2, 3, \ldots$ such that the set of points determining the partition γ_n increases with n, maximum length of the intervals of the partition γ_n decreases to 0 as $n \to \infty$, $S^{\overline{\gamma}n}$ decreases, $s^{\overline{\gamma}n}$ increases and the Riemann integral of f is given by

$$\int_a^b f(x)\,dx = \lim_{n \to \infty} S^{\overline{\gamma}n} = \lim_{n \to \infty} s^{\overline{\gamma}n}. \tag{32.1}$$

Let us write $S_n = S^{\gamma n}$, $s_n = s^{\gamma n}$. Then

$$S_1 \geqslant S_2 \geqslant \ldots S_n \geqslant \ldots \geqslant s_n \geqslant s_{n-1} \geqslant \ldots \geqslant s_2 \geqslant s_1.$$

Let

$$f^0(x) = \lim_{n \to \infty} S_n(x),$$

$$f_0(x) = \lim_{n \to \infty} s_n(x).$$

Since $S_n \geqslant f \geqslant s_n$, it follows that $f^0 \geqslant f \geqslant f_0$. Further f^0 and f_0 are borel functions and

$$\int_{[a,\,b]} f^0\,dL = \lim_{n \to \infty} \int_{[a,\,b]} S_n\,dL = \lim_{n \to \infty} \overline{S}_n,$$

$$\int_{[a,\,b]} f_0\,dL = \lim_{n \to \infty} \int_{[a,\,b]} s_n\,dL = \lim_{n \to \infty} \bar{s}_n.$$

Now Eq. (32.1) implies that

$$\int_{[a,\,b]} f^0\,dL = \int_a^b f(x)\,dx = \int_{[a,\,b]} f_0\,dL$$

and hence

$$\int_{[a,\,b]} (f^0 - f_0)\,dL = 0.$$

Thus f^0, f, f_0 agree almost everywhere. In other words f is measurable with respect to $([a, b]\ \overline{\mathscr{B}}_{[a,\,b]}, L)$ and

$$\int_a^b f(x)\,dx = \int_{[a,\,b]} f\,dL,$$

where $\overline{\mathscr{B}}_{[a,\,b]}$ is the completion of $\mathscr{B}_{[a,\,b]}$ with respect to Lebesgue measure. Further the set $\{x : f^0(x) = f(x) = f_0(x)\}$ is also the set of continuity points of f. Our discussion can now be summarised in the form of a proposition.

Proposition. 32.1. Let f be a Riemann integrable function over the interval $[a, b]$. Then f is integrable with respect to the Lebesgue measure on $[a, b]$ and the two integrals over $[a, b]$ are equal. Further f is continuous a.e. $x(L)$.

Remark. 32.2. Of course, there are Lebesgue integrable functions which are not Riemann integrable. If A and A' are dense borel subsets of $[a, b]$, then χ_A is integrable with respect to the Lebesgue measure and not Riemann integrable.

§33. Riesz Representation Theorem

Let X be a metric space and let $C(X)$ be the space of all bounded real valued continuous functions on X. If μ is a probability measure on $\mathscr{B}_X$, we define

$$\Lambda_\mu (f) = \int f \, d\mu \text{ for } f \in C(X).$$

Then Λ_μ possesses the following properties:

(i) if $f \geqslant 0$ then $\Lambda_\mu (f) \geqslant 0$;

(ii) $\Lambda_\mu (1) = 1$, where 1 also stands for the constant continuous function identically equal to unity;

(iii) $\Lambda_\mu (af + bg) = a\Lambda_\mu (f) + b\Lambda_\mu (g)$, $a, b \in R, f, g, \in C(X)$.

Now there arises the following natural question : if $\Lambda : C(X) \to R$ is a map satisfying the above three properties does there exist a probability measure μ such that $\Lambda = \Lambda_\mu$? We shall investigate this problem in the present section.

Definition. 33.1. A map $\Lambda : C(X) \to R$ is said to be a *normalised non-negative linear functional if*

(i) $\Lambda (1) = 1$, $\Lambda (f) \geqslant 0$ whenever $f \geqslant 0$;

(ii) $\Lambda (af + bg) = a \; \Lambda (f) + b \; \Lambda \; (g)$ for all, $a, b \in R$ and $f, g \in C(X)$.

For any normalised non-negative linear functional Λ on $C(X)$, let

$$\lambda_\Lambda(C) = \inf \{\Lambda(f) : f \in C(X), f \geqslant \chi_C\} \tag{33.1}$$

for every closed set C. We shall now adopt the notations of Section 20.

Proposition. 33.2. Let X be a metric space and let Λ be a normalised non-negative linear functional on $C(X)$. Then the function λ_Λ (defined by Eq. 33.1) is a smooth probability content on the closed subsets of X. (See Definition 20.1)

Proof. It is clear from property (i) and Eq. (33.1) that $0 < \lambda_\Lambda(C) \leqslant 1$ for all $C \in \mathscr{C}_X$ and $\lambda_\Lambda(X) = 1$. If $C_1 \supset C_2$ and $C_1, C_2 \in \mathscr{C}_X$, then $\chi_{C_1} \geqslant \chi_{C_2}$. Hence $\lambda_\Lambda(C_1) \geqslant \lambda_\Lambda(C_2)$. Now suppose that C_1 and C_2 are any two closed sets. Then $\chi_{C_1 \cup C_2} \leqslant \chi_{C_1} + \chi_{C_2}$. If $f \geqslant \chi_{C_1}$ and $g \geqslant \chi_{C_2}$, then $f+g \geqslant \chi_{C_1 \cup C_2}$. Hence

$$\lambda_\Lambda(C_1 \cup C_2) \leqslant \Lambda(f+g) = \Lambda(f) + \Lambda(g).$$

Taking infimum over f and g satisfying the required conditions, successively, we get

$$\lambda_\Lambda(C_1 \cup C_2) \leqslant \lambda_\Lambda(C_1) + \lambda_\Lambda(C_2).$$

Let now C_1, C_2 be two disjoint closed sets. By Proposition 20.2 there exists a function $h \in C(X)$ such that $0 \leqslant h \leqslant 1$, $h(x) = 0$ for all $x \in C_1$, $h(x) = 1$ for all $x \in C_2$. If $f \in C(X)$ and $f \geqslant \chi_{C_1 \cup C_2}$, then $f(1-h) \geqslant \chi_{C_1}$ and $fh \geqslant \chi_{C_2}$. Hence

$$\begin{aligned}\Lambda(f) &= \Lambda(f(1-h)) + \Lambda(fh)\\ &\geqslant \lambda_\Lambda(C_1) + \lambda_\Lambda(C_2).\end{aligned}$$

Now taking infimum over f, we get

$$\lambda_\Lambda(C_1 \cup C_2) \geqslant \lambda_\Lambda(C_1) + \lambda_\Lambda(C_2).$$

This shows that λ_Λ is a probability content on $\mathscr{C}_X$.

Now we shall prove the smoothness of λ_Λ. To this end let $\varepsilon > 0$, $0 < \gamma < 1$ be arbitrary constants and let C be a fixed closed set. By definition we can construct an $f \in C(X)$ such that

$$f \geqslant \chi_C,\ \Lambda(f) \leqslant \lambda_\Lambda(C) + \varepsilon. \tag{33.2}$$

Let

$$\{x : f(x) > \gamma\} = G_\gamma,\ \{x : f(x) \geqslant \gamma\} = C_\gamma.$$

Since $\dfrac{f}{\gamma} \geqslant \chi_{C_\gamma}$ it follows that

$$\Lambda(f) = \gamma \Lambda(f/\gamma) \geqslant \gamma \lambda_\Lambda(C_\gamma). \tag{33.3}$$

Now Eqs. (33.2) and (33.3) imply that

$$\lambda_\Lambda(C_\gamma) \leqslant \frac{1}{\gamma}(\lambda_\Lambda(C) + \varepsilon).$$

Since $C_\gamma = \overline{G}_\gamma \supset G_\gamma \supset C$, we have

$$\inf\{\lambda_\Lambda(\overline{G}),\ G \supset C,\ G \text{ open}\} \leqslant \frac{1}{\gamma}(\lambda_\Lambda(C) + \varepsilon).$$

Letting $\varepsilon \to 0$ and $\gamma \to 1$, we obtain the smoothness of the content λ_Λ. This completes the proof.

Remark. 33.3. By Proposition 20.5 the content λ_Λ can be uniquely extended to a regular finitely additive probability distribution (see Remark 20.6) on the boolean algebra $\mathscr{F}_X$ generated by the class G_x of all open subsets of X. We shall denote it by μ. Thus we arrive at a boolean probability space $(X, \mathscr{F}_X, \mu)$ starting from the linear functional Λ.

In order to proceed further we need the idea of integration on a boolean probability space $(X, \mathscr{F}_X, \mu)$. Consider any partition $\mathscr{P}$ of X into a finite number of disjoint sets $F_1, F_2, \ldots, F_k$ from the boolean algebra $\mathscr{F}_X$. For any bounded real function f on X define the upper and lower sums

$$S(\mathscr{P}) = \sum_{j=1}^{k} \mu(F_j)\, M_j,$$

$$s(\mathscr{P}) = \sum_{j=1}^{k} \mu(F_j)\, m_j,$$

where

$$M_j = \sup \{f(x), x \in F_j\},$$

$$m_j = \inf \{f(x), x \in F_j\}, j=1, 2, \ldots, k.$$

If

$$\inf_{\mathscr{P}} S(\mathscr{P}) = \sup_{\mathscr{P}} s(\mathscr{P}), \tag{33.4}$$

where the infimum and supremum are taken over all partitions of the type mentioned in the beginning then we say that f is integrable with respect to μ and define the *integral* $\int f\, d\mu$ as the common number given by Eq. (33.4). Exactly as in the classical theory of Riemann integration it follows that every bounded function f satisfying the property

$$f^{-1}((a, b]) \in \mathscr{F}_X \text{ for all } a, b \in R,$$

is integrable. In particular, bounded continuous functions on X are integrable. (Indeed, $f^{-1}((a, b]) = f^{-1}((a, b)) \cup f^{-1}(\{b\})$ is the union of an open set and a closed set.) Then the following properties are proved easily.

(i) if a, b are real constants and $f, g \in C(X)$, then
$\int (af+bg)d\mu = a \int f\, d\mu + b \int g\, d\mu$;

(ii) $\int f\, d\mu \geqslant 0$ if $f \geqslant 0$ and $f \in C(X)$;

(iii) $\left|\int f\,d\mu\right| \leqslant \sup_{x \in X} |f(x)|$; $\int 1\,d\mu = 1$;

(iv) if $A, B \in \mathscr{F}_X$, $A \cap B = \varnothing$, $f \in C(X)$, then
$\int_{A \cup B} f\,d\mu = \int_A f\,d\mu + \int_B f\,d\mu$, where $\int_E f\,d\mu$ denotes $\int \chi_E f\,d\mu$ for all $E \in \mathscr{F}_X$.

The reader may note the following: if X is the interval $[\alpha, \beta]$, $\mathscr{F}_X$ is the boolean algebra of all finite disjoint unions of intervals of the form $[a, a]$, $(a, b]$ where a and b vary in $[\alpha, \beta]$ and μ is the Lebesgue measure, the integral defined above is precisely the Riemann integral.

Proposition. 33.4. (Riesz representation theorem). Let X be a metric space and let $C(X)$ be the space of all real bounded continuous functions on X. Let Λ be a normalised non-negative linear functional on X. Then there exists a unique regular (finitely additive) probability distribution μ on the boolean algebra $\mathscr{F}_X$ generated by the class of all open subsets of X, such that

$$\Lambda(f) = \int f\,d\mu \text{ for all } f \in C(X). \tag{33.5}$$

Conversely every regular probability distribution μ on $\mathscr{F}_X$ determines a normalised non-negative linear functional Λ on $C(X)$ by Eq. (33.5).

Proof. From the linear functional Λ we define the smooth probability content λ_Λ by Eq. (33.1) and as mentioned at the beginning of Remark 33.3 we construct the boolean probability space $(X, \mathscr{F}_X, \mu)$. We shall now prove Eq. (33.5). Let $f \in C(X)$, $0 \leqslant f \leqslant 1$. For any positive integer n, let

$$G_i = \{x : f(x) > i/n\},\ i = 0, 1, 2, \ldots, n;$$

Then $G_0 \supset G_1 \supset \ldots \supset G_n = \varnothing$. Let a_i be the continuous function on the unit interval $[0, 1]$, which is 0 in $\left[0, \frac{i-1}{n}\right]$, 1 in $\left[\frac{i}{n}, 1\right]$ and linear in $\left[\frac{i-1}{n}, \frac{i}{n}\right]$. Let

$$f_i(x) = a_i(f(x)),\ x \in X,\ i = 1, 2, \ldots, n.$$

Then we have

$$\frac{1}{n}(a_1(t) + a_2(t) + \ldots + a_n(t)) = t$$

and hence

$$\frac{1}{n}(f_1(x) + f_2(x) + \ldots + f_n(x)) = f(x),$$

$$\frac{1}{n}(\Lambda(f_1) + \Lambda(f_2) + \ldots + \Lambda(f_n)) = \Lambda(f). \tag{33.6}$$

Since $f_i \geqslant \chi_{G_i}$ and for any closed set $C \subset G_i$, $\chi_{G_i} \geqslant \chi_C$ we have $f_i \geqslant \chi_C$ and hence

$$\Lambda(f_i) \geqslant \lambda_\Lambda(C) = \mu(C),$$

where λ_Λ is defined by Eq. (33.1). The regularity of μ now implies that

$$\Lambda(f_i) \geqslant \mu(G_i) \text{ for every } i.$$

Hence Eq. (33.6) implies that

$$\begin{aligned}
\Lambda(f) &\geqslant \frac{1}{n} \sum_{i=1}^{n} \mu(G_i) = \sum_{i=1}^{n} \left(\frac{i}{n} - \frac{i-1}{n}\right) \mu(G_i) \\
&= \sum_{i=1}^{n} \frac{i}{n} [\mu(G_i) - \mu(G_{i+1})] \\
&= \left[\sum_{i=1}^{n-1} \frac{i+1}{n} \mu(G_i - G_{i+1})\right] - \frac{1}{n} \mu(G_1) \\
&\geqslant \left[\sum_{i=1}^{n-1} \int_{G_i - G_{i+1}} f\, d\mu\right] - \frac{1}{n} \mu(G_1) \\
&= \int_{G_1} f\, d\mu - \frac{1}{n} \mu(G_1) \\
&\geqslant \int_X f\, d\mu - \frac{1}{n}.
\end{aligned}$$

Letting $n \to \infty$ we have

$$\Lambda(f) \geqslant \int f\, d\mu \text{ for all } f \in C(X),\ 0 \leqslant f \leqslant 1. \tag{33.7}$$

If $f \geqslant 0$ and $f \in C(X)$ then there exists a positive constant c such that $0 \leqslant cf \leqslant 1$. By inequality (33.7),

$$\Lambda(f) = \frac{1}{c} \Lambda(cf) \geqslant \frac{1}{c} \int cf\, d\mu = \int f\, d\mu.$$

If f is any element in $C(X)$ there exists a constant c_1 such that $f(x) + c_1 \geqslant 0$ for all x. Hence

$$\begin{aligned}
\Lambda(f) &= \Lambda(f + c_1) - c_1 \\
&\geqslant \int (f + c_1)\, d\mu - c_1 = \int f\, d\mu.
\end{aligned}$$

Thus inequality (33.7) holds for all $f \in C(X)$. Changing f to $-f$ we have

$$-\Lambda(f) = \Lambda(-f) \geqslant \int (-f)\, d\mu = -\int f\, d\mu$$

or, equivalently,

$$\Lambda(f) \leqslant \int f\, d\mu \text{ for all } f \in C(X).$$

Combining the two we get Eq. (33.5).

Now suppose that ν is another regular probability distribution on $(X, \mathscr{F}_X)$ such that

$$\Lambda(f) = \int f\, d\nu \text{ for all } f \in C(X).$$

Let C be any closed set. We can find two sequences $\{G_n\}$, $\{H_n\}$ of open sets such that

$$G_1 \supset G_2 \supset \ldots \supset C,$$
$$H_1 \supset H_2 \supset \ldots \supset C,$$
$$\lim_{n \to \infty} \mu(G_n) = \mu(C), \quad \lim_{n \to \infty} \nu(H_n) = \nu(C).$$

Putting $V_n = G_n \cap H_n$ we see that V_n also decreases and

$$\lim_{n \to \infty} \mu(V_n) = \mu(C), \quad \lim_{n \to \infty} \nu(V_n) = \nu(C). \tag{33.8}$$

Now choose a sequence of continuous functions f_n such that $0 \leqslant f_n \leqslant 1$ and

$$f_n(x) = 1 \text{ if } x \in C,$$
$$= 0 \text{ if } x \notin V_n.$$

This can be done because C and V'_n are disjoint closed sets. We have

$$\int f_n\, d\mu = \int_C f_n\, d\mu + \int_{V_n - C} f_n\, d\mu$$
$$= \mu(C) + \int_{V_n - C} f_n\, d\mu. \tag{33.9}$$

We have

$$\int_{V_n - C} f_n\, d\mu \leqslant \mu(V_n - C) = \mu(V_n) - \mu(C).$$

By Eqs. (33.8) and (33.9)

$$\lim_{n \to \infty} \int f_n\, d\mu = \mu(C).$$

Similarly

$$\lim_{n \to \infty} \int f_n\, d\nu = \nu(C).$$

Thus $\mu(C) = \nu(C)$ for all closed sets. Hence regularity implies that $\mu = \nu$. This proves the uniqueness of μ. The converse follows from the definition of the integral with respect to μ. This completes the proof.

Corollary. 33.5. (Riesz representation theorem for the compact case). Let X be a compact metric space and let $C(X)$ be the space of real continuous functions on X. To every normalised non-negative linear functional Λ on $C(X)$ there corresponds a unique probability measure μ on the borel σ-algebra $\mathscr{B}_X$ of X such that

$$\Lambda(f) = \int f \, d\mu \text{ for all } f \in C(X).$$

Conversely, every probability measure on $\mathscr{B}_X$ determines a normalised non-negative linear functional on $C(X)$.

Proof. This follows immediately from Propositions 33.4, 20.8 and the fact that the integral of f with respect to μ outlined in Remark 33.3 is the same as the integral of sections 28 and 29 for continuous functions.

Exercise. 33.6. Let X be a locally compact second countable metric space and let $C_0(X)$ be the set of all continuous functions with compact support (i.e., for every $f \in C_0(X)$, there exists a compact set K, depending on f such that $f(x) = 0$ for all $x \notin K$.) Let $\Lambda : C_0(X) \to R$ be a linear map such that $\Lambda(f) \geqslant 0$ whenever $f \geqslant 0$. For any compact set $K \subset X$, let

$$\lambda(K) = \inf \{\Lambda(f) : f \geqslant \chi_K, f \in C_0(X)\}.$$

Then λ is a compact content. λ satisfies the smoothness property:

$$\lambda(K) = \inf\{\lambda(\bar{G}), G \text{ open}, G \supset K \text{ and } \bar{G} \text{ compact}\}.$$

There exists a unique σ-finite measure μ in $\mathscr{B}_X$ such that $\mu(K) = \lambda(K)$ for all compact sets and

$$\Lambda(f) = \int f \, d\mu \text{ for all } f \in C_0(X).$$

(Hint: Use Corollary 20.12 and Remark 20.13)

Proposition 33.4 and Corollary 33.5 show that the only method of constructing non-negative linear functionals on $C(X)$ is through integration with respect to finitely or countably additive measures on an algebra of sets containing the class of all open sets. It is only natural to raise the following question: What are the linear functionals which are not necessarily non-negative? We shall soon establish under quite general conditions that every 'bounded' linear functional can be expressed as the difference of two non-negative linear functionals.

Definition. 33.7. Let X be a set and let $A(X)$ be a set of bounded real valued functions on X satisfying the following conditions:

(i) if $f, g \in A(X)$ and $a, b \in R$, then the functions $af + bg$, fg and $\max(f, g)$ lie in $A(X)$;

(ii) if $f \in A(X)$ and $\frac{1}{f}$ is bounded then $\frac{1}{f} \in A(X)$;

(iii) the constant function 1 belongs to $A(X)$.

We shall say that $A(X)$ is a *function ring* over X.

Remark. 33.8. As examples of function rings we mention the following:

(i) the space $C(X)$ of all bounded real valued continuous functions on a topological space X;

(ii) the space of all bounded real valued borel functions on any borel space;

(iii) the space of all real valued simple functions on any borel space.

Definition. 33.9. Let $A(X)$ be a function ring over X. A map $\Lambda : A(X) \to R$ is called a *bounded linear functional* if (i) $\Lambda(af+bg) = a\,\Lambda(f)+b\,\Lambda(g)$ for all $a, b \in R$ and $f, g \in A(X)$; (ii) there exists a constant $\alpha > 0$ such that $|\Lambda(f)| \leqslant \alpha \|f\|$ for all $f \in A(X)$, where $\|f\| = \sup\limits_{x \in X} |f(x)|$. The number $\|\Lambda\|$ defined by

$$\|\Lambda\| = \inf \{\alpha : |\Lambda(f)| \leqslant \alpha \|f\| \text{ for all } f \in A(X)\},$$

is called the *norm* of Λ. The number $\|f\|$ mentioned above is also called the *norm* of f. In particular $|\Lambda(f)| \leqslant \|\Lambda\| \, \|f\|$.

Let $A_+(X) = \{f : f \in A(X), f \geqslant 0\}$. The linear functional Λ is said to be *non-negative* if $\Lambda(f) \geqslant 0$ for all $f \in A_+(X)$. To every linear functional Λ we associate a functional $|\Lambda|$ as follows:

$$|\Lambda|(f) = \sup \{|\Lambda(\phi)| + |\Lambda(\psi)| : \phi+\psi = f;\ \phi, \psi \in A_+(X)\} \quad \text{if } f \in A_+(X), \tag{33.10}$$

$$|\Lambda|(f) = |\Lambda|(f^+) + |\Lambda|(f^-) \text{ if } f \in A(X), \tag{33.11}$$

where

$$f^+ = \max(f, 0),\ f^- = \max(-f, 0). \tag{33.12}$$

Let

$$\Lambda^+(f) = \frac{|\Lambda|(f) + \Lambda(f)}{2}, \tag{33.13}$$

$$\Lambda^-(f) = \frac{|\Lambda|(f) - \Lambda(f)}{2}. \tag{33.14}$$

Then $|\Lambda|$, Λ^+ and Λ^- are called the *variation*, *positive part* and *negative part* of Λ respectively.

Proposition. 33.10. Let $A(X)$ be a function ring over X and let Λ be any bounded linear functional on it. Then $|\Lambda|$, Λ^+ and Λ^- are non-negative linear functionals on $A(X)$ and $\Lambda = \Lambda^+ - \Lambda^-$, $|\Lambda| = \Lambda^+ + \Lambda^-$. Further $\Lambda^+(f) \geqslant \Lambda(f)$ for all $f \in A_+(X)$. If Λ_1 is any non-negative linear functional such that $\Lambda_1(f) \geqslant \Lambda(f)$ for all $f \in A_+(X)$, then $\Lambda_1(f) \geqslant \Lambda^+(f)$ for all $f \in A_+(X)$.

Proof. Because of condition (i) in Definition 33.7, f^+ and f^- defined by Eq. (33.12) belong to $A_+(X)$. By Eqs. (33.10) and (33.11) we have

$$|\Lambda|(f) \leqslant 4 \,\|\Lambda\|\, \|f\| \text{ for all } f \in A(X);$$

$$|\Lambda|(cf) = c\,|\Lambda|(f) \text{ if } c \in R \text{ and } f \in A(X).$$

Let $f, g \in A_+(X)$ and let $f + g \geqslant c > 0$. Let $\phi_1, \phi_2, \phi_3, \phi_4 \in A_+(X)$, $\phi_1 + \phi_2 = f$, $\phi_3 + \phi_4 = g$. Then

$$\sum_{i=1}^{4} |\Lambda(\phi_i)| = \left|\Lambda\left(\sum_{i\,:\,\Lambda(\phi_i)\geqslant 0} \phi_i\right)\right| + \left|\Lambda\left(\sum_{i\,:\,\Lambda(\phi_i)< 0} \phi_i\right)\right|$$

$$\leqslant |\Lambda|(f+g).$$

First taking supremum over ϕ_1, ϕ_2 and then over ϕ_3, ϕ_4 we get

$$|\Lambda|(f) + |\Lambda|(g) \leqslant |\Lambda|(f+g). \tag{33.15}$$

For any $\varepsilon > 0$, choose $\phi, \psi \in A_+(X)$ such that $\phi + \psi = f + g$ and

$$|\Lambda|(f+g) \leqslant |\Lambda(\phi)| + |\Lambda(\psi)| + \varepsilon. \tag{33.16}$$

Since $f + g \geqslant c > 0$, we have by conditions (i) and (ii) of Definition 33.7,

$$f = \frac{\phi f}{f+g} + \frac{\psi f}{f+g},$$

$$g = \frac{\phi g}{f+g} + \frac{\psi g}{f+g}. \tag{33.17}$$

and

$$|\Lambda(\phi)| \leqslant \left|\Lambda\left(\frac{\phi f}{f+g}\right)\right| + \left|\Lambda\left(\frac{\phi g}{f+g}\right)\right|$$

$$|\Lambda(\psi)| \leqslant \left|\Lambda\left(\frac{\psi f}{f+g}\right)\right| + \left|\Lambda\left(\frac{\psi g}{f+g}\right)\right|$$

From Eqs. (33.16), (33.17) and the definition of $|\Lambda|$, we get

$$|\Lambda|(f+g) \leqslant \varepsilon + |\Lambda|(f) + |\Lambda|(g).$$

Since ε is arbitrary this together with Eq. (33.15) shows that

$$|\Lambda|(f+g) = |\Lambda|(f) + |\Lambda|(g). \tag{33.18}$$

Now let $f, g \in A_+(X)$ be arbitrary. Then $1+f \geqslant 1$, $1+f+g \geqslant 1$ and the above result shows that

$$\begin{aligned} |\Lambda|(1+f+g) &= |\Lambda|(1) + |\Lambda|(f+g) \\ &= |\Lambda|(1+f) + |\Lambda|(g) \\ &= |\Lambda|(1) + |\Lambda|(f) + |\Lambda|(g). \end{aligned}$$

Since $|\Lambda|$ takes only finite values, Eq. (33.18) holds for $f, g \in A_+(X)$. If f and g are arbitrary,

$$f+g = (f+g)^+ - (f+g)^- = f^+ + g^+ - f^- - g^-$$

and hence

$$(f+g)^+ + f^- + g^- = (f+g)^- + f^+ + g^+.$$

Since both sides are sums of elements in $A_+(X)$, Eq. (33.11) implies that $|\Lambda|$ is a linear functional. Hence Λ^+, Λ^- defined by Eqs. (33.13) and (33.14) are also linear functionals. If $f \in A_+(X)$, Eq. (33.10) implies that

$$|\Lambda|(f) \geqslant |\Lambda(f)|.$$

Hence Λ^+ and Λ^- are non-negative linear functionals. Adding and subtracting Eqs. (33.13) and (33.14) successively we get

$$|\Lambda| = \Lambda^+ + \Lambda^- \text{ and } \Lambda = \Lambda^+ - \Lambda^-.$$

To prove the last part consider any non-negative linear functional Λ_1 such that $\Lambda_1(f) \geqslant \Lambda(f)$ for all $f \in A_+(X)$. Let $\varepsilon > 0$ be arbitrary and let $f \in A_+(X)$. We choose ϕ, ψ in $A_+(X)$ such that

$$\phi + \psi = f,\ \Lambda(\phi) \geqslant 0,\ \Lambda(\psi) \leqslant 0 \text{ and}$$
$$|\Lambda|(f) \leqslant \Lambda(\phi) - \Lambda(\psi) + \varepsilon.$$

Then

$$\begin{aligned} \Lambda^+(f) &= \frac{|\Lambda|(f) + \Lambda(f)}{2} \\ &\leqslant \tfrac{1}{2}\{\Lambda(\phi) - \Lambda(\psi) + \epsilon + \Lambda(\phi) + \Lambda(\psi)\} \\ &= \Lambda(\phi) + \frac{\varepsilon}{2} \\ &\leqslant \Lambda_1(\phi) + \frac{\varepsilon}{2} \\ &\leqslant \Lambda_1(\phi) + \Lambda_1(\psi) + \frac{\varepsilon}{2} = \Lambda_1(f) + \frac{\varepsilon}{2}. \end{aligned}$$

The arbitrariness of ε implies that $\Lambda^+(f) \leqslant \Lambda_1(f)$ for all f in $A_+(X)$. This completes the proof.

Exercise. 33.11. Let $A^*(X)$ denote the space of all bounded linear functionals on the function ring $A(X)$. If $\Lambda_1, \Lambda_2 \in A^*(X)$ we write $\Lambda_1 \geqslant \Lambda_2$ if $\Lambda_1 - \Lambda_2$ is a non-ngeative linear functional. Let

$$\Lambda_1 \vee \Lambda_2 = \Lambda_1 + (\Lambda_2 - \Lambda_1)^+,$$
$$\Lambda_1 \wedge \Lambda_2 = \Lambda_1 - (\Lambda_1 - \Lambda_2)^+.$$

Then (i) $\Lambda_1 \vee \Lambda_2 \geqslant \Lambda_1, \Lambda_2$. If $\Lambda \geqslant \Lambda_1$ and $\Lambda \geqslant \Lambda_2$ then $\Lambda \geqslant \Lambda_1 \vee \Lambda_2$; (ii) $\Lambda_1 \wedge \Lambda_2 \leqslant \Lambda_1, \Lambda_2$. If $\Lambda \leqslant \Lambda_1$ and $\Lambda \leqslant \Lambda_2$ then $\Lambda \leqslant \Lambda_1 \wedge \Lambda_2$. (In other words $\Lambda_1 \vee \Lambda_2$ is the least upperbound and $\Lambda_1 \wedge \Lambda_2$ is the greatest lower bound of Λ_1 and Λ_2 in the ordering $\geqslant$.)

Remark. 33.12. Let $(X, \mathscr{B})$ be a borel space and let $\mu: \mathscr{B} \to [0, \infty]$ be a map with the following properties:

(i) $\sup\limits_{E \in \mathscr{B}} |\mu(E)| < \infty$

(ii) if $E = \bigcup\limits_{i=1}^{\infty} E_i$ and $E_1, E_2, \ldots$ are disjoint, then $\mu(E) = \sum\limits_{1}^{\infty} \mu(E_i)$, where the infinite series on the right hand side converges absolutely.

Then μ is called a *totally finite signed measure* on $(X, \mathscr{B})$. For any $E \in \mathscr{B}$, let

$$|\mu|(E) = \sup\{|\mu(A)| + |\mu(B)| : A, B \in \mathscr{B}, A \cap B = \emptyset, A \cup B = E\};$$

$$\mu^+(E) = \frac{|\mu|(E) + \mu(E)}{2}$$

$$\mu^-(E) = \frac{|\mu|(E) - \mu(E)}{2}.$$

A slight modification of the proof of Proposition 33.10 shows that $|\mu|$, μ^+ and μ^- are totally finite measures and $\mu = \mu^+ - \mu^-$, $|\mu| = \mu^+ + \mu^-$. If μ_1, μ_2 are two totally finite signed measures we define

$$\mu_1 \vee \mu_2 = \mu_1 + (\mu_2 - \mu_1)^+,$$
$$\mu_1 \wedge \mu_2 = \mu_1 - (\mu_1 - \mu_2)^+.$$

We say that $\mu_1 \geqslant \mu_2$ if $\mu_1 - \mu_2$ is a measure. Then $\geqslant$ is a partial ordering. Further (i) $\mu_1 \vee \mu_2 \geqslant \mu_1, \mu_2$. If $\mu \geqslant \mu_i$, $i=1, 2$ then $\mu \geqslant \mu_1 \vee \mu_2$; (ii) $\mu_1 \wedge \mu_2 \leqslant \mu_1, \mu_2$. If $\mu \leqslant \mu_i$, $i=1, 2$ then $\mu \leqslant \mu_1 \wedge \mu_2$. (In other words $\mu_1 \vee \mu_2$ is the least upperbound and $\mu_1 \wedge \mu_2$ is the greatest lowerbound of μ_1 and μ_2 in the ordering $\geqslant$.)

Exercise. 33.13. For any totally finite signed measure μ on a borel space $(X, \mathscr{B})$, let $\|\mu\| = |\mu|(X)$. Then $\|c\mu\| = |c|\,\|\mu\|$, $\|\mu_1 + \mu_2\| \leqslant \|\mu_1\| + \|\mu_2\|$ and $\|\mu\| = 0$ if and only if $\|\mu\| = 0$.

If $\{\mu_n\}$ is a sequence of totally finite signed measures such that $\lim_{m, n\to\infty} \| \mu_m - \mu_n \| = 0$, then there exists a μ such that $\| \mu_n - \mu \| \to 0$.

(In other words the space of all totally finite signed measures is a *complete normed linear space* or equivalently a *Banach space*. This name is in honour of the Polish mathematician S. Banach who created the subject of functional analysis.)

Remark. 33.14. Remark 33.12 and Exercise 33.13 are summarised by saying that the space of all totally finite signed measures on a borel space $(X, \mathscr{B})$ is a *Banach lattice*. A *lattice* is a partially ordered set with an ordering $\geqslant$ such that any two elements have a unique maximum and a unique minimum under $\geqslant$.

§34. Some Integral Inequalities

In this section we shall prove a series of basic inequalities which lead to the construction of many function spaces. Such spaces constitute the foundations of modern functional analysis.

Proposition. 34.1. If $0 \leqslant p \leqslant 1$, $q=1-p$ and $x>0$, then $e^{px} \leqslant pe^x + q$. If $p>0$, equality is obtained if and only if $x=0$.

Proof. We have

$$\begin{aligned} e^{px} &= 1+px+p^2 \frac{x^2}{2!} + \dots \\ &\leqslant 1+px+p\frac{x^2}{2!}+\dots+p\frac{x^n}{n!}+\dots \\ &\leqslant q+pe^x. \end{aligned}$$

This completes the proof.

Proposition. 34.2. If $a\geqslant 0$, $b \geqslant 0$, $0<\alpha<1$, $\alpha+\beta=1$, then

$$a^\alpha b^\beta \leqslant \alpha a + \beta b. \tag{34.1}$$

Equality is attained if and only if $a=b$.

Proof. If a, b or $a-b$ equals zero the above inequality is trivial. So we may assume that $a>b>0$. Dividing both sides of Eq. (34.1) by b, we see that it is enough to prove that

$$\left(\frac{a}{b}\right)^\alpha \leqslant \alpha\left(\frac{a}{b}\right)+\beta$$

If we put $x=\log \frac{a}{b}$, $p=\alpha$, $q=\beta$ in the preceding proposition the proof is complete.

Proposition. 34.3. (Holder's inequality). Let $p>1$, $q>1$ and $\frac{1}{p}+\frac{1}{q}=1$. Let $(X, \mathscr{B}, \mu)$ be a σ-finite measure space and let f, g be complex borel functions such that

$$\int |f|^p \, d\mu < \infty, \int |g|^q \, d\mu < \infty.$$

Then fg is integrable over X and

$$\left| \int fg \, d\mu \right| \leqslant \left(\int |f|^p \, d\mu \right)^{1/p} \left(\int |g|^q \, d\mu \right)^{1/q}. \tag{34.2}$$

Proof. Putting

$$\|f\|_p = \left(\int |f|^p \, d\mu \right)^{1/p}, \tag{34.3}$$

$$\|g\|_q = \left(\int |g|^q \, d\mu \right)^{1/q}, \tag{34.4}$$

$$a = \frac{|f|^p}{\|f\|_p^p}, \quad b = \frac{|g|^q}{\|g\|_q^q},$$

$$\alpha = \frac{1}{p}, \quad \beta = \frac{1}{q}$$

in inequality (34.1) we get

$$\frac{|fg|}{\|f\|_p \, \|g\|_q} \leqslant \frac{1}{p} \frac{|f|^p}{\|f\|_p^p} + \frac{1}{q} \frac{|g|^q}{\|g\|_q^q}.$$

Integrating both sides with respect to μ, we have

$$\frac{\int |fg| \, d\mu}{\|f\|_p \, \|g\|_q} \leqslant 1,$$

which is stronger than inequality (34.2). The proof is complete.

Remark. 34.4. (Schwarz's inequality.) Putting $p=q=2$ in Holder's inequality and changing g to its complex conjugate we have

$$\left| \int f\bar{g} \, d\mu \right| \leqslant \|f\|_2 \, \|g\|_2.$$

Equality is attained if and only if $g=cf$ for some constant c, if $\|f\|_2 < \infty$, $\|g\|_2 < \infty$.

Proposition. 34.5. Let $(X, \mathscr{B}, \mu)$ be a σ-finite measure space and let f, g be complex borel functions on X such that $\|f\|_p < \infty$, $\|g\|_p < \infty$ (See Eq. 34.3) for some $p>0$. Then $\|f+g\|_p < \infty$.

Proof. We have $|f+g| \leqslant |f| + |g| \leqslant 2 \max(|f|, |g|)$. Hence $|f+g|^p \leqslant 2^p \max(|f|^p, |g|^p) \leqslant 2^p (|f|^p + |g|^p)$. This shows that $|f+g|^p$ is integrable and the proof is complete.

Definition. 34.6. Let $(X, \mathscr{B}, \mu)$ be a σ-finite measure space. The space $L_p(\mu)$ is defined for $0<p<\infty$ by

$$L_p(\mu) = \{ f : \int |f|^p \, d\mu < \infty \}$$

where f stands for an equivalence class of complex valued functions with respect to μ. If $p=\infty$,

$$L_\infty(\mu) = \{f : \operatorname*{ess\,sup}_X |f| < \infty\}.$$

For any $f \in L_p(\mu)$, its *norm* $\|f\|_p$ is defined by Eq. (34.3), if $1 \leqslant p < \infty$. Otherwise $\|f\|_\infty = \operatorname*{ess\,sup}_X |f|$.

Exercise. 34.7. If $f \in L_1(\mu)$, $g \in L_\infty(\mu)$, then

$$\|fg\|_1 \leqslant \|f\|_1 \, \|g\|_\infty$$

Proposition. 34.8. (Minkowski's inequality.) Let $p \geqslant 1$. Then for any $f, g \in L_p(\mu)$,

$$\|f+g\|_p \leqslant \|f\|_p + \|g\|_p. \tag{34.5}$$

Proof. If $p=1$ or ∞ the proof is trivial. Otherwise, we have

$$\int |f+g|^p \, d\mu \leqslant \int |f| \, |f+g|^{p-1} \, d\mu + \int |g| \, |f+g|^{p-1} \, d\mu. \tag{34.6}$$

Let $q = \dfrac{p}{p-1}$. Then $\dfrac{1}{p} + \dfrac{1}{q} = 1$. By Proposition 34.5, f, g and $f+g$ belong to $L_p(\mu)$. Now applying Holder's inequality to the pairs of functions $|f|$, $|f+g|^{p-1}$ and $|g|$, $|f+g|^{p-1}$ we obtain from inequality (34.6)

$$\|f+g\|_p^{\,p} = \int |f+g|^p d\mu \leqslant \|f\|_p \, \|f+g\|_p^{\,p/q} + \|g\|_p \, \|f+g\|_p^{\,p/q}.$$

This implies the inequality (34.5) and completes the proof.

Corollary. 34.9. For $p \geqslant 1$, $L_p(\mu)$ is a vector space with the norm $\|.\|_p$ which satisfies

(i) $\|f\|_p = 0$ if and only if $f=0$;
(ii) $\|cf\|_p = |c| \, \|f\|_p$ for any constant c;
(iii) $\|f+g\|_p \leqslant \|f\|_p + \|g\|_p$.

In other words $L_p(\mu)$ is a normed linear space for every $p \geqslant 1$.

Proposition. 34.10. (Riesz-Fischer theorem). Let $(X, \mathscr{B}, \mu)$ be a σ-finite measure space. Suppose $f_n \in L_p(\mu)$, $n=1, 2, \ldots$ and $p \geqslant 1$. If $\lim_{m, n \to \infty} \|f_m - f_n\|_p = 0$ as $n \to \infty$, then there exists an $f \in L_p(\mu)$ such that $\|f_n - f\|_p \to 0$ as $n \to \infty$.

Proof. Because of Minkowski's inequality it follows that $d(f, g) = \|f-g\|_p$ is a metric in $L_p(\mu)$. Under this metric $\{f_n\}$ is a Cauchy sequence. To prove the proposition it is therefore enough to show that a subsequence of $\{f_n\}$ converges to a limit. By going to a subsequence, if necessary, we may assume without loss of generality that

$$\|f_n - f_{n+1}\|_p < \frac{1}{2^n}, \quad n=1, 2, \ldots.$$

By Minkowski's inequality,

$$\left\| \sum_{k=1}^{n} |f_k - f_{k+1}| \right\|_p \leqslant \sum_{1}^{\infty} \|f_k - f_{k+1}\|_p \leqslant 1.$$

Let

$$g = \sum_{1}^{\infty} |f_k - f_{k+1}|.$$

By monotone convergence theorem,

$$\|g\|_p = \lim_{n\to\infty} \left\| \sum_{1}^{n} |f_k - f_{k+1}| \right\|_p \leqslant 1.$$

Hence g is finite a.e. (μ). Hence the infinite series $\sum_{n=1}^{\infty} (f_n - f_{n-1})$, where $f_0=0$, converges absolutely a.e. (μ) and the sum upto n terms of this series is always dominated by the function $|f_1|+g$ which lies in $L_p(\mu)$. The sum upto n terms of this series is nothing but f_n. Let $\lim_{n\to\infty} f_n = f$ a.e. (μ). By Lebesgue dominated convergence theorem $\lim_{n\to\infty} \|f_n - f\|_p = 0$. This completes the proof in the case $p<\infty$. The case $p=\infty$ is left to the reader as an exercise.

Remark. 34.11. Corollary 34.9 and Riesz-Fischer theorem together imply that $L_p(\mu)$ is a complete normed linear space or a Banach space under the norm $\|.\|_p$.

Exercise. 34.12. If the σ-algebra $\mathcal{B}$ is generated by a countable family $\mathcal{E}$ of sets then $L_p(\mu)$ is a separable metric space under the metric d defined by $d(f, g) = \|f-g\|_p, f, g \in L_p(\mu)$. (Hints: (i) simple functions are dense in $L_p(\mu)$; (ii) simple functions with rational values are dense in $L_p(\mu)$; (iii) the boolean algebra generated by $\mathcal{E}$ is countable, (iv) Proposition 18.7).

Proposition. 34.13. If μ is a probability measure and $f \in L_p(\mu)$ for some $p>1$, then $f \in L_{p_1}(\mu)$ for every $1 \leqslant p_1 \leqslant p$ and $\|f\|_{p_1} \leqslant \|f\|_p$.

Proof. Since $f \in L_p(\mu)$, it follows that $|f|^{p_1} \in L_{\frac{p}{p_1}}(\mu)$ and the constant function 1 belongs to $L_{p'}(\mu)$ for every $p' \geqslant 0$. Hence by applying Holder's inequality to the functions $|f|^{p_1}$ and 1 by considering them as elements of $L_{\frac{p}{p_1}}(\mu)$ and $L_{\frac{p}{p-p_1}}(\mu)$ respectively we have

$$\int |f|^{p_1} \, d\mu \leqslant \left(\int |f|^p \, d\mu \right)^{\frac{p_1}{p}}.$$

Raising both sides to the $\frac{1}{p_1}$th power we get the required result.

Proposition. 34.14. (Chebyshev's inequality.) Let $f \in L_p(\mu)$ where $p \geqslant 1$. Then for any $a>0$,

$$\mu\{x : |f(x)| \geqslant a\} \leqslant \left(\frac{\|f\|_p}{a}\right)^p.$$

Proof. We have

$$\begin{aligned} \int |f|^p \, d\mu &\geqslant \int_{\{x:\, |f(x)| \geqslant a\}} |f|^p \, d\mu \\ &\geqslant a^p \int_{\{x:\, |f(x)| \geqslant a\}} 1 \, d\mu \\ &= a^p \, \mu\{x : |f(x)| \geqslant a\}. \end{aligned}$$

This completes the proof.

Definition. 34.15. Let $(X, \mathscr{B}, \mu)$ be a probability space. A family of random variables $\{f_\alpha\}$ with values in a metric space Y is said to be *mutually independent* if for any finite number of the α's, say $\alpha_1, \alpha_2, \ldots, \alpha_k$ and borel sets $E_1, E_2, \ldots, E_k$ in X, the events $f_{\alpha_i}^{-1}(E_i)$, $i=1, 2, \ldots, k$ are mutually independent, i.e.,

$$\mu\left(\bigcap_{i=1}^{k} f_{\alpha_i}^{-1}(E_i)\right) = \prod_{i=1}^{k} \mu\left(f_{\alpha_i}^{-1}(E_i)\right).$$

Exercise. 34.16. If $\{f_\alpha\}$ are mutually independent random variables on a probability space $(X, \mathscr{B}, \mu)$ with values in a metric space Y and $\{\phi_\alpha\}$ is a family of borel maps from Y into another metric space Z, then $\{\phi_\alpha \circ f_\alpha\}$ is a family of mutually independent random variables.

Exercise. 34.17. If $f_1, f_2, \ldots, f_k$ are mutually independent, real (or complex) valued random variables on a probability space $(X, \mathscr{B}, \mu)$ and Ef_j, $j=1, 2, \ldots, k$ exist, then $Ef_1 f_2 \ldots f_k$ exists and equals $Ef_1 \cdot Ef_2 \ldots Ef_k$. (Hint: It is enough to prove for two random variables. Since this is true for simple random variables it holds for non-negative random variables. Since any random variable can be split into positive and negative parts the general case goes through.)

Definition. 34.18. Let $(X, \mathscr{B}, \mu)$ be a probability space and let $f_1, f_2, \ldots, f_k$ be complex valued random variables. Then for

$$\mathbf{r} = \begin{pmatrix} r_1 \\ r_2 \\ \cdot \\ \cdot \\ \cdot \\ r_k \end{pmatrix}, \quad r_i \text{ is a non-negative integer for } i=1, 2, \ldots, k,$$

the number $\mathbf{E}f_1^{r_1} f_2^{r_2} \ldots f_k^{r_k}$ (if it exists) is called the $\mathbf{r}$th *moment* of the random vector

$$\mathbf{f} = \begin{pmatrix} f_1 \\ f_2 \\ \cdot \\ \cdot \\ \cdot \\ f_k \end{pmatrix}.$$

If $f_1, f_2 \in L_2(\mu)$, then by Schwarz's inequality and Proposition 34.13, $\mathbf{E}f_1 \bar{f_2}$, $\mathbf{E}f_1$, $\mathbf{E}f_2$ exist and the number

$$\mathbf{cov}\,(f_1, f_2) = \mathbf{E}f_1 \bar{f_2} - \mathbf{E}f_1\, \mathbf{E}\bar{f_2}$$

is called the *covariance between f_1 and f_2*. If $f_1 = f_2 = f$, $\mathbf{cov}\,(f, f) = \mathbf{E}\,|f|^2 - |\mathbf{E}f|^2$ is called the *variance* of f and denoted by $\mathbf{V}(f)$. (It may be noted that $\mathbf{cov}\,(f_1, f_2) = \mathbf{E}(f_1 - \mathbf{E}f_1)\,\overline{(f_2 - \mathbf{E}f_2)}$. In particular $\mathbf{V}(f) \geqslant 0$. $\mathbf{V}(f) = 0$ if and only if f is a constant a.e.)

Exercise. 34.19. For constants $c_1, c_2, \ldots, c_n$ and complex valued random variables $f_1, f_2, \ldots, f_n$ in $L_2(\mu)$,

$$\mathbf{V}(c_1 f_1 + c_2 f_2 + \ldots + c_n f_n) = \sum_{i,j} c_i \bar{c}_j\, \mathbf{cov}\,(f_i, f_j).$$

In particular, the covariance matrix $((\sigma_{ij}))$ defined by $\sigma_{ij} = \mathbf{cov}\,(f_i, f_j)$ is positive semi-definite. If the covariance matrix is singular then there exist constants $c, c_1, c_2, \ldots, c_n$ such that

$$c_1 f_1 + c_2 f_2 + \ldots + c_n f_n = c \text{ a.e. } (\mu).$$

Remark. 34.20. If f, g are complex valued random variables in $L_2(\mu)$, then $(\mathbf{V}(f))^{1/2}$ is called the *standard deviation* of f, and denoted by $\sigma(f)$. The number

$$\mathbf{r}(f, g) = \frac{\mathbf{cov}\,(f, g)}{\sigma(f)\,\sigma(g)}$$

is called the *correlation coefficient* between f and g. It follows from Schwarz's inequality that $|\mathbf{r}(f, g)| \leqslant 1$ and equality is attained if and only if there exist constants α, β such that $f = \alpha g + \beta$ a.e. (μ). Thus the correlation coefficient may be used as a measure of ' linear dependence ' of one random variable on another.

Exercise. 34.21. (Kolmogorov's inequality). Let $(X, \mathscr{B}, \mu)$ be a probability space and let $f_1, f_2, \ldots, f_n$ be mutually independent real valued random variables with $\mathbf{E}f_i = m_i$, $\mathbf{V}(f_i) = \sigma_i^2$, $i = 1, 2, \ldots, n$.

If $S_k = f_1+f_2+\ldots+f_k$, $M_k = m_1+m_2+\ldots+m_k$, $v_k^2 = \sigma_1^2+\sigma_2^2+\ldots+\sigma_k^2$, $k = 1, 2, \ldots, n$, then

$$\mu\left\{\frac{|S_k - M_k|}{v_n} \leqslant t,\ k = 1, 2, \ldots, n\right\} \geqslant 1 - t^{-2}.$$

(Hint: Carry out the proof exactly as in Proposition 11.3 without any change.)

Exercise. 34.22. Prove Proposition 11.4 and Corollaries 11.5 and 11.6 after replacing simple random variables by real valued random variables on a probability space $(X, \mathscr{B}, \mu)$. Hence deduce the strong law of large numbers: if $f_1, f_2, \ldots$ is a sequence of independent real valued random variables on $(X, \mathscr{B}, \mu)$ such that

$$\mathbf{E}f_i = m_i,\ \mathbf{V}(f_i) = \sigma_i^2 \text{ and } \sum_{k=1}^{\infty} k^{-2}\,\sigma_k^2 < \infty,$$

then

$$\lim_{n\to\infty} \frac{f_1+f_2+\ldots+f_n - (m_1+m_2+\ldots+m_n)}{n} = 0 \text{ a.e. } (\mu).$$

In particular, if $\mathbf{E}f_i = m$, $\mathbf{V}(f_i) = \sigma^2$ for all i,

$$\lim_{n\to\infty} \frac{f_1 + \ldots + f_n}{n} = m \text{ a.e. } (\mu).$$

We shall now prove an inequality concerning integrals of convex functions, which is frequently used in many problems of probability and statistics. To this end we introduce a few definitions.

Definition. 34.23. A set $E \subset R^k$ is said to be *convex* if $\mathbf{x}, \mathbf{y} \in E$, $0 \leqslant p, q \leqslant 1$, $p + q = 1$ imply that $p\mathbf{x} + q\mathbf{y} \in E$. A real valued function ϕ defined on E is said to be *convex* if

$$\phi(p\mathbf{x} + q\mathbf{y}) \leqslant p\phi(\mathbf{x}) + q\phi(\mathbf{y})$$

for all $0 \leqslant p, q \leqslant 1$, $p + q = 1$, $\mathbf{x}, \mathbf{y} \in E$.

Exercise. 34.24. If E is an open interval in R and ϕ is a real valued twice differentiable function such that $\phi''(x) \geqslant 0$ for all $x \in E$, then ϕ is convex on E.

Exercise. 34.25. If E is an open convex subset of R^k and ϕ is a real valued twice differentiable function with continuous second order derivatives on E and the $k \times k$ matrix $\left(\left(\frac{\partial^2\phi}{\partial x_i\,\partial x_j}\right)\right)$ is positive semi-definite on E, then ϕ is convex on E. (Hint: use Taylor expansion upto second degree.)

Proposition. 34.26. Let ϕ be a real valued convex function on a convex set $E \subset R^k$. Let $p_i \geqslant 0$, $i = 1, 2, \ldots, n$, $\sum_i p_i = 1$ and let $\mathbf{x}^i \in E$, $i = 1, 2, \ldots, n$. Then

$$\phi\left(\sum_{i=1}^{n} p_i \mathbf{x}^i\right) \leqslant \sum_{i=1}^{n} p_i \phi(\mathbf{x}^i) \tag{34.7}$$

Proof. Without loss of generality we may assume that $p_i > 0$ for every i. If $n=2$, inequality (34.7) holds by the definition of convexity. We shall now prove for a general n assuming that Eq. (34.7) holds up to $n-1$. We have

$$\sum_{i=1}^{n} p_i \mathbf{x}^i = (1-p_n)\left[\sum_{i=1}^{n-1} \frac{p_i}{1-p_n} \mathbf{x}^i\right] + p_n \mathbf{x}^n.$$

Hence by induction hypothesis

$$\phi\left(\sum_{i=1}^{n} p_i \mathbf{x}^i\right) \leqslant (1-p_n)\, \phi\left(\sum_{i=1}^{n-1} \frac{p_i}{1-p_n} \mathbf{x}^i\right) + p_n \phi(\mathbf{x}^n)$$

$$\leqslant \sum_{i=1}^{n-1} p_i \phi(\mathbf{x}^i) + p_n \phi(\mathbf{x}^n).$$

This completes the proof.

Proposition. 34.27. Let $K \subset R^k$ be a compact convex set and let ϕ be a real valued continuous convex function on K. If μ is a probability measure on the borel subsets of K, then

$$\phi\left(\int_K \mathbf{x}\, d\mu\right) \leqslant \int_K \phi(\mathbf{x})\, d\mu$$

where $\int_K \mathbf{x}\, d\mu$ is the column vector whose ith coordinate is $\int_K x_i\, d\mu$, $i=1, 2, \ldots, k$.

Proof. Since K is compact and convex it follows that $\int_K \mathbf{x}\, d\mu \in K$ for every probability measure μ. Since ϕ is continuous it follows that ϕ is uniformly continuous. Let $\varepsilon > 0$ be arbitrary. Then there exists a $\delta > 0$ such that

$$|\phi(\mathbf{x}) - \phi(\mathbf{y})| < \varepsilon \text{ if } \mathbf{x}, \mathbf{y} \in K, \|\mathbf{x} - \mathbf{y}\| < \delta,$$

where $\|\mathbf{x}\|$ is the usual Euclidean norm $(\sum_i x_i^2)^{1/2}$. We can divide K into disjoint borel sets E_j, $j=1, 2, \ldots, n$ such that

$$K = \bigcup_j E_j; \quad \text{diameter } (E_j) < \delta.$$

Let $\mathbf{x}^j \in E_j, j=1, 2, \ldots, n$ be n points. Then

$$\left| \int_K \phi(\mathbf{x}) d\mu - \sum_j \phi(\mathbf{x}^j) \mu(E_j) \right|$$
$$\leqslant \sum_j \int_{E_j} |\phi(\mathbf{x}) - \phi(\mathbf{x}^j)| \, d\mu(\mathbf{x}) \leqslant \varepsilon \sum \mu(E_j) = \varepsilon. \tag{34.8}$$

Further

$$\| \int_K \mathbf{x} \, d\mu - \sum_j \mu(E_j) \mathbf{x}^j \|$$
$$\leqslant \sum_j \int_{E_j} \| \mathbf{x} - \mathbf{x}^j \| \, d\mu \leqslant \delta \sum_j \mu(E_j) = \delta.$$

Hence

$$|\phi(\int_K \mathbf{x} \, d\mu) - \phi(\sum_j \mu(E_j) \mathbf{x}^j)| < \varepsilon. \tag{34.9}$$

From inequalities (34.8), (34.9) and Proposition 34.26 we get

$$\phi(\int_K \mathbf{x} \, d\mu) \leqslant \varepsilon + \phi(\sum_j \mu(E_j) \mathbf{x}^j)$$
$$\leqslant \varepsilon + \sum_j \phi(\mathbf{x}^j) \mu(E_j)$$
$$\leqslant 2\varepsilon + \int_K \phi(\mathbf{x}) \, d\mu.$$

Since ε is arbitrary the proof is complete.

Proposition. 34.28. (Jensen's inequality.) Let $E \subset R^k$ be a convex set of the form $E = \bigcup_i K_i$, where $K_1 \subset K_2 \subset \ldots$ is an increasing sequence of compact convex sets. Let μ be a probability measure on the borel subsets of E such that $\int \| \mathbf{x} \| \, d\mu < \infty$. If ϕ is a real valued continuous convex function on E and ϕ is integrable with respect to μ, then

$$\phi(\int_E \mathbf{x} \, d\mu) \leqslant \int_E \phi(\mathbf{x}) \, d\mu.$$

Proof. We observe that $\lim_{n\to\infty} \mu(K_n) = \mu(E) = 1$. The integrability conditions imply that

$$\lim_{n\to\infty} \frac{\int_{K_n} \mathbf{x} \, d\mu}{\mu(K_n)} = \int_E \mathbf{x} \, d\mu$$

$$\lim_{n\to\infty} \frac{\int_{K_n} \phi(\mathbf{x}) \, d\mu}{\mu(K_n)} = \int \phi(\mathbf{x}) \, d\mu$$

Hence by Proposition 34.27 and the continuity of ϕ,

$$\phi(\int_E \mathbf{x} \, d\mu) = \lim_{n\to\infty} \phi(\mu(K_n)^{-1} \int_{K_n} \mathbf{x}(d\mu))$$
$$\leqslant \lim_{n\to\infty} \mu(K_n)^{-1} \int_{K_n} \phi(\mathbf{x}) \, d\mu = \int_E \phi(\mathbf{x}) d\mu.$$

This completes the proof.

Remark. 34.29. Let $E \subset R^k$ be a convex set of the form $\bigcup_i K_i$ where $\{K_n\}$ is an increasing sequence of compact convex sets. Let ϕ be a real valued continuous convex function on E. Let $(X, \mathscr{B}, P)$ be a probability space and let $f_1, f_2, \ldots, f_k$ be real valued random variables on X such that the map

$$x \to \mathbf{f}(x) = \begin{pmatrix} f_1(x) \\ f_2(x) \\ \cdot \\ \cdot \\ \cdot \\ f_k(x) \end{pmatrix}$$

takes values in E. If $\mathbf{E}\,|\phi(\mathbf{f})| < \infty$ and $\mathbf{E}\,\|\mathbf{f}\| < \infty$, where

$$\|\mathbf{f}\| = (\Sigma f_i^2)^{\frac{1}{2}}, \text{ then}$$

$$\phi(\mathbf{Ef}) \leqslant \mathbf{E}\,\phi(\mathbf{f}), \tag{34.10}$$

where $\mathbf{Ef}$ is the column vector whose ith coordinate is $\mathbf{E}f_i$. This follows immediately from the preceding proposition if we put $P\mathbf{f}^{-1} = \mu$ and use Proposition 29.9. For convex functions ϕ Jensen's inequality is used in the form (34.10) frequently in applications.

Corollary. 34.30. If ϕ is a real valued function in the interval (a, b) such that $\phi''(x) \geqslant 0$ and ϕ'' is continuous then

$$\phi(\mathbf{E}f) \leqslant \mathbf{E}\,\phi(f)$$

for any random variable f with values in (a, b), defined on a probability space $(X, \mathscr{B}, P)$.

Remark. 34.31. Soon we shall establish that every convex function defined on an open convex subset of R^k is continuous. This would imply that we can drop the assumption of continuity of ϕ in Proposition 34.28 and Remark 34.29 if E is an open convex subset of R^k.

Proposition. 34.32. Let f be a monotonic increasing convex function defined on an interval $[a, b)$. Then $\lim_{x \to a} f(x) = f(a)$.

Proof. We write

$$x = \frac{y-x}{y-a}\,a + \frac{x-a}{y-a}\,y, \quad a < x < y.$$

The convexity of f implies that

$$f(x) \leqslant \frac{y-x}{y-a} f(a) + \frac{x-a}{y-a} f(y).$$

Hence

$$\frac{f(x)-f(a)}{x-a} \leqslant \frac{f(y)-f(a)}{y-a}.$$

If $f(x) \not\to f(a)$ as $x \to a$ then it follows that $f(y)-f(a)=\infty$. This is a contradiction and hence the proof is complete.

Proposition. 34.33. Let ϕ be a real valued convex function on an open convex set $G \subset R^k$ such that ϕ is bounded on bounded sets. If $\{\mathbf{x}_n\}$ is any sequence of points in G such that $\mathbf{x}_n \to \mathbf{x}$ as $n \to \infty$ and $\mathbf{x} \in G$, then

$$\overline{\lim_{n\to\infty}} \; \phi(\mathbf{x}_n) \leqslant \phi(\mathbf{x}). \tag{34.11}$$

Proof. Let $\bar{S}(\mathbf{x}, \rho)=\{\mathbf{y}: \|\mathbf{y}-\mathbf{x}\| \leqslant \rho\}$ be the closed sphere with centre $\mathbf{x}$ and radius ρ under the Euclidean norm. Since G is open and $\mathbf{x} \in G$ there exists an $a>0$ such that $\bar{S}(\mathbf{x}, \rho) \subset G$ for all $0 \leqslant \rho < a$. Let $f(\rho)= \sup \{\phi(\mathbf{y}), \mathbf{y} \in \bar{S}(\mathbf{x}, \rho)\}$. The convexity of ϕ implies that

$$f(\rho) = \sup \{\phi(\mathbf{y}), \|\mathbf{y}-\mathbf{x}\| = \rho\}.$$

(i.e., supremum over the closed sphere=supremum over the boundary.) Further $f(\rho)$ is monotonic increasing in $[0, a)$ and convex. Indeed, if $0 \leqslant \rho_1 < \rho < a$ any point $\mathbf{y}$ on $\bar{S}(\mathbf{x}, \rho)$ is of the form $p\boldsymbol{\xi}+q\boldsymbol{\eta}$ where $\boldsymbol{\xi} \in \bar{S}(\mathbf{x}, \rho_1)$, $\boldsymbol{\eta} \in \bar{S}(\mathbf{x}, \rho_2)$. Hence

$$\begin{aligned}\phi(\mathbf{y}) &\leqslant p\,\phi(\boldsymbol{\xi}) + q\,\phi(\boldsymbol{\eta})\\ &\leqslant p f(\rho_1)+q f(\rho_2).\end{aligned}$$

Taking supremum over $\mathbf{y}$ we have

$$f(\rho) \leqslant p\, f(\rho_1) + q\, f(\rho_2).$$

By the preceding proposition $\lim_{\rho\to 0} f(\rho) = f(0) = \phi(\mathbf{x})$. This proves inequality (34.11).

Proposition. 34.34. Let ϕ be a real valued convex function on an open convex set $G \subset R^k$ such that ϕ is bounded on bounded sets. Then ϕ is continuous.

Proof. Let $\mathbf{x} \in G$ and let $\mathbf{x}_n \in G$, $\lim_{n\to\infty} \mathbf{x}_n=\mathbf{x}$ and $\lim_{n\to\infty} \phi(\mathbf{x}_n)<\phi(\mathbf{x})$.

Then we may assume without loss of generality that there exists an $a>0$ such that

$$\phi(\mathbf{x}_n) < \phi(\mathbf{x}) - a \text{ for all } n.$$

Since G is open we can select a point $\mathbf{y}_n$ on the line joining $\mathbf{x}$ and $\mathbf{x}_n$ such that $\| \mathbf{x} - \mathbf{y}_n \| = \gamma > 0$ and $\mathbf{y}_n \in G$, where $\gamma > 0$ is a suitable constant. Then

$$\mathbf{x} = \frac{\| \mathbf{x}_n - \mathbf{x} \|}{\| \mathbf{x}_n - \mathbf{y}_n \|} \mathbf{y}_n + \frac{\| \mathbf{x} - \mathbf{y}_n \|}{\| \mathbf{x}_n - \mathbf{y}_n \|} \mathbf{x}_n.$$

Then

$$\phi(\mathbf{x}) \leqslant \frac{\| \mathbf{x}_n - \mathbf{x} \|}{\| \mathbf{x}_n - \mathbf{y}_n \|} \phi(\mathbf{y}_n) + \frac{\| \mathbf{x} - \mathbf{y}_n \|}{\| \mathbf{x}_n - \mathbf{y}_n \|} \phi(\mathbf{x}_n).$$

Hence

$$\begin{aligned} a &< \phi(\mathbf{x}) - \phi(\mathbf{x}_n) \\ &\leqslant \frac{\| \mathbf{x}_n - \mathbf{x} \|}{\| \mathbf{x}_n - \mathbf{y}_n \|} [\phi(\mathbf{y}_n) - \phi(\mathbf{x}_n)]. \end{aligned}$$

The right hand side tends to zero as $n \to \infty$. This is a contradiction. Hence Proposition 34.33 implies the continuity of ϕ. This completes the proof.

CHAPTER FIVE

Measures on Product Spaces

§35. Transition Measures and Fubini's Theorem

We shall examine how measures can be constructed on product spaces out of what are called 'transition measures'. The product measures turn out to be special cases of such a construction. Later we shall see how integration in the product space is reduced to successive integration over the marginal spaces. When the transition measures happen to be transition probability measures they acquire a practical significance of great value and form the foundations of the study of Markov processes.

Definition. 35.1. Let $(X_i, \mathscr{B}_i)$, $i=1, 2$ be borel spaces. If $(X, \mathscr{B})$ is a borel space and f is a borel map from $(X_1 \times X_2, \mathscr{B}_1 \times \mathscr{B}_2)$ into $(X, \mathscr{B})$ we say that f is a *borel map of two variables* (x_1, x_2), $x_i \in X_i$, $i=1, 2$. If $X=R$ or the extended real line and $\mathscr{B}$ is its borel σ-algebra we say that f is a borel function of two variables (x_1, x_2). For any fixed $x_1 \in X_1$, we write f^{x_1} for the map from X_2 into X, defined by

$$f^{x_1}(x_2) = f(x_1, x_2), \; x_2 \in X_2.$$

Similarly, for any fixed $x_2 \in X_2$, we write

$$f_{x_2}(x_1) = f(x_1, x_2), \; x_1 \in X_1.$$

f^{x_1} and f_{x_2} are called *sections* of f by x_1 and x_2 respectively. If $E \subset X_1 \times X_2$ we write

$$E^{x_1} = \{x_2 : (x_1, x_2) \in E\},$$
$$E_{x_2} = \{x_1 : (x_1, x_2) \in E\}.$$

E^{x_1} and E_{x_2} are called *sections* of the set E by x_1 and x_2 respectively.

Exercise. 35.2. Sections of subsets of $X_1 \times X_2$ satisfy the following properties:

(i) $(\bigcup_i E_i)^{x_1} = \bigcup_i E_i^{x_1}$;

(ii) $(\bigcap_i E_i)^{x_1} = \bigcap_i E_i^{x_1}$;

(iii) $(E')^{x_1} = (E^{x_1})'$;

(iv) $\chi_E^{x_1} = \chi_{E^{x_1}}$

Proposition. 35.3. Let f be a borel map of two variables from $(X_1 \times X_2, \mathscr{B}_1 \times \mathscr{B}_2)$ into a borel space $(X, \mathscr{B})$. Then the section f^{x_1} is a borel map from $(X_2, \mathscr{B}_2)$ into $(X, \mathscr{B})$. In particular, if $E \in \mathscr{B}_1 \times \mathscr{B}_2$, the section $E^{x_1} \in \mathscr{B}_2$.

Proof. For any $B \subset X$, we have

$$\chi_{(f^{x_1})^{-1}(B)} = \chi^{x_1}_{f^{-1}(B)} = \chi_{[f^{-1}(B)]^{x_1}}.$$

To prove the measurability of f^{x_1}, it is therefore enough to prove that E^{x_1} is measurable whenever $E \in \mathscr{B}_1 \times \mathscr{B}_2$. Let

$$\mathscr{L} = \{E: E \in \mathscr{B}_1 \times \mathscr{B}_2,\ E^{x_1} \in \mathscr{B}_2 \text{ for every } x_1 \in X_1\}.$$

(See Remark 13.10). If $E_n \in \mathscr{L}$ and E_n increases or decreases to E, then $E_n{}^{x_1}$ increases or decreases accordingly to E^{x_1} for every $x_1 \in X_1$. Hence $\mathscr{L}$ is a monotone class. If $E = B_1 \times B_2$, $B_i \in \mathscr{B}_i$, $i=1, 2$ then

$$\begin{aligned} E^{x_1} &= B_2 \text{ if } x_1 \in B_1, \\ &= \emptyset \text{ if } x_1 \notin B_1. \end{aligned}$$

Hence every borel rectangle $B_1 \times B_2 \in \mathscr{L}$. Hence finite disjoint unions of borel rectangles belong to $\mathscr{L}$. Since finite disjoint unions of borel rectangles constitute a boolean algebra, Proposition 14.4 implies that $\mathscr{L} = \mathscr{B}_1 \times \mathscr{B}_2$. This completes the proof.

Definition. 35.4. Let $(X_i, \mathscr{B}_i)$, $i=1, 2$ be borel spaces. A map $\lambda : X_1 \times \mathscr{B}_2 \to [0, \infty]$ is called a *transition measure* if

(i) for every fixed $B \in \mathscr{B}_2$, $\lambda(x_1, B)$ is a borel function of x_1;

(ii) for every fixed $x_1 \in X_1$, $\lambda(x_1, B)$ is a σ-finite measure as a function of B.

λ is called a *σ-finite transition measure* if $X_2 = \bigcup_{i=1}^{\infty} B_i$ and $\lambda(x_1, B_i) < \infty$ for all $x_1 \in X_1$ and every $i=1, 2, \ldots$. It is called a *transition probability* if $\lambda(x_1, X_2) \equiv 1$, $x_1 \in X_1$.

Example. 35.5. Let μ be a σ-finite measure on $(X_2, \mathscr{B}_2)$. Let $\lambda(x_1, B) = \mu(B)$ for all $x_1 \in X_1$, $B \in \mathscr{B}_2$. Then λ is a transition measure.

Example. 35.6. Let $f(x, y)$ be a non-negative continuous function in $[0, 1] \times [0, 1]$. For any borel set $B \subset [0, 1]$, let

$$\lambda(x, B) = \int_B f(x, y)\, dy$$

where dy denotes integration with respect to the Lebesgue measure. Then λ is a transition measure.

Example. 35.7. Let

$$P = \begin{pmatrix} p_{11} & p_{12} & \cdots & p_{1k} \\ p_{21} & p_{22} & \cdots & p_{2k} \\ \cdots & \cdots & \cdots & \cdots \\ p_{k1} & p_{k2} & \cdots & p_{kk} \end{pmatrix}$$

be a $k \times k$ matrix of non-negative numbers such that each row adds up to unity, i.e., $\sum_j p_{ij} = 1$ for each i. Let $X_1 = X_2 = \{1, 2, \ldots, k\}$. For any $i \in X_1$, $B \subset X_2$, let

$$\lambda(i, B) = \sum_{j \in B} p_{ij}.$$

Then λ is a transition probability on $(X_1 \times X_2, \mathscr{B}_1 \times \mathscr{B}_2)$. ($P$ is called the associated *transition probability matrix.*)

Remark. 35.8. Transition probabilities have the following interpretation: there are two systems whose states are described by the points of X_1 and X_2. The statistical behaviour of the elementary outcomes of the second system depends on the state of the first system. If the first system is in the state x_1 then the probability that the state of the second system is in a set B is described by the transition probability $\lambda(x_1, B)$.

As a special case we can consider the following situation. Suppose communication of messages takes place between two stations S_1 and S_2. The messages being transmitted from S_1 constitute the set X_1. The messages received at S_2 constitute the set X_2. If the same message $x_1 \in X_1$ is transmitted repeatedly the received message need not be the same always. In such a case we say that there is noise during transmission. If x_1 is transmitted the probability of the received message being in the set B is described by a transition probability $\lambda(x_1, B)$. We say that λ is the transition probability of the ' communication channel' between S_1 and S_2.

Proposition. 35.9. Let f be a non-negative borel function of two variables on $(X_1 \times X_2, \mathscr{B}_1 \times \mathscr{B}_2)$ and let $\lambda(x_1, B)$ be a σ-finite transition measure on $X_1 \times \mathscr{B}_2$. Then $\int f(x_1, x_2)\, \lambda(x_1, dx_2)$ is a borel function on $(X_1, \mathscr{B}_1)$. (Here dx_2 indicates integration with respect to the measure $\lambda(x_1, .)$.)

Proof. Since non-negative functions are limits of monotonically increasing sequences of non-negative simple functions an application of monotone convergence theorem shows that it is enough to prove the result when f is simple. This in turn implies that it is enough to prove it when f is a characteristic function. Since λ is a σ-finite transition measure there exist disjoint sets $B_1, B_2, \ldots$ in $\mathscr{B}_2$ such that

$$X_2 = \bigcup_j B_j,\ \lambda(x_1, B_j) < \infty \text{ for all } j=1, 2, \ldots, \text{ and for all } x_1 \in X_1.$$

If $f = \chi_E$, $E \in \mathscr{B}_1 \times \mathscr{B}_2$, then by Proposition 35.3 and monotone convergence theorem

$$\int \chi_E(x_1, x_2)\, \lambda(x_1, dx_2) = \sum_{i=1}^{\infty} \int \chi_E(x_1, x_2)\, \chi_{B_i}(x_2)\, \lambda(x_1, dx_2).$$

(Here we treat x_1 as fixed while integrating with respect to $\lambda(x_1, \ldots)$ in the variable x_2). Thus it is enough to prove that for all $E \in \mathscr{B}_1 \times \mathscr{B}_2$, the function

$$\int \chi_E(x_1, x_2)\, \chi_{B_i}(x_2)\, \lambda(x_1, dx_2) \tag{35.1}$$

is borel on $(X_1, \mathscr{B}_1)$. Let

$$\mathscr{L}_i = \{E : E \in \mathscr{B}_1 \times \mathscr{B}_2,\ \int \chi_E(x_1, x_2)\, \chi_{B_i}(x_2)\, \lambda(x_1, dx_2) \text{ is borel}\}.$$

(See Remark 13.10!) If $E = A \times B$ is a borel rectangle then the function (35.1) is $\chi_A(x_1)\, \lambda(x_1, BB_i)$. Hence $E \in \mathscr{L}_i$. Further $\mathscr{L}_i$ is closed under finite disjoint unions and hence contains the boolean algebra of all finite disjoint unions of borel rectangles. If $E_n \in \mathscr{L}_i$ and E_n increases or decreases monotonically to E, then by dominated convergence theorem

$$\lim_{n \to \infty} \int \chi_{E_n}(x_1, x_2)\, \chi_{B_i}(x_2)\, \lambda(x_1, dx_2)$$
$$= \int \chi_E(x_1, x_2)\, \chi_{B_i}(x_2)\, \lambda(x_1, dx_2).$$

Hence $\mathscr{L}_i$ is a monotone class. By Proposition 14.4, $\mathscr{L}_i = \mathscr{B}_1 \times \mathscr{B}_2$ for all i. This completes the proof.

Definition. 35.10. Let $(X_i, \mathscr{B}_i)$, $i = 1, 2$ be borel spaces. A σ-finite transition measure λ on $X_1 \times \mathscr{B}_2$ is said to be *uniformly σ-finite* if there exist sequences $\{A_n\}$ and $\{B_n\}$ of disjoint sets such that

(i) $\bigcup_n A_n = X_1$, $A_n \in \mathscr{B}_1$, $n = 1, 2, \ldots$;

(ii) $\bigcup_n B_n = X_2$, $B_n \in \mathscr{B}_2$, $n = 1, 2, \ldots$;

(iii) $\sup_{x_1 \in A_m} \lambda(x_1, B_n) < \infty$ for all $m = 1, 2, \ldots$ and $n = 1, 2, \ldots$

Proposition. 35.11. Let $(X_i, \mathscr{B}_i)$, $i = 1, 2$ be borel spaces and let λ be a uniformly σ-finite transition measure on $X_1 \times \mathscr{B}_2$. If μ is a σ-finite measure on $(X_1, \mathscr{B}_1)$ and $E \in \mathscr{B}_1 \times \mathscr{B}_2$, the number

$$\nu(E) = \int \left[\int \chi_E(x_1, x_2)\, \lambda(x_1, dx_2) \right] \mu(dx_1) \tag{35.2}$$

is well-defined. Further ν is a σ-finite measure on $\mathscr{B}_1 \times \mathscr{B}_2$. If f is a non-negative borel function on $(X_1 \times X_2, \mathscr{B}_1 \times \mathscr{B}_2)$, then

$$\int f\, d\nu = \int \left[\int f(x_1, x_2)\, \lambda(x_1, dx_2) \right] \mu(dx_1). \tag{35.3}$$

Proof. It follows from Proposition 35.9 that ν is well-defined. The countable additivity of ν is an immediate consequence of monotone convergence theorem. Since λ is uniformly σ-finite there exist countable partitions $\{A_n\}$ and $\{B_n\}$ of X_1 and X_2 respectively satisfying properties (i) to (iii) of Definition 35.10. Since μ is σ-finite there exists a partition $\{C_n\}$ of X_1 such that $\mu(C_n) < \infty$ for each n. Hence $\{(A_i C_j) \times B_k\}$, $i=1, 2, \ldots$; $j=1, 2, \ldots$; $k=1, 2, \ldots$ is a countable partition of $X_1 \times X_2$ into borel rectangles. Further

$$\begin{aligned} \nu((A_i C_j) \times B_k) &= \int_{A_i C_j} \lambda(x_1, B_k)\, \mu(dx_1) \\ &\leqslant \left[\sup_{x_1 \in A_i} \lambda(x_1, B_k) \right] \mu(C_j) < \infty. \end{aligned}$$

This shows that ν is a σ-finite measure.

If $f = \chi_E$, Eq. (35.3) is same as Eq. (35.2). Thus Eq. (35.3) holds for all non-negative simple functions. By monotone convergence theorem, Eq. (35.3) holds for all non-negative borel functions. This completes the proof.

Remark. 35.12. If λ is a transition probability and μ is a probability measure then ν defined by Eq. (35.2) is a probability measure on the product borel space $(X_1 \times X_2, \mathscr{B}_1 \times \mathscr{B}_2)$.

Remark. 35.13. If $\lambda(x_1, B) = \lambda(B)$ is independent of x_1, where λ is a σ-finite measure on $\mathscr{B}_2$, then the measure ν defined by Eq. (35.2) has the property

$$\nu(A \times B) = \mu(A)\, \lambda(B) \text{ for all } A \in \mathscr{B}_1,\ B \in \mathscr{B}_2. \tag{35.4}$$

If ν' is another measure such that for any borel rectangle $A \times B$,

$$\nu'(A \times B) = \mu(A)\, \lambda(B).$$

then ν and ν' agree on all borel rectangles and hence on the boolean algebra of their finite disjoint unions. Hence $\nu = \nu'$ on the σ-algebra $\mathscr{B}_1 \times \mathscr{B}_2$. Thus we have the following result: given σ-finite measures μ, λ on the borel spaces $(X_1, \mathscr{B}_1)$ and $(X_2, \mathscr{B}_2)$ respectively, there exists a unique σ-finite measure ν on the product space $(X_1 \times X_2, \mathscr{B}_1 \times \mathscr{B}_2)$ which satisfies Eq. (35.4). This measure ν is called the *product* of the two measures μ and λ and denoted by $\mu \times \lambda$.

If $(X_i, \mathscr{B}_i)$, $i=1,2,\ldots, k$ are borel spaces and μ_i is a σ-finite measure on $\mathscr{B}_i$ for every i we can, by repeating the above procedure, construct a unique σ-finite measure $\nu=\mu_1\times\mu_2\times\ldots\times\mu_k$ on the product borel space $(X_1\times X_2\times\ldots\times X_k, \mathscr{B}_1\times\mathscr{B}_2\times\ldots\times\mathscr{B}_k)$, satisfying the condition

$$\nu(B_1\times B_2\times\ldots\times B_k) = \prod_{i=1}^{k} \mu_i(B_i).$$

The measure ν is called the product of the measures μ_i, $i=1, 2, \ldots, k$. If $X_i=R$, $\mathscr{B}_i=\mathscr{B}_R$ and μ_i is the Lebesgue measure on R for every $i=1, 2, \ldots, k$ then the product measure ν is called the *Lebesgue measure in* R^k. Its completion is also called the Lebesgue measure.

The product measure $\nu=\mu\times\lambda$ has the following property: for any non-negative borel function $f(x_1, x_2)$ on $(X_1\times X_2, \mathscr{B}_1\times\mathscr{B}_2)$

$$\begin{aligned}\int f\, d\nu &= \int \left[\int f(x_1, x_2)\, d\lambda(x_2)\right] d\mu(x_1)\\ &= \int \left[\int f(x_1, x_2)\, d\mu(x_1)\right] d\lambda(x_2).\end{aligned}$$

This is one form of what is classifically known as Fubini's theorem.

Proposition. 35.14. (Generalised Fubini's theorem). Let $(X_i, \mathscr{B}_i)$, $i = 1, 2$ be borel spaces. Let λ be a uniformly σ-finite transition measure on $\mathscr{B}_1$. If ν is the measure determined by λ and μ on $\mathscr{B}_1\times\mathscr{B}_2$ by Eq. (35.2) then a borel function f on $(X_1\times X_2, \mathscr{B}_1\times\mathscr{B}_2)$ is integrable with respect to ν if and only if

(i) $\int |f(x_1, x_2)|\, \lambda(x_1, dx_2) < \infty$ a.e. $x_1(\mu)$;

(ii) $\int [\int |f(x_1, x_2)|\, \lambda(x_1, dx_2)]\, \mu(dx_1) < \infty$.

In such a case

$$\int f\, d\nu = \int \left[\int f(x_1, x_2)\, \lambda(x_1, dx_2)\right] \mu(dx_1).$$

Proof. By definition f is integrable with respect to ν if and only if $\int |f|\, d\nu < \infty$. By Proposition 35.11 conditions (i) and (ii) are necessary and sufficient. Since $f=f^+-f^-$, $f^+\leqslant|f|$, $f^-\leqslant|f|$ and f^+ and f^- are non-negative the last equation follows from the validity of Eq. (35.3) for f^+ and f^- separately.

Corollary. 35.15. (Fubini's theorem) If μ and λ are σ-finite measures on $(X_1, \mathscr{B}_1)$ and $(X_2, \mathscr{B}_2)$ respectively, a borel function f on $(X_1\times X_2, \mathscr{B}_1\times\mathscr{B}_2)$ is integrable with respect to $\mu\times\lambda$ if and only if

(i) $\int |f(x_1, x_2)|\, \lambda(dx_2) < \infty$ a.e. $x_1\ (\mu)$;

(ii) $\int [\int |f(x_1, x_2)|\, \lambda(dx_2)]\, \mu(dx_1) < \infty$.

In such a case

$$\begin{aligned}\int f(x_1, x_2)\, (\mu\times\lambda)\, (dx_1\, dx_2) &= \int \left[\int f(x_1, x_2)\, \lambda(dx_2)\right] \mu(dx_1)\\ &= \int \left[\int f(x_1, x_2)\, \mu(dx_1)\right] \lambda(dx_2).\end{aligned}$$

Proof. The first part follows immediately from the preceding proposition if we put $\lambda(x_1, B)=\lambda(B)$ for all $B\in\mathscr{B}_2$, $x_1\in X_1$. The second part follows from interchanging the roles of λ and μ and X_1 and X_2, i.e., considering $\mu(x_2, B)=\mu(B)$, $B\in\mathscr{B}_1$, $x_2\in X_2$ as a transition measure and observing that the measure spaces $(X_1\times X_2, \mathscr{B}_1\times\mathscr{B}_2, \mu\times\lambda)$ and $(X_2\times X_1, \mathscr{B}_2\times\mathscr{B}_1, \lambda\times\mu)$ are isomorphic through the map $(x_1, x_2)\to(x_2, x_1)$.

Corollary. 35.16. Under the notations of Proposition 35.14 any borel function f on $(X_1\times X_2, \mathscr{B}_1\times\mathscr{B}_2)$ satisfies the equation

$$f(x_1, x_2)=0 \text{ a.e. } (x_1, x_2)\ (\nu)$$

if and only if

$$f(x_1, x_2)=0 \text{ a.e. } x_2\ (\lambda(x_1,\ldots)), \text{ a.e. } x_1\ (\mu).$$

In particular, when $\boldsymbol{\nu}=\mu\times\lambda$ where μ and λ are σ-finite measures on $(X_1, \mathscr{B}_1)$ and $(X_2, \mathscr{B}_2)$ respectively,

$$f(x_1, x_2)=0, \text{ a.e. } (x_1, x_2)\ (\mu\times\lambda)$$

if and only if

$$f(x_1, x_2)=0, \text{ a.e. } x_2\ (\lambda), \text{ a.e. } x_1\ (\mu).$$

Proof. This follows from the generalised Fubini's theorem and the fact that $f=0$ if and only if $|f|=0$.

Proposition. 35.17. Let $(X_i, \mathscr{B}_i)$, $i=1, 2$ be borel spaces and let μ and λ be σ-finite measures on $\mathscr{B}_1$ and $\mathscr{B}_2$ respectively. Then

$$\begin{aligned}(\mu\times\lambda)\ (E) &= \int \mu(E_{x_2})\ \lambda(dx_2)\\ &= \int \lambda(E^{x_1})\ \mu(dx_1),\end{aligned} \tag{35.5}$$

for all $E\in\mathscr{B}_1,\times\mathscr{B}_2$, where E^{x_1} and E_{x_2} are sections of E by x_1 and x_2 respectively.

Proof. In Proposition 35.11, put $\lambda(x_1, B)=\lambda(B)$ for all x_1. Then Eq. (35.2) and Corollary 35.15 yield Eq. (35.5) and that completes the proof.

Exercise. 35.18. If $(X_i, \mathscr{B}_i, \mu_i)$, $i=1, 2,\ldots, k$ are σ-finite measure spaces and $T_i: X_i\to X_i$ is a borel map such that $\mu_i\ T_i^{-1}=\nu_i$ for each i, then $(\mu_1\times\mu_2\times\ldots\times\mu_k)\ T^{-1}=\nu_1\times\nu_2\times\ldots\times\nu_k$ where T is the map which sends $(x_1, x_2, \ldots, x_k)$ to $(T_1x_1, T_2x_2, \ldots, T_kx_k)$. In particular, if μ_i is invariant under T_i, i.e., $\mu_i\ T_i^{-1}=T_i$ for every i, then $\mu_1\times\mu_2\times\ldots\times\mu_k$ is invariant under T. If L is the Lebesgue measure in R^k then L is invariant under all translations $T_{\mathbf{a}}: \mathbf{x}\to\mathbf{x}+\mathbf{a}$, $\mathbf{a}\in R^k$.

Exercise. 35.19. Let $(X_i, \mathscr{B}_i)$, $i=1, 2, 3, 4$ be borel spaces. If λ_1 and λ_2 are transition probabilities on $X_1\times\mathscr{B}_2$ and $X_2\times\mathscr{B}_3$ respectively then

$$(\lambda_1 \circ \lambda_2)\ (x_1, B)=\int \lambda_2(x_2, B)\ \lambda_1(x_1, dx_2)$$

is a transition probability on $X_1\times\mathscr{B}_3$. If λ_3 is a transition probability on $X_3\times\mathscr{B}_4$, then $(\lambda_1 \circ \lambda_2) \circ \lambda_3=\lambda_1 \circ (\lambda_2 \circ \lambda_3)$.

Exercise. 35.20. Let X be the finite set $\{1, 2, \ldots, k\}$ and let $\mathscr{B}$ be the algebra of all subsets of X. Let λ_1 and λ_2 be transition probabilities on $X \times \mathscr{B}$ determined by transition probability matrices P_1 and P_2 respectively (See Example 35.7). Then the transition probability matrix associated with $\lambda_1 \circ \lambda_2$ is the matrix $P_1 P_2$.

Exercise. 35.21. Let $(X_i, \mathscr{B}_i)$, $i=1, 2, \ldots$ be a sequence of borel spaces and let λ_i be a transition probability on $X_i \times \mathscr{B}_{i+1}$, $i=1, 2, \ldots$. Let μ be a probability measure on $(X_1, \mathscr{B}_1)$. For any $E \in \mathscr{B}_1 \times \mathscr{B}_2 \times \ldots \times \mathscr{B}_n$, let

$$\mu_n(E) = \int [\ldots[\int [\int \chi_E(x_1, x_2, \ldots, x_n)\, \lambda_{n-1}(x_{n-1}, dx_n)] \lambda_{n-2}(x_{n-2}, dx_{n-1})]\ldots]\, \lambda_1(x_1, dx_2)]\, \mu(dx_1).$$

Then $\{\mu_n\}$ is a consistent family of probability measures on the sequence of spaces $\{X_1 \times X_2 \times \ldots \times X_n\}$. (Hence by Proposition 27.4 and Example 27.6 there exists a measure $\tilde{\mu}$ on $(\prod_{i=1}^{\infty} X_i, \prod_{i=1}^{\infty} \mathscr{B}_i)$ such that $\tilde{\mu}\, \pi_n^{-1} = \mu_n$, $n=1, 2, \ldots$, where π_n is the projection from $\prod_{i=1}^{\infty} X_i$ onto $(X_1, \times X_2 \times \ldots \times X_n)$. The measure space $(\prod_{i=1}^{\infty} X_i, \prod_{i=1}^{\infty} \mathscr{B}_i, \tilde{\mu})$ is usually called a *discrete time Markov process* with initial distribution μ at time 1 and transition probability λ_n at time n. (The name Markov process is in honour of the Russian mathematician A. Markov who first investigated them.)

Remark. 35.22. One can interpret Exercise 35.21 as follows: Consider a sequence of statistical experiments where the elementary outcomes of the nth experiment belong to the sample space X_n with a σ-algebra $\mathscr{B}_n$ of events. Outcomes of the first experiment occur according to the distribution μ_1. If in the first n experiments the outcomes are $x_1, x_2, \ldots, x_n$ the outcome of the $n+1$-th experiment is distributed according to the probability measure $\lambda_n(x_n, \ldots)$. The statistical behaviour of the outcomes of the $n+1$th experiment depends only on the outcome at the n-th experiment when the outcomes of the first n experiments are known. This is what is known as the *Markov character* of the process.

Remark. 35.23. Let $(X, \mathscr{B})$ be a borel space and let $\{\lambda_t, t>0\}$ be a family of transition probabilities on $X \times \mathscr{B}$ such that

$$\lambda_t \circ \lambda_s = \lambda_{t+s}, \; t \geqslant 0, \; s \geqslant 0. \tag{35.6}$$

Then $\{\lambda_t, t>0\}$ is called a one parameter semi group of transition probabilities. Eq. (35.6) is called the *Chapman-Kolmogorov equation.* Let μ be a probability measure on $\mathcal{B}$. For any $t_1<t_2<\ldots<t_n$, let $\mu_{t_1 t_2 \ldots t_n}$ be the probability measure defined on $(X \times X \times \ldots \times X, \mathcal{B}\times\mathcal{B}\times\ldots\times\mathcal{B})$ by

$$\mu_{t_1 t_2 \ldots t_n}(E)=\int \chi_E(x_1, x_2, \ldots, x_n)\, \lambda_{t_n-t_{n-1}}(x_{n-1}, dx_n)$$
$$\lambda_{t_{n-1}-t_{n-2}}(x_{n-2}, dx_{n-1}) \ldots \lambda_{t_2-t_1}(x_1, dx_2)\, \lambda_{t_1}(x_0, dx_1)\, \mu(dx_0)$$
$$\text{if } t_1>0.$$
$$\mu_{t_1 t_2 \ldots t_n}(E)= \int \chi_E(x_1, x_2, \ldots, x_n)\, \lambda_{t_n-t_{n-1}}(x_{n-1}, dx_n)$$
$$\lambda_{t_{n-1}-t_{n-2}}(x_{n-2}, dx_{n-1}) \ldots \lambda_{t_2-t_1}(x_1, dx_2)\, \mu(dx_1)$$
$$\text{if } t_1=0.$$

Then the family of probability measures $\{\mu_{t_1 t_2 \ldots t_n}\}$ is consistent and hence by Proposition 27.8 and Example 27.10 determines a probability measure $\tilde{\mu}$ on the product space $(\prod_{t\geqslant 0} X_t, \prod_{t\geqslant 0} \mathcal{B}_t)$, $X_t=X$ for all $t\geqslant 0$, whose finite dimensional distributions are $\mu_{t_1 t_2 \ldots t_n}$.

As a particular example one may consider $X=R$, $\mathcal{B}=\mathcal{B}_R$ and

$$\lambda_t(x, E)=\frac{1}{\sqrt{2\pi t}}\int_E e^{-\frac{1}{2t}(x-y)^2}\, dy, \quad t>0.$$

This is known as the transition probability of the *standard brownian motion process.*

§36. Convolution of Probability Measures on R^n

Let λ, μ be probability measures on the borel σ-algebra of R^n. Consider the product probability measure $\lambda\times\mu$ in R^{2n}. Then for any borel set $E\subset R^n$, we have by Fubini's theorem

$$\begin{aligned}\lambda\times\mu \{(\mathbf{x}, \mathbf{y}) : \mathbf{x}+\mathbf{y}\in E\} &= \int \chi_E (\mathbf{x}+\mathbf{y})\, d(\lambda\times\mu)\\ &= \int [\int \chi_E (\mathbf{x}+\mathbf{y})\, d\lambda(\mathbf{x})]\, d\mu(\mathbf{y})\\ &= \int \lambda(E-\mathbf{y})\, d\mu(\mathbf{y}) = \int \mu(E-\mathbf{x})\, d\lambda(\mathbf{x}). \qquad (36.1)\end{aligned}$$

where $E-\mathbf{y}=\{\mathbf{z}-\mathbf{y}, \mathbf{z}\in R^n\}$. Now we can describe this in the language of probability as follows: if λ and μ are distributions of independent R^n valued random variables $\mathbf{f}$ and $\mathbf{g}$, then the distribution of their sum $\mathbf{f}+\mathbf{g}$ is given by Eq. (36.1). The measure defined by Eq. (36.1) is usually denoted by $\lambda * \mu$. Thus

$$\begin{aligned}(\lambda * \mu)(E) &= \int \lambda(E-\mathbf{y})\, d\mu(\mathbf{y})\\ &= \int \mu(E-\mathbf{x})\, d\lambda(\mathbf{x}) = (\mu * \lambda)(E),\end{aligned}$$

is the distribution of $\mathbf{f}+\mathbf{g}$. $\lambda * \mu$ is called the *convolution* of λ and μ.

Exercise. 36.1. For three probability measures λ, μ, ν in R^n,

$$(\lambda * \mu) * \nu = \lambda * (\mu * \nu),$$

$$(p\lambda + q\mu) * \nu = p(\lambda * \nu) + q(\mu * \nu) \text{ for } 0 \leqslant p \leqslant 1,\ 0 \leqslant q \leqslant 1,\ p+q=1.$$

Exercise. 36.2. A probability measure λ is said to be *degenerate* at a point $\mathbf{x}$ if $\lambda(\{\mathbf{x}\}) = 1$. If $\lambda * \mu$ is degenerate at a point $\mathbf{x}$, then λ and μ are degenerate at points $\mathbf{y}$ and $\mathbf{z}$ such that $\mathbf{x} = \mathbf{y} + \mathbf{z}$. (Hint: Use Corollary 35.16.)

Remark. 36.3. If L denotes the Lebesgue measure in R and f is any real valued borel function on R^n, we shall write $\int_E f(\mathbf{x}) d\mathbf{x}$ or $\int_E f(x_1, x_2, \ldots, x_n)\, dx_1\, dx_2 \ldots dx_n$ for $\int_E f\, dL$. A probability distribution μ on R^n is said to have *density function* f if

$$\mu(E) = \int_E f(\mathbf{x})\, d\mathbf{x} \text{ for all borel sets } E \subset R^n.$$

Since $\mu(E) \geqslant 0$ for all E, it follows that $f(\mathbf{x}) \geqslant 0$ a.e. $\mathbf{x}(L)$. Further

$$\int f(\mathbf{x})\, d\mathbf{x} = 1.$$

If μ, ν are probability distributions with density functions f and g respectively then

$$(\mu \times \nu)\,(E \times F) = \mu(E)\, \nu(F) = \int_{E \times F} f(\mathbf{x})\, g(\mathbf{y})\, d\mathbf{x}\, d\mathbf{y}.$$

Hence for finite disjoint unions of borel rectangles we have the relation

$$(\mu \times \nu)\,(A) = \int_A f(\mathbf{x})\, g(\mathbf{y})\, d\mathbf{x}\, d\mathbf{y}.$$

Thus the above relation extends to all borel sets $A \subset R^{2n}$. In other words $f(\mathbf{x})\, g(\mathbf{y})$ is the density function of $\mu \times \nu$. An application of Fubini's theorem shows that

$$\begin{aligned}(\mu * \nu)\,(E) &= \int \chi_E\,(\mathbf{x}+\mathbf{y})\, f(\mathbf{x})\, g(\mathbf{y})\, d\mathbf{x}\, d\mathbf{y} \\ &= \int \left[\int \chi_E\,(\mathbf{x}+\mathbf{y})\, g(\mathbf{y})\, d\mathbf{y} \right] f(\mathbf{x})\, d\mathbf{x}.\end{aligned}$$

Since L is invariant under translations (See Exercise 35.18), we have

$$\int \chi_E\,(\mathbf{x}+\mathbf{y})\, g(\mathbf{y})\, d\mathbf{y} = \int \chi_E\,(\mathbf{y})\, g(\mathbf{y}-\mathbf{x})\, d\mathbf{y}.$$

Hence (by Fubini's theorem again!)

$$(\mu * \nu)\,(E) = \int_E \left[\int g(\mathbf{x}-\mathbf{y})\, f(\mathbf{x})\, d\mathbf{x} \right] d\mathbf{y}.$$

Thus $\mu * \nu$ has density function $\int g(\mathbf{y}-\mathbf{x})\, f(\mathbf{x})\, d\mathbf{x}$. We denote this by $g * f$. Since $\mu * \nu = \nu * \mu$ it follows that

$$(f * g)\,(\mathbf{x}) = \int f(\mathbf{x}-\mathbf{y})\, g(\mathbf{y})\, d\mathbf{y} = \int g(\mathbf{x}-\mathbf{y})\, f(\mathbf{y})\, d\mathbf{y} = (g * f)\,(\mathbf{x}) \quad \text{a.e. } \mathbf{x}(L). \tag{36.2}$$

Exercise. 36.4. Let μ_i be a multivariate normal distribution with mean vector $\mathbf{m}_i$ and covariance matrix Σ_i, $i = 1, 2$. Then $\mu_1 * \mu_2$ is the multivariate normal distribution with mean vector $\mathbf{m}_1 + \mathbf{m}_2$ and covariance matrix $\Sigma_1 + \Sigma_2$.

Exercise. 36.5. Let $\mu_{\alpha, \beta}$ be the probability measure on the real line with density function

$$f_{\alpha, \beta}(x) = e^{-\alpha x} \frac{x^{\beta-1}}{\Gamma(\beta)} \quad \text{if } x > 0,$$
$$= 0 \qquad\qquad \text{if } x \leqslant 0,$$

where $\alpha > 0$, $\beta > 0$. Then

$$(f_{\alpha, \beta} * f_{\alpha, \beta'})(x) = f_{\alpha, \beta+\beta'}(x) \text{ for all } \alpha>0, \beta>0, \beta'>0.$$

($\mu_{\alpha, \beta}$ is known as the *gamma distribution* with parameters α and β. If $\alpha=1/2$, $\beta=n/2$, $\mu_{\alpha, \beta}$ is called the *Chi-squared distribution* with n *degrees of freedom*. Any random variable which obeys this distribution is called a χ^2_n random variable. If χ^2_m and χ^2_n are independent random variables, then $\chi^2_m + \chi^2_n$ is a χ^2_{m+n} random variable.)

Exercise. 36.6. Let μ, ν be probability measures on the real line such that $\mu(\mathbb{Z}) = \nu(\mathbb{Z}) = 1$, where $\mathbb{Z}$ is the set of all integers. Suppose

$$\mu(\{j\}) = p_j, \ \nu(\{j\}) = q_j, \ j = 0, \pm 1, \pm 2, \ldots.$$

Then $(\mu * \nu)(\mathbb{Z}) = 1$ and

$$(\mu * \nu)(\{j\}) = \sum_k p_{j-k}\, q_k = \sum_k q_{j-k}\, p_k, \ j = 0, \pm 1, \pm 2, \ldots.$$

In particular, if μ and ν are Poisson distributions with parameters α and β respectively, $\mu * \nu$ is a Poisson distribution with parameter $\alpha + \beta$.

Exercise. 36.7. Consider a sequence of independent binomial trials with probability of success equal to p. Let X_k be the number of trial at which the k-th success occurs. Then

$$P\{X_k=n\} = \binom{n-1}{k-1} p^k q^{n-k}, \ n=k, k+1, \ldots.$$

The distribution of X_k is the k-fold convolution of the distribution of X_1. (The distribution of X_1 is called the *geometric distribution.* The distribution of X_k is called the *Pascal distribution.*)

Exercise. 36.8. If f, g are independent real valued random variables with distributions μ, ν respectively find the distribution of the random variable fg.

§37. The Lebesgue Measure on R^n

We recall that the Lebesgue measure on R^n is the n-fold product of Lebesgue measure on the real line. We know from Exercise 35.18 that the Lebesgue measure on R^n is invariant under all translations. In R^n we shall use the notation

$$E+\mathbf{a} = \{\mathbf{x}+\mathbf{a}, \mathbf{x} \in E\},$$
$$E-\mathbf{a} = \{\mathbf{x}-\mathbf{a}, \mathbf{x} \in E\} \text{ for all } E \subset R^n, \mathbf{a} \in R^n.$$

We now have the following characterisation of Lebesgue measure.

Proposition. 37.1. Let μ be a σ-finite measure on the borel σ-algebra of R^n such that

(i) $\mu\{\mathbf{x}: 0 < x_i \leqslant 1, i = 1, 2, ..., n\} = 1$

(ii) $\mu(E) = \mu(E+\mathbf{a})$ for all borel sets E and all $\mathbf{a} \in R^n$.

Then μ is the Lebesgue measure in R^n.

Proof. From condition (ii) it follows that the measure of the rectangle $\{\mathbf{x} : a_i < x_i \leqslant a_i + b_i, i = 1, 2, ..., n\}$ is independent of the vector $\mathbf{a}$ whose i-th coordinate is a_i for every i. Let $b_i = \frac{1}{k}$, $i=1, 2, ..., n$, where k is a positive integer. Then the unit cube $\{\mathbf{x}: 0<x_i \leqslant 1, i=1, 2, ..., n\}$ can be written as a disjoint union of k^n cubes of the form $\{\mathbf{x}: a_i<x_i \leqslant a_i+\frac{1}{k}, i=1, 2, ..., n\}$ which have equal measure. Thus every cube with sides $1/k$ has measure $1/k^n$. Since every rectangle with rational side lengths can be written as a disjoint union of cubes, each of which has side length $1/k$ for some positive integer k, it follows that every such rectangle has measure equal to the product of the side lengths. This means that μ coincides with Lebesgue measure on all sets which are finite disjoint unions of rectangles with rational side lengths. Since such sets generate the borel σ-algebra it follows that μ is the Lebesgue measure.

Proposition. 37.2. Let L be the Lebesgue measure in R^n and let T be a non-singular linear transformation on R^n. Then $L(T(E)) = |\det T|\, L(E)$ for all borel sets E.

Proof. Let $L'(E)=L(T(E))$. Then the non-singularity of T implies that L' is also a σ-finite measure. Further

$$L'(E+\mathbf{a})=L(TE+T\mathbf{a})=L(TE)=L'(E).$$

Hence L' is invariant under all translations. Further $L'(E)\neq 0$ whenever E is open. In particular $L'(E)\neq 0$ when E is the cube $\{\mathbf{x} : 0<x_i<1,\ i=1, 2, \ldots, n\}$. Hence by the preceding proposition there exists a constant $c(T)$ such that $L'(E)=c\ (T)\ L(E)$. Thus

$$L(TE)=c(T)\ L(E), \text{ for all borel sets } E, \tag{37.1}$$

for every non-singular linear transformation T. For any two non-singular linear transformations T_1, T_2 we have $T_1T_2(E)=T_1(T_2(E))$. Hence repeated application of Eq. (37.1) shows that

$$c(T_1T_2)=c(T_1)\ c(T_2) \tag{37.2}$$

for all non-singular linear transformations T_1 and T_2. If T is an orthogonal linear transformation, we put for E the open unit ball in R^n with origin as centre in Eq. (37.1). Then $TE=E$ and Eq. (37.1) shows that $c(T)=1$ for all orthogonal linear transformations. Any non-singular linear transformation T can be written as O_1DO_2 where O_1, O_2 are orthogonal transformations and D is a diagonal transformation whose matrix is diagonal with positive elements along the diagonal. Then by Eq. (37.2), $c(T)=c(D)$. Now for any $\mathbf{x}$,

$$D\mathbf{x}=\begin{pmatrix} d_1x_1 \\ d_2x_2 \\ \cdot \\ \cdot \\ \cdot \\ d_nx_n \end{pmatrix},\ d_i>0,\ i=1, 2, \ldots, n.$$

Such a transformation changes a rectangle with sides $l_1, l_2, \ldots, l_n$ into a rectangle with sides $l_1d_1, l_2d_2, \ldots, l_nd_n$. If E is a rectangle with sides $l_1, l_2, \ldots, l_n$,

$$l_1d_1\ l_2d_2 \ldots l_nd_n=L(D(E))=c(D)L(E)=c(D)l_1l_2\ldots l_n.$$

Hence $c(D)=d_1d_2\ldots d_n=|\det D|$. Since $|\det O|=1$ for any orthogonal matrix O, it follows that $c(T)=|\det T|$. This completes the proof.

Exercise. 37.3. If T is a singular linear transformation $L(T(R^n))=0$. Hence Proposition 37.2 holds even when T is singular.

Proposition. 37.4. (Change of variable formula) Let Ω be an open subset of R^n and let the Lebesgue measure be restricted to Ω.

Let $T : \Omega \to R^n$ be a homeomorphism of Ω into R^n, defined by

$$T(\mathbf{x}) = \begin{pmatrix} y_1(\mathbf{x}) \\ y_2(\mathbf{x}) \\ \cdot \\ \cdot \\ \cdot \\ y_n(\mathbf{x}) \end{pmatrix}, \quad \mathbf{x} = \begin{pmatrix} x_1 \\ x_2 \\ \cdot \\ \cdot \\ \cdot \\ x_n \end{pmatrix} \in \Omega.$$

and let $\frac{\partial y_i}{\partial x_j}$ be continuous on Ω for all i and j. Let

$$j(T, \mathbf{x}) = \det \left(\left(\frac{\partial y_i}{\partial x_j} \right) \right), \; 1 \leqslant i, j \leqslant n. \tag{37.3}$$

Then for any non-negative borel function f on the open set $T(\Omega)$,

$$\int_{T\Omega} f(\mathbf{y}) \, d\mathbf{y} = \int_{\Omega} f(T\mathbf{x}) \, |j(T, \mathbf{x})| \, d\mathbf{x} \tag{37.4}$$

where $d\mathbf{x}$, $d\mathbf{y}$ denote integration with respect to Lebesgue measure.

Proof. For any $\mathbf{x} \in R^n$, let

$$|\mathbf{x}| = max\,(|x_i|, i=1, 2, \ldots, n).$$

For any linear transformation V of R^n into R^n, let

$$|V| = \sup_{|\mathbf{x}|=1} |V\mathbf{x}|.$$

Then $|\mathbf{x}|$ and $|V|$ have all the properties of a 'norm' as $\mathbf{x}$ varies in R^n and V over the class of linear transformations. Indeed

(i) $|\mathbf{x} + \mathbf{y}| \leqslant |\mathbf{x}| + |\mathbf{y}|$,
(ii) $|c\mathbf{x}| = |c| \, |\mathbf{x}|$, c scalar,
(iii) $|\mathbf{x}| = 0$ if and only if $\mathbf{x} = 0$,
(iv) $|V_1 + V_2| \leqslant |V_1| + |V_2|$,
(v) $|cV| = |c| \, |V|$, c scalar,
(vi) $|V| = 0$ if and only if $V = 0$,
(vii) $|V_1 V_2| \leqslant |V_1| \, |V_2|$,
(viii) $|V\mathbf{x}| \leqslant |V| \, |\mathbf{x}|$.

Let

$$\mathcal{J}(T, \mathbf{x}) = \left(\left(\frac{\partial y_i}{\partial x_j} \right) \right), \; 1 \leqslant i, j \leqslant n,$$

denote the *Jacobian matrix* of the homeomorphism T. Then $\mathcal{J}(T, \mathbf{x})$ is a non singular matrix for all $\mathbf{x} \in \Omega$. It may be noted that for a homeomorphism T_1 from the open set $\Omega_1 \subset R^n$ onto the open set $\Omega_2 \subset R^n$ and a homeomorphism T_2 from Ω_2 onto the open set $\Omega_3 \subset R^n$,

$$\mathcal{J}(T_2 T_1, \mathbf{x}) = \mathcal{J}(T_2, T_1 \mathbf{x})\, \mathcal{J}(T_1, \mathbf{x}) \text{ for all } \mathbf{x} \in \Omega_1. \tag{37.5}$$

Indeed, this is a consequence of the standard formula for the differential coefficient of a function of functions. (This is known as a *cocycle equation.* Such an equation appears in several fields of mathematics. It is a starting point of what is currently known as cohomology theory.) If T is a linear transformation defined by a matrix V, then $\mathcal{J}(T, \mathbf{x}) = V$. Let now $A \subset \Omega$ be a compact rectangle. Let

$$\omega(\delta) = \sup_{\substack{|\mathbf{x} - \mathbf{y}| < \delta \\ \mathbf{x},\, \mathbf{y} \in A}} |\mathcal{J}(T, \mathbf{x}) - \mathcal{J}(T, \mathbf{y})|. \tag{37.6}$$

Here the matrix $\mathcal{J}(T, \mathbf{x})$ is considered as a linear transformation of R^n. The continuity of $\mathcal{J}(T, \mathbf{x})$ in Ω and hence its uniform continuity in A implies that

$$\lim_{\delta \to 0} \omega(\delta) = 0.$$

Putting $\Omega_1 = \Omega_3 = T\Omega$, $\Omega_2 = \Omega$, and $T_1 = T^{-1}$, $T_2 = T$ in Eq. (37.5) we have

$$I = \mathcal{J}(T, T^{-1}\mathbf{x})\, \mathcal{J}(T^{-1}, \mathbf{x}), \tag{37.7}$$

where I denotes the identity matrix of order n. Let

$$\sup_{\mathbf{x} \in A} |\mathcal{J}(T, \mathbf{x})^{-1}| = m. \tag{37.8}$$

If

$$K(\mathbf{a}, \delta) = \left\{\mathbf{x} : -\frac{\delta}{2} < x_i - a_i \leqslant \frac{\delta}{2} \text{ for all } i\right\},$$

then

$$L(K(\mathbf{a}, \delta)) = \delta^n.$$

If now the cube $K(\mathbf{a}, \delta) \subset A$, we have by Taylor's theorem in several variables

$$(T\mathbf{x})_i - (T\mathbf{a})_i = y_i(\mathbf{x}) - y_i(\mathbf{a})$$

$$= \sum_{j=1}^{n} \frac{\partial y_i}{\partial x_j} (\mathbf{a} + \theta_i (\mathbf{x} - \mathbf{a}))\, (x_j - a_j),$$

for some $0 < \theta_i < 1$. Hence

$$T\mathbf{x} - T\mathbf{a} = \mathcal{J}(T, \mathbf{a})\, (\mathbf{x} - \mathbf{a}) + \mathbf{v}(\mathbf{x}),$$

where Eq. (37.6) implies

$$|\mathbf{v}(\mathbf{x})| \leqslant n \left(\max_{i,j} \left| \frac{\partial y_i}{\partial x_j} (\mathbf{a} + \theta_i (\mathbf{x} - \mathbf{a})) - \frac{\partial y_i}{\partial x_j} (\mathbf{a}) \right| \right) |\mathbf{x} - \mathbf{a}|$$

$$\leqslant n\, \omega(\delta)\delta, \text{ for all } \mathbf{x} \in K(\mathbf{a}, \delta).$$

Thus, by Eq. (37.8),

$$|\mathcal{J}(T, \mathbf{a})^{-1} T\mathbf{x} - \mathcal{J}(T, \mathbf{a})^{-1} T\mathbf{a}| \leqslant |\mathcal{J}(T, \mathbf{a})^{-1}| \, |\mathbf{v}(\mathbf{x})| + |\mathbf{x} - \mathbf{a}|$$
$$\leqslant mn\, \omega(\delta)\delta + \frac{\delta}{2} = \frac{\delta}{2}(1 + 2\, mn\, \omega(\delta)),$$
$$\text{for all } \mathbf{x} \in K(\mathbf{a}, \delta).$$

Hence

$$\mathcal{J}(T, \mathbf{a})^{-1} T\mathbf{x} \in \overline{K(\mathcal{J}(T, \mathbf{a})^{-1} T\mathbf{a}, \delta(1 + 2\, mn\, \omega(\delta)))}.$$
$$\text{for all } \mathbf{x} \in K(\mathbf{a}, \delta).$$

Hence

$$L(\mathcal{J}(T, \mathbf{a})^{-1} T(K(\mathbf{a}, \delta)) \leqslant [\delta\,(1+2\, mn\, \omega(\delta))]^n$$
$$= [1+2\, mn\, \omega(\delta)]^n\, L(K(\mathbf{a}, \delta)).$$

By Proposition 37.2, the left hand side of the above inequality is $|j(T, \mathbf{a})|^{-1} L(T(K(\mathbf{a}, \delta)))$, where $j(T, \mathbf{x})$ is given by Eq. (37.3). Thus

$$L(T(K(\mathbf{a}, \delta))) \leqslant [1+2mn\, \omega(\delta)]^n\, |j(T, \mathbf{a})|\, L(K(\mathbf{a}, \delta)).$$

Let A_1 be the maximal left open right closed rectangle contained in the compact rectangle A with which we started. Then for any $\varepsilon > 0$, we can divide A_1 into a finite number of disjoint cubes of the form $K(\mathbf{a}^{(i)}, \delta)$, where $\delta < \varepsilon$. Hence

$$L(TA_1) = \sum_i L(T(K(\mathbf{a}^{(i)}, \delta))$$
$$\leqslant [1+2mn\, \omega(\delta)]^n \sum_i |j(T, \mathbf{a}^{(i)})|\, L(K(\mathbf{a}^{(i)}, \delta)).$$

Since $|j(T, \mathbf{x})|$ is continuous, we obtain by letting $\varepsilon \to 0$ and hence $\delta \to 0$

$$L(TA_1) \leqslant \int_{A_1} |j(T, \mathbf{x})|\, d\mathbf{x} \qquad (37.9)$$

(where the right hand side is Riemann integral of $|j(T, .)|$ and hence coincides with the Lebesgue integral). Thus (37.9) holds for every bounded left open right closed rectangle A_1 whose closure is completely contained in Ω. Hence for all borel sets $E \subset \Omega$,

$$L(T(E)) \leqslant \int_E |j(T, \mathbf{x})|\, d\mathbf{x}.$$

This can be rewritten as

$$\int_{T\Omega} \chi_E\,(T^{-1}\mathbf{x})\, d\mathbf{x} \leqslant \int_\Omega \chi_E\,(\mathbf{x})\, |j(T, \mathbf{x})|\, d\mathbf{x},\ E \subset \Omega.$$

Hence for any non-negative borel function f on Ω,

$$\int_{T\Omega} f(T^{-1}\mathbf{x})\, d\mathbf{x} \leqslant \int_\Omega f(\mathbf{x})\, |j(T, \mathbf{x})|\, d\mathbf{x}.$$

Hence for any non-negative borel function g on $T\Omega$, applying the above inequality for the homeomorphisms T and T^{-1} successively, we get

$$\int_{T\Omega} g(\mathbf{x})\, d\mathbf{x} = \int_{T\Omega} (g \circ T)\, (T^{-1}\mathbf{x})\, d\mathbf{x} \leqslant \int_{\Omega} (g \circ T)\, (\mathbf{x})\, |j(T, \mathbf{x})|\, d\mathbf{x}$$
$$\leqslant \int_{T\Omega} g(\mathbf{x})\, |j(T, T^{-1}\mathbf{x})|\, |j(T^{-1}, \mathbf{x})|\, d\mathbf{x}. \qquad (37.10)$$

By (37.7),

$$|j(T, T^{-1}\mathbf{x})\, j(T^{-1}, \mathbf{x})| = 1.$$

Hence (37.10) implies that

$$\int_{T\Omega} g(\mathbf{x})\, d\mathbf{x} = \int_{\Omega} (g \circ T)\, (\mathbf{x})\, |j(T, \mathbf{x})|\, d\mathbf{x}.$$

This completes the proof.

Corollary. 37.5. If T is a mapping from an open interval I_1 onto another open interval I_2 such that $T'(x) \neq 0$ for all $x \in I_1$, then

$$\int_{I_2} g(x)\, dx = \int_{I_1} g(Tx)\, |T'(x)|\, dx,$$

for any non-negative borel function g on I_2.

Example. 37.6. For any positive definite real matrix Σ of order $n \times n$ and any column vector $\mathbf{m}$, we shall show that

$$\int_{R^n} \exp\left\{-\tfrac{1}{2}\, (\mathbf{x}-\mathbf{m})'\, \Sigma^{-1}\, (\mathbf{x}-\mathbf{m})\right\} d\mathbf{x} = (\sqrt{2\pi})^n\, (\det \Sigma)^{\frac{1}{2}}, \qquad (37.11)$$

where the prime sign denotes transpose. Indeed, since the Lebesgue measure is invariant under translations it follows that the left hand side of Eq. (37.11) is independent of $\mathbf{m}$. (See Exercise 35.18). Put $\mathbf{m}=0$. Since Σ is positive definite we can write $\Sigma = T'T$, where T is a non-singular matrix. Making the transformation $T\mathbf{x} = \mathbf{y}$ we see from Proposition 37.4 that the left hand side of Eq. (37.11) is equal to

$$|\det T| \int_{R^n} \exp\left\{-\tfrac{1}{2}(y_1^2+y_2^2+\ldots+y_n^2)\right\} dy_1\, dy_2 \ldots dy_n$$
$$= |\det T|\, \left(\int_R e^{-\frac{1}{2}y^2}\, dy\right)^n$$
$$= |\det T|\, (\sqrt{2\pi})^n = |\det \Sigma|^{\frac{1}{2}}\, (\sqrt{2\pi})^n.$$

Example. 37.7. Let $Q_k(\mathbf{x}) = \sum_{i=1}^{n} x_i^{2k}$. We shall evaluate the Lebesgue measure of the set

$$E = \{\mathbf{x} : Q_k\, (A\mathbf{x}) \leqslant 1\},$$

where A is a non-singular linear transformation of R^n onto itself.

First of all we observe that as $\mathbf{x}$ varies in E, the point $A\mathbf{x}$ varies in he set

$$F = \{\mathbf{x} : Q_k\, (\mathbf{x}) \leqslant 1\}.$$

By Proposition 37.2,

$$L(F)=L(AE)=|\det A|\,L(E). \tag{37.12}$$

Let

$$v_n=L\{\mathbf{x}: Q_k(\mathbf{x})\leqslant 1\}. \tag{37.13}$$

The Lebesgue measure in R^n is the product of the Lebesgue measures in R^{n-1} and R. The section of F by $x_n=x$ is given by

$$F_x=\{(x_1, x_2, \ldots, x_{n-1}) : x_1^{2k}+x_2^{2k}+\ldots+x_{n-1}^{2k}\leqslant 1-x^{2k}\}.$$

Thus $F_x=\varnothing$ if $|x|>1$. If $|x|<1$, divide both sides by $1-x^{2k}$ in the above inequality and observe from Eq. (37.12) that

$$L(F_x)=v_{n-1}\,(1-x^{2k})^{(n-1)/2k}.$$

By Proposition 35.17,

$$\begin{aligned} v_n=L(F)&=v_{n-1}\int_{-1}^{+1}(1-x^{2k})^{(n-1)/2k}\,dx.\\ &=2v_{n-1}\int_0^1(1-y)^{\frac{n-1}{2k}}\,y^{\frac{1}{2k}-1}\,\frac{dy}{2k}\\ &=\frac{v_{n-1}}{k}\,\frac{\Gamma\left(\frac{n-1}{2k}+1\right)\Gamma\left(\frac{1}{2k}\right)}{\Gamma\left(\frac{n}{2k}+1\right)}. \end{aligned}$$

This recurrence relation shows that

$$v_n=\frac{\left[\Gamma\left(\frac{1}{2k}\right)\right]^n}{k^n\,\Gamma\left(\frac{n}{2k}+1\right)}.$$

Thus

$$L\{\mathbf{x}: Q_k(A\mathbf{x})\leqslant 1\}=\frac{|\det A|^{-1}}{k^n}\,\frac{\left[\Gamma\left(\frac{1}{2k}\right)\right]^n}{\Gamma\left(\frac{n}{2k}+1\right)}.$$

In particular, the volume of the unit ball in R^n is given by

$$L\{\mathbf{x}: \sum_i x_i^2\leqslant 1\}=\frac{\pi^{n/2}}{\Gamma\left(\frac{n}{2}+1\right)}.$$

Exercise. 37.8. The Lebesgue measure of the standard polyhedron

$$\left\{\mathbf{x}: x_i\geqslant 0,\ i=1, 2, \ldots, n,\ \sum_{i=1}^n x_i\leqslant 1\right\}\ \text{is}\ \frac{1}{n!}.$$

Exercise. 37.9. If C is the closed convex hull generated by the origin 0 and n linearly independent vectors $\mathbf{v}_1, \mathbf{v}_2, \ldots, \mathbf{v}_n$ in R^n then the Lebesgue measure of C is $\frac{1}{n!} | \det \Gamma |$, where Γ is the matrix whose jth column is $\mathbf{v}_j$ for all $1 \leqslant j \leqslant n$. (Hint: the image of the standard polyhedron under the linear transformation Γ is C.) This result is known as Gauss' formula. It may be noted that

$$C = \left\{ \sum_{i=1}^{n} p_i \mathbf{v}_i, p_i \geqslant 0 \text{ for all } i \text{ and } \sum_{i=1}^{n} p_i = 1 \right\}.$$

Proposition. 37.10. Let $\Omega \subset R^n$ be an open domain and let $\{T_t, t \in R\}$ be a family of homeomorphisms of Ω onto itself such that $T_t T_s = T_{t+s}$ for all $t, s \in R$. (Such a family is known as a *one parameter group of homeomorphisms.*) For each $\mathbf{x} \in \Omega$, let $T_t \mathbf{x} = \mathbf{y}(t, \mathbf{x})$. Let y_i, $\frac{\partial y_i}{\partial x_j}$, $\frac{\partial^2 y_i}{\partial_t \partial x_j}$ exist and be continuous on $R \times \Omega$. Then T_t preserves the Lebesgue measure L in Ω for all t if and only if

$$\sum_{i=1}^{n} \frac{\partial^2 y_i}{\partial t \partial x_i} (0, \mathbf{x}) = 0 \text{ for all } \mathbf{x} \in \Omega .$$

Proof. Since $T_t T_s = T_{t+s}$, we have $T_0{}^2 = T_0$ and hence $T_0 = I$, where I is the identity transformation. Hence $\mathbf{y}(0, \mathbf{x}) = \mathbf{x}$ or equivalently $y_i(0, \mathbf{x}) = x_i$ for all $1 \leqslant i \leqslant n$. It follows from Proposition 37.4 that T_t preserves L for all t if and only if

$$|j(T_t, \mathbf{x})| = 1 \text{ for all } t \in R, \mathbf{x} \in \Omega , \tag{37.14}$$

where $j(T_t, \mathbf{x}) = \det \left(\left(\frac{\partial y_i}{\partial x_j} (t, \mathbf{x}) \right) \right)$. Let us write

$$a(t, \mathbf{x}) = j(T_t, \mathbf{x}).$$

Then by Eq. (37.5),

$$a(t+s, \mathbf{x}) = a(s, T_t \mathbf{x}) \, a(t, \mathbf{x}).$$

Differentiating both sides with respect to s at $s=0$, we obtain

$$\frac{\partial a}{\partial t} (t, \mathbf{x}) = \frac{\partial a}{\partial s} (0, T_t \mathbf{x}) a(t, \mathbf{x}). \tag{37.15}$$

Since a is continuous in $R \times \Omega$ it follows that Eq. (37.14) will be fulfilled if and only if $j(T_t, \mathbf{x})$ is independent of t or equivalently,

$$\frac{\partial a}{\partial t} (t, \mathbf{x}) = 0 \text{ for all } t \in R, \mathbf{x} \in \Omega .$$

Since T_t is a homeomorphism, Eq. (37.15) implies that the above condition will be fulfilled if and only if

$$\frac{\partial \alpha}{\partial s}(0, \mathbf{x})=0 \text{ for all } \mathbf{x}\in \Omega .$$

The standard rule for differentiating a determinant shows that

$$\frac{\partial \alpha}{\partial s}(s, \mathbf{x}) = \sum_i \begin{vmatrix} \frac{\partial y_1}{\partial x_1} & \frac{\partial y_1}{\partial x_2} & & \frac{\partial y_1}{\partial x_n} \\ \cdots & \cdots & \cdots & \cdots \\ \frac{\partial^2 y_i}{\partial s\, \partial x_1} & \frac{\partial^2 y_i}{\partial s\, \partial x_2} & \cdots & \frac{\partial^2 y_i}{\partial s\, \partial x_n} \\ \cdots & \cdots & \cdots & \cdots \\ \frac{\partial y_n}{\partial x_1} & \frac{\partial y_n}{\partial x_2} & \cdots & \frac{\partial y_n}{\partial x_n} \end{vmatrix}$$

where in the i-th term under summation the i-th row of the determinant alone is differentiated with respect to s. At $s=0$, $\frac{\partial y_i}{\partial x_j}=\delta_{ij}$, where $\delta_{ij}=1$ or 0 according as $i=j$ or $i\neq j$. Hence

$$\frac{\partial \alpha}{\partial s}(0, \mathbf{x})= \sum_{i=1}^{n} \frac{\partial^2 y_i}{\partial s\, \partial x_i}(0, \mathbf{x}).$$

This completes the proof.

Remark. 37.11. We shall now see an illustration of the preceding proposition in classical mechanics. Let $\Omega \subset R^{2n}$ be an open subset. We shall represent any point in R^{2n} by $(\mathbf{x}, \mathbf{y})$, where $\mathbf{x}$ and $\mathbf{y}$ are column vectors in R^n whose i-th coordinates are x_i and y_i respectively. Let H be a function on Ω with continuous second order derivatives. Suppose $\{T_t, t\in R\}$ is a one parameter group of homeomorphisms of Ω such that

$$T_t(\mathbf{x}, \mathbf{y})=(\mathbf{q}(t, \mathbf{x}, \mathbf{y}), \mathbf{p}(t, \mathbf{x}, \mathbf{y})),$$

where

$$\mathbf{q}=\begin{pmatrix} q_1 \\ q_2 \\ \cdot \\ \cdot \\ \cdot \\ q_n \end{pmatrix}, \mathbf{p}=\begin{pmatrix} p_1 \\ p_2 \\ \cdot \\ \cdot \\ \cdot \\ p_n \end{pmatrix}.$$

Suppose the q_i's and p_i's satisfy the differential equations:

$$\frac{dq_i}{dt} = \frac{\partial H}{\partial p_i}(\mathbf{q}, \mathbf{p}),$$
$$\frac{dp_i}{dt} = -\frac{\partial H}{\partial q_i}(\mathbf{q}, \mathbf{p}) \tag{37.16}$$

The system of equations (37.16) is called a *Hamiltonian system* with *Hamiltonian H*. We have from (37.16) and the fact that $\mathbf{q}(0, \mathbf{x}, \mathbf{y}) = \mathbf{x}$ and $\mathbf{p}(0, \mathbf{x}, \mathbf{y}) = \mathbf{y}$,

$$\begin{aligned}\frac{\partial^2 q_i}{\partial t\, \partial x_i}(0, \mathbf{x}, \mathbf{y}) &= \frac{\partial^2 q_i}{\partial x_i\, \partial t}(0, \mathbf{x}, \mathbf{y}) \\ &= \frac{\partial^2 H}{\partial x_i\, \partial y_i}(\mathbf{x}, \mathbf{y}) \\ \frac{\partial^2 p_i}{\partial t\, \partial y_i}(0, \mathbf{x}, \mathbf{y}) &= \frac{\partial^2 p_i}{\partial y_i\, \partial t}(0, \mathbf{x}, \mathbf{y}) \\ &= -\frac{\partial^2 H}{\partial y_i\, \partial x_i}(0, \mathbf{x}, \mathbf{y}).\end{aligned}$$

Thus

$$\sum_i \frac{\partial^2 q_i}{\partial t\, \partial x_i} + \sum_i \frac{\partial^2 p_i}{\partial t\, \partial y_i} = 0, \text{ when } t = 0.$$

In other words the condition of Proposition 37.10 is fulfilled. Hence T_t preserves the Lebesgue measure in Ω. This result, namely, that a *dynamical* group of transformations induced by a Hamiltonian H depending only on positions and momenta preserves the Lebesgue measure in the *phase space* Ω, is known in classical mechanics as *Liouville's theorem.* We have paid special attention to this result because this is the starting point of modern statistical mechanics and ergodic theory.

Exercise. 37.12. If a random variable f has standard normal distribution then f^2 has χ_1^2 distribution. (Hint: use Corollary 37.5). If $\mathbf{f}$ has n-dimensional multivariate normal distribution with mean vector $\mathbf{m}$ and non-singular covariance matrix Σ and $\Sigma = CC'$, then the vector $\mathbf{g} = C^{-1}(\mathbf{f} - \mathbf{m})$ has a multivariate normal distribution with mean vector $\mathbf{0}$ and covariance matrix I. Hence the random variable

$$(\mathbf{f} - \mathbf{m})'\, \Sigma^{-1}\, (\mathbf{f} - \mathbf{m}) = g_1^2 + g_2^2 + \dots + g_n^2$$

has a distribution which is an n-fold convolution of χ_1^2 distribution, i.e., χ_n^2 distribution. (See Exercise 36.5.)

§38. The Convolution Algebra L_1 (R^n)

Let $L_1(R^n)$ denote the space of all real (or complex) borel functions on R^n which are integrable with respect to the Lebesgue measure over R^n. If $f, g \in L_1(R^n)$ it follows from Fubini's theorem that $f(\mathbf{x})\, g(\mathbf{y})$ belongs to $L_1(R^{2n})$. Again Fubini's theorem and translation invariance of Lebesgue measure imply that

$$\int f(\mathbf{x})\, g(\mathbf{y})\, d\mathbf{x}\, d\mathbf{y} = \int f(\mathbf{x}-\mathbf{y})\, g(\mathbf{y})\, d\mathbf{x}\, d\mathbf{y}.$$

Further $\int f(\mathbf{x}-\mathbf{y})\, g(\mathbf{y})\, d\mathbf{y}$ is finite almost everywhere and belongs to $L_1(R^n)$. As in Section 36 we write

$$(f * g)\,(\mathbf{x}) = \int f(\mathbf{x}-\mathbf{y})\, g(\mathbf{y})\, d\mathbf{y}.$$

Then $f * g$ is the *convolution* of f and g. An application of Fubini's theorem yields immediately the following result.

Proposition. 38.1. If $f, g \in L_1(R^n), f * g$ is defined and belongs to $L_1(R^n)$. Further

(i) $f * g = g * f$,
(ii) $(f * g) * h = f * (g * h)$,
(iii) $(af + bg) * h = af * h + bg * h$ for all scalars a and b,
(iv) $\|f * g\|_1 \leqslant \|f\|_1 \; \|g\|_1$ in $L_1(R^n)$,
(v) if $\|f_n - f\|_1 \to 0$, $\|g_n - g\|_1 \to 0$ in $L_1(R^n)$, then $\|f_n * g_n - f * g\|_1 \to 0$ as $n \to \infty$.

Remark. 38.2. The above proposition together with Riesz-Fischer theorem is summarised by saying that under convolution, $L_1(R^n)$ is a *Banach algebra.*

§39. Approximation in L_p Spaces with respect to Lebesgue Measure in R^n.

We shall construct a rich family of smooth functions approximating borel functions in various ways. This turns out to be the starting point of the theory of distributions or generalised functions formulated by L. Schwartz and S. L. Sobolev.

We shall denote by L_p (R^n) the space

$$\{f : \int |f(\mathbf{x})|^p\, d\mathbf{x} < \infty\},$$

where f stands for any equivalence class of functions with respect to the Lebesgue measure.

Proposition. 39.1. For any borel set $E \subset R^n$ of finite Lebesgue measure the function $L(E \Delta (E+\mathbf{x}))$ is uniformly continuous in the variable $\mathbf{x}$.

Proof. This is done exactly along the same lines as in the proof of Proposition 21.2.

Corollary. 39.2. (Lebesgue's theorem.) For any function $f \in L_p(R^n)$, $p \geqslant 1$, the function

$$\int |f(\mathbf{x}+\mathbf{y}) - f(\mathbf{x})|^p \, d\mathbf{x}$$

is uniformly continuous in $\mathbf{y}$. In particular,

$$\lim_{\mathbf{y}\to\mathbf{0}} \int |f(\mathbf{x}+\mathbf{y}) - f(\mathbf{x})|^p \, d\mathbf{x} = 0.$$

Proof. If $f = \chi_E$, where $L(E) < \infty$, the corollary holds because of Proposition 39.1. Hence for any simple function in $L_p(R^n)$ the same result holds because of Minkowski inequality. Since simple functions are dense in $L_p(R^n)$, we can for any $\varepsilon > 0$, choose a simple function s such that $\|f - s\|_p < \varepsilon$. For any $\mathbf{y} \in R^n$ let us write $T_\mathbf{y}$ for the map

$$(T_\mathbf{y} f)(\mathbf{x}) = f(\mathbf{x}+\mathbf{y}). \tag{39.1}$$

If $f \in L_p(R^n)$ it follows from the translation invariance of the Lebesgue measure that $T_\mathbf{y} f \in L_p(R^n)$ and $\|T_\mathbf{y} f\|_p = \|f\|_p$. Hence

$$\|T_\mathbf{y} f - f\|_p \leqslant \|T_\mathbf{y} f - T_\mathbf{y} s\|_p + \|T_\mathbf{y} s - s\|_p + \|s - f\|_p$$
$$\leqslant 2\varepsilon + \|T_\mathbf{y} s - s\|_p.$$

Hence

$$\lim_{\mathbf{y}\to\mathbf{0}} \|T_\mathbf{y} f - f\|_p \leqslant 2\varepsilon.$$

Since ε is arbitrary $\lim_{\mathbf{y}\to\mathbf{0}} \|T_\mathbf{y} f - f\|_p = 0$. Since

$$\|T_{\mathbf{y}_1} f - T_{\mathbf{y}_2} f\| = \|T_{\mathbf{y}_1 - \mathbf{y}_2} f - f\|_p,$$

the required uniform continuity follows immediately and the proof is complete.

Remark. 39.3. Consider the Banach space $L_p(R^n)$. A map $T : L_p(R^n) \to L_p(R^n)$ is said to be a Banach space automorphism if

(i) $T(af + bg) = a\,Tf + b\,Tg$ for all f, $g \in L_p(R^n)$ and scalars a, b;

(ii) $\|Tf\|_p = \|f\|_p$;

(iii) T is onto.

All such automorphisms form a group A. For each $\mathbf{y} \in R^n$, let $T_\mathbf{y} : L_p(R^n) \to L_p(R^n)$ be the map defined by Eq. (39.1). Then $T_\mathbf{y}$ is in A for every $\mathbf{y}$. Further $T_{\mathbf{y}_1} T_{\mathbf{y}_2} = T_{\mathbf{y}_1 + \mathbf{y}_2}$ for all $\mathbf{y}_1, \mathbf{y}_2 \in A$ and $\lim_{\mathbf{y}\to\mathbf{0}}$

$\| T_{\mathbf{y}} f - f \|_p = 0$ for all $f \in L_p(R^n)$. In other words the map $\mathbf{y} \to T_{\mathbf{y}}$ is a homomorphism from the additive group R^n into the group A with the property that the map $\mathbf{y} \to T_{\mathbf{y}} f$ from R^n into $L_p(R^n)$ is continuous. One says that the map $\mathbf{y} \to T_{\mathbf{y}}$ is a *continuous representation* of the group R^n in the group A.

We shall now proceed with the task of constructing smooth functions.

Proposition. 39.4. Let

$$\begin{aligned} f_0(t) &= e^{1/t} \text{ if } t < 0, \\ &= 0 \quad \text{if } t \geqslant 0. \end{aligned} \tag{39.2}$$

Then f_0 is infinitely differentiable at all points of the real line.

Proof. It is left to the reader.

Proposition. 39.5. Let ϕ be the function defined on R^n by

$$\phi(\mathbf{x}) = \frac{f_0(\|\mathbf{x}\|^2 - 1)}{\int f_0(\|\mathbf{x}\|^2 - 1)\, d\mathbf{x}}, \quad \mathbf{x} \in R^n$$

where $\|\mathbf{x}\|^2 = \sum_{i=1}^{n} x_i^2$ and f_0 is given by Eq. (39.2). Then ϕ is well defined and ϕ possesses the following properties:

(i) $\phi(\mathbf{x}) \geqslant 0$,

(ii) $\phi(\mathbf{x}) = 0$ if $\|\mathbf{x}\| \geqslant 1$,

(iii) $\int \phi(\mathbf{x})\, d\mathbf{x} = 1$,

(iv) ϕ is infinitely differentiable at all points of R^n.

Proof. Since $f_0(\|\mathbf{x}\|^2 - 1)$ vanishes outside the unit ball and is strictly positive in the open unit ball it follows that

$$0 < \int f_0(\|\mathbf{x}\|^2 - 1)\, d\mathbf{x} < \infty$$

and hence ϕ is well defined. Since ϕ is the composition of infinitely differentiable functions f and $\|\mathbf{x}\|^2 - 1$, it follows that ϕ is infinitely differentiable. Properties (i) to (iii) are obvious and the proof is complete.

For any borel function f and any $\varepsilon > 0$ we define

$$f_\varepsilon(\mathbf{x}) = \int f(\mathbf{x} - \varepsilon \mathbf{y})\, \phi(\mathbf{y})\, d\mathbf{y}. \tag{39.3}$$

where ϕ is as in Proposition 39.5. Since ϕ vanishes outside the unit ball the integration in Eq. (39.3) is performed only over the region $\|\mathbf{y}\| < 1$. As $\mathbf{y}$ varies over the unit ball $\mathbf{x} - \varepsilon \mathbf{y}$ varies over a ball of radius ε and centre $\mathbf{x}$. Thus $f_\varepsilon(\mathbf{x})$ is the *weighted average* of the values of f in a sphere of radius ε with centre $\mathbf{x}$. As $\varepsilon \to 0$ we can therefore expect f_ε to approximate f. We shall soon make this precise.

Definition. 39.6. Let $\Omega \subset R^n$ be an open subset. A borel function f on Ω is said to be *locally integrable* if, for every compact set $K \subset \Omega$, $\int_K |f(\mathbf{x})| \, d\mathbf{x} < \infty$. The function f is said to have *support contained in* K if $f(\mathbf{x}) = 0$ for all $\mathbf{x} \notin K$. In such a case we shall say that f has *compact support*. For any compact $K \subset R^n$, we shall write

$$K^\varepsilon = \{\mathbf{x} : d(\mathbf{x}, K) < \varepsilon\},$$

where $d(\mathbf{x}, K) = \inf \{ \|\mathbf{x} - \mathbf{y}\|, \mathbf{y} \in K\}$. K^ε is called the ε-neighbourhood of K.

Remark. 39.7. Let f be a locally integrable function on an open set $\Omega \subset R^n$, with support contained in a compact set $K \subset \Omega$. Then there exists a $\delta > 0$ such that $K^\delta \subset \Omega$. Then f_ε defined by Eq. (39.3) is meaningful for $\varepsilon < \delta$. Indeed, it is enough to know the values of f in an ε-neighbourhood of the point $\mathbf{x}$ in order to calculate $f_\varepsilon(\mathbf{x})$. If the support of f is contained in K, we can define $\chi_K f$ in R^n and write $f_\varepsilon = (\chi_k f)_\varepsilon$. In such a case the support of f_ε is contained in $K^\varepsilon \subset \Omega$. Thus f_ε may be considered as a function with compact support in Ω for all $\varepsilon < \delta$.

Proposition. 39.8. Let f be a locally integrable function on R^n. Then f_ε defined by Eq. (39.3) is infinitely differentiable. If f has support contained in a compact set K, f_ε has support contained in an ε-neighbourhood of K. If $\sup_x |f(x)| \leqslant c$ then $\sup_x |f_\varepsilon(x)| \leqslant c$.

Proof. By change of variables and Proposition 37.2.

$$f_\varepsilon(\mathbf{x}) = \varepsilon^{-n} \int f(\mathbf{y}) \, \phi\left(\frac{1}{\varepsilon}(\mathbf{x} - \mathbf{y})\right) d\mathbf{y}.$$

Since the support of ϕ is contained in a compact set and $0 \leqslant \phi \leqslant 1$, and f is locally integrable f_ε is well defined. Since any derivative of ϕ has compact support, repeated application of mean value theorem and Lebesgue's dominated convergence theorem shows that f_ε is infinitely differentiable. The second part of the theorem is obvious.

Proposition. 39.9. Let $f \in L_p(R^n)$ where $p \geqslant 1$. Then f_ε defined by Eq. (39.3) is infinitely differentiable. Further $f_\varepsilon \in L_p(R^n)$ and

$$\lim_{n \to \infty} \|f_\varepsilon - f\|_p = 0.$$

Proof. Since $f \in L_p(R^n)$ it follows from Holder's inequality that f is locally integrable. By Proposition 39.8, f_ε is well-defined and infinitely differentiable. Since ϕ is a probability density we have from Proposition 34.13 and Fubini's theorem

$$\begin{aligned}\int |f_\varepsilon(\mathbf{x}) - f(\mathbf{x})|^p \, d\mathbf{x} &= \int \left| \int [f(\mathbf{x} - \varepsilon \mathbf{y}) - f(\mathbf{x})] \, \phi(\mathbf{y}) \, d\mathbf{y} \right|^p d\mathbf{x} \\ &\leqslant \int \left[\int |f(\mathbf{x} - \varepsilon \mathbf{y}) - f(\mathbf{x})|^p \, \phi(\mathbf{y}) \, d\mathbf{y} \right] d\mathbf{x} \\ &\leqslant \int \left[\int |f(\mathbf{x} - \varepsilon \mathbf{y}) - f(\mathbf{x})|^p \, d\mathbf{x} \right] \phi(\mathbf{y}) \, d\mathbf{y}. \end{aligned} \tag{39.4}$$

By Minkowski's inequality, $\int |f(\mathbf{x} - \varepsilon \mathbf{y}) - f(\mathbf{x})|^p \, d\mathbf{x}$ is bounded in $\mathbf{y}$. Further by Corollary 39.2 this tends to 0 as $\varepsilon \to 0$. Hence by Lebesgue dominated convergence theorem the right hand side of (39.4) tends to 0 as $\varepsilon \to 0$. This completes the proof.

Exercise. 39.10. Let $\Omega \subset R^n$ be an open set and let $L_p(\Omega)$ denote the L_p space of functions with respect to the Lebesgue measure restricted to Ω. For any $f \in L_p(\Omega)$, there exists a sequence $\{f_n\}$ of infinitely differentiable functions with compact support such that $\lim_{n \to \infty} \|f_n - f\|_p = 0$. (Hint: use Remark 39.7.)

Exercise. 39.11. Let $\Omega \subset R^n$ be an open set and let f be a real valued continuous function on Ω. Then there exists a sequence $\{f_n\}$ of infinitely differentiable functions with compact support such that

$$\lim_{n \to \infty} \sup_{\mathbf{x} \in K} |f_n(\mathbf{x}) - f(\mathbf{x})| = 0$$

for every compact $K \subset \Omega$.

Remark. 39.12. Let $\Omega \subset R^n$ be an open set and let $C_0^\infty(\Omega)$ denote the space of all real valued infinitely differentiable functions on Ω with compact support. Then $C_0^\infty(\Omega)$ is a vector space. For any $\psi \in C_0^\infty(\Omega)$, let

$$\sup_{\mathbf{x} \in \Omega} \left| \frac{\partial^{|\mathbf{r}|} \psi}{\partial x_1^{r_1} \partial x_2^{r_2} \ldots \partial x_n^{r_n}} (\mathbf{x}) \right| = \|\psi\|_{\mathbf{r}}$$

where

$$\mathbf{r} = \begin{pmatrix} r_1 \\ r_2 \\ \cdot \\ \cdot \\ \cdot \\ r_n \end{pmatrix}, \quad |\mathbf{r}| = r_1 + r_2 + \ldots + r_n$$

is a column vector of non-negative integers. We define a topology on $C_0^\infty(\Omega)$ by declaring a sequence $\psi_1, \psi_2, \ldots$ to converge to ψ if and only if there exists a compact set K such that all the ψ_n's and ψ have their support in K and

$$\lim_{n\to\infty} \| \psi_n - \psi \|_{\mathbf{r}} = 0$$

for every non-negative integral vector $\mathbf{r}$. Then $C_0^\infty(\Omega)$ becomes a *topological vector space*. A map $\Lambda : C_0^\infty(\Omega) \to R$ is called a *generalised function* or *distribution* (in the sense of Schwarz and Sobolev) if

(i) $\Lambda(a_1\psi_1 + a_2\psi_2) = a_1\Lambda(\psi_1) + a_2\,\Lambda(\psi_2)$ for all $a_1, a_2 \in R;\ \psi_1, \psi_2 \in C_0^\infty(\Omega)$;

(ii) $\Lambda(\psi_n) \to 0$ whenever $\psi_n \to 0$ (in the topology described above.)

Any locally integrable function f on Ω may be considered as a generalised function through the map Λ_f defined by

$$\Lambda_f(\psi) = \int f\psi d\mathbf{x},\ \psi \in C_0^\infty(\Omega).$$

It is easy to check that Λ_f is indeed a generalised function. If f_1 is another locally integrable function such that the generalised functions Λ_f and Λ_{f_1} are identical then

$$\int (f - f_1)\psi\, d\mathbf{x} = 0, \text{ for all } \psi \in C_0^\infty(\Omega). \tag{39.5}$$

By Propositions 39.8, 39.9 and Exercise 39.10 it is clear that the characteristic function χ_K of any compact set $K \subset \Omega$ can be approximated in measure by a sequence $\psi_n \in C_0^\infty(\Omega)$ which is uniformly bounded. Hence Eq. (39.5) implies that

$$\int_K (f - f_1) d\mathbf{x} = 0 \text{ for all compact } K \subset \Omega.$$

Hence $f = f_1$ a.e. $\mathbf{x}$. Thus there is a one to one correspondence between the ordinary function f and the generalised function Λ_f.

If μ is a σ-finite measure in Ω with finite μ measure for every compact $K \subset \Omega$, then Λ_μ defined by

$$\Lambda_\mu(\psi) = \int \psi d\mu,\ \psi \in C_0^\infty(\Omega)$$

is also a generalised function. Thus measures are generalised functions.

Now suppose f is a function on Ω such that $\dfrac{\partial f}{\partial x_i}$ is defined on Ω and both f and $\dfrac{\partial f}{\partial x_i}$ are continuous. Let us see what is the relation between the generalised function Λ_f and $\Lambda_{\partial f/\partial x_i}$. We have, integrating by parts,

$$\Lambda_{\frac{\partial f}{\partial x_i}}(\psi) = \int \frac{\partial f}{\partial x_i} \psi \, d\mathbf{x}$$

$$= - \int f \frac{\partial \psi}{\partial x_i} \, d\mathbf{x}$$

$$= - \Lambda_f \left(\frac{\partial \psi}{\partial x_i} \right) \text{ for all } \psi \in C_0^{\infty}(\Omega).$$

We imitate this equation for all generalised functions. To every generalised function Λ on $C_0^{\infty}(\Omega)$ we define its *derivative* $D_i \Lambda$ by

$$(D_i \Lambda)(\psi) = -\Lambda \left(\frac{\partial \psi}{\partial x_i} \right).$$

Then $D_i \Lambda$ is a generalised function. Repeating this procedure we can construct more generalised functions by the equations

$$\left(D_1^{r_1} \, D_2^{r_2} \ldots D_n^{r_n} \, \Lambda \right) (\psi)$$

$$= (-1)^{r_1 + r_2 + \ldots + r_n} \Lambda \left(\frac{\partial^{|\mathbf{r}|} \psi}{\partial_{x_1}^{r_1} \, \partial_{x_2}^{r_2} \ldots \partial_{x_n}^{r_n}} \right).$$

This enables one to develop a calculus of generalised functions and formulate differential equations in terms of generalised functions. This yields a far-reaching generalisation of the classical differential calculus. The aim of the present remark is just to indicate to the reader how measure theory has played a role in leading to new ideas in functional analysis. The interested reader should consult [11], [19], [20].

Exercise. 39.13. Let $\Omega \subset R^n$ be an open subset and let Λ be a generalised function such that

$$\Lambda(\psi) \geqslant 0 \text{ whenever } \psi \geqslant 0 \text{ and } \psi \in C_0^{\infty}(\Omega).$$

Then there exists a σ-finite measure μ on the borel σ-algebra of Ω such that

$$\Lambda(\psi) = \int \psi \, d\mu, \text{ for all } \psi \in C_0^{\infty}(\Omega)$$

and $\mu(K) < \infty$ for all compact $K \subset \Omega$.

CHAPTER SIX

Hilbert Space and Conditional Expectation

§40. Elementary Properties of Banach Spaces

In Section 34 we introduced the spaces $L_p(\mu)$ for $p \geqslant 1$ and mentioned that they are Banach spaces. Among these the space $L_2(\mu)$ has a special role to play in our subject. In fact all statistical problems which centre around 'extrapolation' or 'prediction' are based on the fundamental properties of the space $L_2(\mu)$. To elaborate on this theme we shall develop the barest minimum of the theory of Banach and Hilbert spaces.

Definition. 40.1. Let $\mathscr{X}$ be a set satisfying the following properties:

(i) $\mathscr{X}$ is a real or complex vector space;

(ii) to every element $x \in \mathscr{X}$ there is a non-negative number $\| x \|$, called the *norm* of x such that (*a*) $\| x \| = 0$ if and only if $x = 0$; (*b*) $\| ax \| = |a| \, \| x \|$ for all scalars a; (*c*) $\| x + y \| \leqslant \| x \| + \| y \|$ for all x, y;

(iii) under the metric d defined by $d(x, y) = \| x - y \|$, $\mathscr{X}$ is a complete metric space.

Then $\mathscr{X}$ is called a *Banach space* or simply a *B-space*.

Remark. 40.2. By Remark 34.11, $L_p(\mu)$ is a B-space for any σ-finite measure μ and any $1 \leqslant p \leqslant \infty$. By Exercise 33.13 the space of all totally finite signed measures on a borel space $(X, \mathscr{B})$ is a B-space. We shall denote this B-space by $M(X)$.

Exercise. 40.3. Let $(X, \mathscr{B})$ be a borel space and let $B(X)$ be the space of all bounded (real or complex) borel functions on X. For any $f \in B(X)$, let $\| f \| = \sup_{x \in X} |f(x)|$. With this norm $B(X)$ is a B-space.

Exercise. 40.4. Let X be a compact metric space and let $C(X)$ be the space of all (real or complex) continuous functions on X. For any $f \in C(X)$ let $\|f\|$ be as in the preceding exercise. Then $C(X)$ is a B-space.

The wealth of examples indicated above shows that it is worth studying B-spaces as a separate subject. Indeed, B-spaces constitute an important part of the subject of modern functional analysis.

Definition. 40.5. A sequence $\{x_n\}$, $n=1, 2, \ldots$ in a B-space $\mathscr{X}$ is said to converge to an element $x \in \mathscr{X}$ if $\lim_{n \to \infty} \|x_n - x\| = 0$ as $n \to \infty$. In such a case we write $x_n \to x$ as $n \to \infty$ or $\lim_{n \to \infty} x_n = x$.

Definition. 40.6. If $\mathscr{X}$, $\mathscr{Y}$ are B-spaces, an *operator* from $\mathscr{X}$ to $\mathscr{Y}$ is a map A from $\mathscr{X}$ into $\mathscr{Y}$ such that

(i) $A(ax_1 + bx_2) = a\, Ax_1 + b\, Ax_2$ for all $x_1, x_2 \in \mathscr{X}$ and all scalars a, b;

(ii) $\lim_{n \to \infty} Ax_n = 0$ for every sequence $\{x_n\}$ in $\mathscr{X}$ such that $\lim_{n \to \infty} x_n = 0$.

If $\mathscr{Y}$ is the B-space of scalars (namely real line or complex plane) with absolute value as norm, an operator from $\mathscr{X}$ to $\mathscr{Y}$ is called a *linear functional*.

Proposition. 40.7. For any operator A from a B-space $\mathscr{X}$ into a B-space $\mathscr{Y}$, let

$$\|A\| = \sup_{\|x\| = 1} \|Ax\|.$$

Then $\|A\| < \infty$ and $\|Ax\| \leqslant \|A\| \ \|x\|$ for all x.

Proof. Suppose $\|A\| = \infty$. Then there exists a sequence $\{x_n\}$ in $\mathscr{X}$ such that $\|x_n\| = 1$ for $n = 1, 2, \ldots$ and $\lim_{n \to \infty} \|Ax_n\| = \infty$. Let $x'_n = \dfrac{x_n}{\|Ax_n\|}$. Then $\|Ax'_n\| = 1$ for all n and $\lim_{n \to \infty} x'_n = 0$. This is a contradiction. This proves the first part. Since

$$\frac{\|Ax\|}{\|x\|} = \left\| A \frac{x}{\|x\|} \right\| \leqslant \|A\| \text{ when } x \neq 0,$$

the second part follows trivially and the proof is complete.

Remark. 40.8. If A is a map from a B-space $\mathscr{X}$ into another B-space $\mathscr{Y}$ satisfying condition (i) of Definition 40.6 and $\|Ax\| \leqslant c\|x\|$ for all x and some fixed positive constant c, then A is an operator.

Remark. 40.9. Let $\mathscr{X}$, $\mathscr{Y}$ be B-spaces and let $\mathscr{A}(\mathscr{X}, \mathscr{Y})$ be the space of all operators from $\mathscr{X}$ to $\mathscr{Y}$. Then for any $A, B \in \mathscr{A}(\mathscr{X}, \mathscr{Y})$ and scalars a, b, $aA + bB$ is defined by $(aA + bB)x = aAx + bBx$. This makes $\mathscr{A}(\mathscr{X}, \mathscr{Y})$ a vector space. Further $\|A\|$ satisfies all the properties of a norm, and with this norm $\mathscr{A}(\mathscr{X}, \mathscr{Y})$ is a B-space. If $\mathscr{Y}$ is the B-space of scalars, the B-space $\mathscr{A}(\mathscr{X}, \mathscr{Y})$ is denoted by $\mathscr{X}^*$ and called the *dual* of $\mathscr{X}$.

By Proposition 33.10 and Riesz representation theorem it follows that for any compact metric space X, the dual of the B-space $C(X)$ is the space $M(X)$ of totally finite signed measures on the borel σ-algebra of X. If $f \in C(X)$ and $\mu \in M(X)$ the associated linear functional is given by

$$\mu(f) = \int f \, d\mu.$$

If $\mathscr{X} = \mathscr{Y}$, we write $\mathscr{A}(\mathscr{X})$ instead of $\mathscr{A}(\mathscr{X}, \mathscr{Y})$. In this case for any A, $B \in \mathscr{A}(\mathscr{X})$ we can define multiplication by $(AB)\,x = A(Bx)$. This makes $\mathscr{A}(\mathscr{X})$ an algebra and further $\|AB\| \leqslant \|A\| \, \|B\|$. We say that $\mathscr{A}(\mathscr{X})$ is a Banach algebra or B-algebra.

Remark. 40.10. It may be noted that the spaces $B(X)$ and $C(X)$ of Exercises 40.3 and 40.4 respectively are also B-algebras under pointwise multiplication. The B-space $L_1(R^n)$ is a B-algebra under convolution. The space $M(R^n)$ also becomes a B-algebra under convolution, when the notion of convolution between probability measures is extended in a natural manner to totally finite signed measures. Once again we remark that the wealth of such examples makes it worthwhile to study the subject of B-algebras. When the multiplication operation is commutative one has a commutative B-algebra. A deep investigation of such algebras was made by the Russian mathematician I. M. Gelfand. See [16].

Exercise. 40.11. Let $(X_i, \mathscr{B}_i)$, $i = 1, 2, 3$ be borel spaces. Let $P : (X_1 \times \mathscr{B}_2) \to [0, 1]$ be a transition probability. Let $B(X_i)$, $M(X_i)$ be the B-spaces of bounded borel functions and totally finite signed measures respectively on $(X_i, \mathscr{B}_i)$. We define operators A_P, B_P by

$$(A_P f)(x_1) = \int f(x_2) \, P(x_1, dx_2), \; f \in B(X_2),$$
$$(B_P \mu)(E) = \int P(x_1, E) \, d\mu(x_1), \; \mu \in M(X_1), \; E \in \mathscr{B}_2.$$

Then A_P is an operator from $B(X_2)$ into $B(X_1)$. B_P is an operator from $M(X_1)$ into $M(X_2)$. Further

$$\|A_P\| \leqslant 1, \; \|B_P\| \leqslant 1.$$

If Q is a transition probability on $X_2 \times \mathscr{B}_3$ and $P \circ Q$ is the composition of P and Q, then

$$A_P A_Q = A_{P \circ Q}, \; B_Q B_P = B_{P \circ Q}.$$

In particular, if $\{P_t, t > 0\}$ is a one parameter semi-group of transition probabilities on $X \times \mathscr{B}$, where $(X, \mathscr{B})$ is a borel space (See Remark 35.23) and $A_t = A_{P_t}$, $B_t = B_{P_t}$, then

$$A_t A_s = A_{t+s}, \; B_t B_s = B_{t+s} \text{ for all } t, s > 0.$$

(This shows that the study of Markov processes in continuous time is intimately connected with the properties of one parameter semi groups of operators on B-spaces.)

Definition. 40.12. A *subspace* S of a B-space $\mathscr{X}$ is a subset of $\mathscr{X}$ satisfying the following properties: (i) S is closed in the topology of $\mathscr{X}$ induced by the metric which is defined by the norm; (ii) S is a vector subspace of $\mathscr{X}$.

For any $x \in \mathscr{X}$ and any subspace $S \subset \mathscr{X}$, the distance $d(x, S)$ is defined by

$$d(x, S) = \inf \{ \| x - y \|, y \in S \}.$$

If there exists an $x_0 \in S$ such that $d(x, S) = \| x - x_0 \|$, we say that x_0 is a *best approximator* in S for the element x.

Remark. 40.13. If $x \in \mathscr{X}$ and S is a subspace it is quite possible that a best approximator does not exist, and even if it exists, it need not be unique. However, in the case of Hilbert spaces we shall show that a unique best approximator exists in every subspace for every element. $L_2(\mu)$ is such a space and that is the reason why $L_2(\mu)$ gets singled out among all $L_p(\mu)$.

Definition. 40.14. Let $\mathscr{X}$, $\mathscr{Y}$ be B-spaces. A sequence $\{A_n\}$ of operators in $\mathscr{A}(\mathscr{X}, \mathscr{Y})$ is said to *converge strongly* to an operator A in $\mathscr{A}(\mathscr{X}, \mathscr{Y})$ if

$$\lim_{n \to \infty} A_n x = Ax \text{ for every } x \in X.$$

Remark. 40.15. It is quite obvious that if $\mathscr{X} = \mathscr{Y}$ and $\lim_{n \to \infty} A_n = A$, $\lim_{n \to \infty} B_n = B$ exist in the strong sense then $\lim_{n \to \infty} (A_n + B_n) = A + B$. It is a theorem that $A_n B_n$ converges strongly to AB but its proof is outside the scope of this book.

§41. Projections in a Hilbert Space

Definition. 41.1. Let $\mathscr{H}$ be a B-space and let (x, y) be a scalar for every $x, y \in \mathscr{H}$ satisfying the following properties:

(i) the norm $\| x \|$ of any element x in $\mathscr{H}$ is given by

$$\| x \|^2 = (x, x) \geqslant 0;$$

(ii) $(ax_1+bx_2, y)=a(x_1, y)+b(x_2, y)$ for all $x, y \in \mathcal{H}$ and all scalars a, b;

(iii) $(x, y)=\overline{(y, x)}$ for all $x, y \in \mathcal{H}$.

Then $\mathcal{H}$ is called a *Hilbert space*. The scalar (x, y) is called the *inner product* betwen x and y. (The name Hilbert space is in honour of the German mathematician D. Hilbert who studied operators on such spaces in connection with the theory of integral equations.)

Example. 41.2. The simplest example of a Hilbert space is the space R^n with the inner product

$$(\mathbf{x}, \mathbf{y})=\sum_{i=1}^{n} x_i y_i, \quad \mathbf{x}, \mathbf{y} \in R^n,$$

where $\mathbf{x}$, $\mathbf{y}$ stand for n-dimensional column vectors whose i-th element is x_i, $i=1, 2, \ldots, n$.

Consider the case $n=2$. The angle θ between the line joining the origin $\mathbf{0}$ and the point $\mathbf{x}$ and the line joining $\mathbf{0}$ and the point $\mathbf{y}$ is given by the formula

$$\cos \theta=\frac{(\mathbf{x}, \mathbf{y})}{\|\mathbf{x}\| \|\mathbf{y}\|},$$

where $\|\mathbf{x}\|=(\mathbf{x}, \mathbf{x})^{\frac{1}{2}}$. The distance between $\mathbf{x}$ and $\mathbf{y}$ is given by $(\mathbf{x}-\mathbf{y}, \mathbf{x}-\mathbf{y})^{\frac{1}{2}}$. In particular, the two lines are perpendicular to each other if and only if $(\mathbf{x}, \mathbf{y})=0$. Thus the importance of the inner product lies in the fact that, it expresses at once the two fundamental ideas of geometry, namely, angle and distance.

Example. 41.3. Consider the space $L_2(\mu)$, where $(X, \mathcal{B}, \mu)$ is a σ-finite measure space. Let

$$(f, g)=\int f\bar{g}\, d\mu \text{ for all } f, g \in L_2(\mu).$$

Then (f, g) is an inner product which makes $L_2(\mu)$ a Hilbert space.

Throughout the rest of this section let $\mathcal{H}$ denote a fixed Hilbert space with inner product (x, y) between any two elements $x, y \in \mathcal{H}$.

Proposition. 41.4. For any $x, y, z \in \mathcal{H}$, the following holds:

(i) $\|x-y\|^2 = \|x\|^2 + \|y\|^2 - 2 \operatorname{Re}(x, y)$, where Re stands for the real part;

(ii) $\|x-y\|^2 + \|x-z\|^2 = 2\left(\left\|x - \frac{y+z}{2}\right\|^2 + \left\|y - \frac{y+z}{2}\right\|^2\right)$;

(iii) $|(x, y)| \leqslant \|x\| \, \|y\|$.

Proof. The first two identities follow from properties (i), (ii) and (iii) of inner product in Definition 41.1. To prove inequality (iii) of the proposition we first rewrite it as

$$\left|\left(\frac{x}{\|x\|}, \frac{y}{\|y\|}\right)\right| \leqslant 1 \text{ for } x \neq 0, y \neq 0$$

and observe that it is enough to prove in the case $\|x\| = \|y\| = 1$. For any scalar t, we have

$$0 \leqslant \|x - ty\|^2 = \|x\|^2 + |t|^2 \|y\|^2 - 2 \operatorname{Re}[\bar{t}(x, y)]$$
$$= 1 + |t|^2 - 2 \operatorname{Re}[\bar{t}(x, y)].$$

Putting $t = (x, y)$, we get $0 \leqslant 1 - |(x, y)|^2$. This completes the proof.

Remark. 41.5. Inequality (iii) is known as *Schwarz inequality.* If x, y, z are imagined to constitute a triangle ABC, then identity (ii) of the proposition is the analogue of the classical Apollonius identity $AB^2 + AC^2 = 2(AD^2 + BD^2)$, where D is the midpoint of BC. Hence we shall call it the *Apollonius identity.*

Proposition. 41.6. If $\lim_{n\to\infty} x_n = x$ and $\lim_{n\to\infty} y_n = y$, where x_n, y_n, x, y are in $\mathcal{H}$ and $n = 1, 2, \ldots$ then

$$\lim_{n\to\infty} (x_n, y_n) = (x, y).$$

Proof. This follows immediately from the boundedness of $\|x_n\|$ and $\|y_n\|$ in n and the inequality

$$|(x_n, y_n) - (x, y)| \leqslant |(x_n - x, y_n)| + |(x, y_n - y)|$$
$$\leqslant \|x_n - x\| \, \|y_n\| + \|x\| \, \|y_n - y\|.$$

(Here we have applied Schwarz inequality.)

Proposition. 41.7. Let $S \subset \mathcal{H}$ be any subset. Then the set

$$S^\perp = \{x : (x, y) = 0 \text{ for every } y \in S\}$$

is a subspace. (It is called the *orthogonal complement* of S.)

Proof. This is an immediate consequence of Definition 41.1 and Proposition 41.6.

Proposition. 41.8. (Projection theorem.) Let $S \subset \mathcal{H}$ be a subspace and let $x \in \mathcal{H}$ be arbitrary. Then there exists a unique best approximator $P^S x$ in S for x. The map $x \to P^S x$ is an operator on $\mathcal{H}$. P^S satisfies the following properties:

(i) $x - P^S x \in S^\perp$ for all x,

(ii) $(P^S x, y) = (x, P^S y) = (P^S x, P^S y)$ for all $x, y \in \mathcal{H}$

(iii) $(P^S)^2 = P^S$

(iv) every element x can be uniquely decomposed as $x = y + z$, where $y \in S$, $z \in S^{\perp}$. Infact $y = P^S x$.

(v) $P^{S^{\perp}} = I - P^S$

(vi) $\|x\|^2 = \|P^S x\|^2 + \|x - P^S x\|^2$ for all x.

Proof. Let x be fixed. Choose a sequence $y_n \in S$ such that

$$\lim_{n\to\infty} \|x - y_n\| = d(x, S) = \delta, \text{ say.}$$

Consider the triple x, y_m, y_n and apply Apollonius identity. Then

$$\|x - y_m\|^2 + \|x - y_n\|^2 - 2\left\|x - \frac{y_m + y_n}{2}\right\|^2 = \tfrac{1}{2}\|y_m - y_n\|^2.$$

Since $y_m, y_n \in S, \dfrac{y_m + y_n}{2} \in S$ and hence

$$\left\|x - \frac{y_m + y_n}{2}\right\|^2 \geqslant \delta^2.$$

Thus

$$\overline{\lim_{m,\,n\to\infty}} \frac{1}{2}\|y_m - y_n\|^2 \leqslant \delta^2 + \delta^2 - 2\delta^2 = 0.$$

The completeness of $\mathcal{H}$ implies that y_n converges to a limit $P^S x$ as $n \to \infty$. Since S is closed $P^S x \in S$. Further

$$\delta = \lim_{n\to\infty} \|x - y_n\| = \|x - P^S x\|.$$

Thus $P^S x$ is a best approximator for x in S. If $x' \in S$ and $\|x - x'\| = \delta$, apply Apollonius identity to the triple x, x', $P^S x$. Then

$$0 \geqslant \delta^2 + \delta^2 - 2\left\|x - \frac{x' + P^S x}{2}\right\|^2 = \frac{1}{2}\|x' - P^S x\|^2.$$

Hence $P^S x = x'$. This proves the uniqueness of the best approximator.

To prove (i) consider the identities

$$\|x - P^S x - ty\|^2 = \|x - P^S x\|^2 + t^2\|y\|^2 - 2t\,\mathrm{Re}(x - P^S x, y),$$
$$\|x - P^S x - ity\|^2 = \|x - P^S x\|^2 + t^2\|y\|^2 - 2t\,\mathrm{Im}(x - P^S x, y),$$

where y is an arbitrary element in S, t is a real parameter, Re and Im denote the real and imaginary parts respectively. Since $P^S x$ is the best approximator for x in S, the qudratic forms in t on the right hand sides of the above two identities attain their minimum when $t = 0$. Differentiating them at $t = 0$, we obtain

$$0 = \mathrm{Re}(x - P^S x, y) = \mathrm{Im}(x - P^S x, y).$$

This proves (i).

To prove (ii) we note that for any x, y,

$$(P^S x, y) = (P^S x, P^S y) + (P^S x, y - P^S y) = (P^S x, P^S y),$$
$$(x, P^S y) = (x - P^S x, P^S y) + (P^S x, P^S y) = (P^S x, P^S y),$$

because $P^S x$, $P^S y \in S$ and $x - P^S x$, $y - P^S y \in S^\perp$.

To prove (iii) we note that whenever $x \in S$, the best approximator for x in S is x itself. Since $P^S x \in S$, we have $P^S(P^S x) = P^S x$.

To prove (iv), let $x = y + z$, where $y \in S$, $z \in S^\perp$. Then $y + z = P^S x + (x - P^S x)$ and $y - P^S x = x - P^S x - z$ is an element which belongs to S and $S^\perp$. Hence $\| y - P^S x \|^2 = (y - P^S x, y - P^S x) = 0$. This proves (iv). Properties (v) and (vi) follow from (iv).

We shall now prove that P^S is an operator. Let $x, y \in \mathcal{H}$ and let a, b be scalars. Then we have two decompositions

$$\begin{aligned} ax + by &= [P^S(ax+by)] + [ax + by - P^S(ax+by)] \\ &= [a\, P^S x + b\, P^S y] + [a(x - P^S x) + b(y - P^S y)], \end{aligned}$$

where the first element is in S and the second in $S^\perp$. By (iv) we have

$$P^S(ax+by) = a\, P^S x + b\, P^S y.$$

Now (vi) implies that

$$\| P^S x \| \leqslant \| x \| \text{ for all } x.$$

This shows that P^S is an operator and completes the proof.

Definition. 41.9. The operator P^S is called the *projection* on S. An operator A on $\mathcal{H}$ is said to be *self adjoint* if $(Ax, y) = (x, Ay)$ for all x, y.

Proposition. 41.10. For any subspace $S \subset \mathcal{H}$, P^S is a self adjoint operator such that $(P^S)^2 = P^S$. Conversely, if A is a self adjoint operator on $\mathcal{H}$ such that $A^2 = A$, then there exists a subspace S such that $A = P^S$.

Proof. The first part is simply a restatement of properties (ii) and (iii) of the operator P^S in Proposition 41.8. We shall prove the converse. Let

$$S = \{x : Ax = x\}.$$

Since A is an operator S is a subspace. Consider the decomposition $x = Ax + (x - Ax)$. Since $A(Ax) = Ax$, $Ax \in S$. Further for any $y \in S$,

$$\begin{aligned} (x - Ax, y) &= (x, y) - (Ax, y) \\ &= (x, y) - (x, Ay) = 0. \end{aligned}$$

Thus $x-Ax \in S^{\perp}$. By property (iv) of Proposition 41.8, $Ax=P^{S}x$. This completes the proof.

Definition. 41.11. Let $\{S_\alpha, \alpha \in T\}$ be a family of subspaces of $\mathcal{H}$. By $\bigvee_\alpha S_\alpha$ we shall denote the smallest subspace containing $\bigcup_\alpha S_\alpha$. By $\bigwedge_\alpha S_\alpha$ we shall denote the subspace $\bigcap_\alpha S_\alpha$, which is also the set theoretic intersection. We say that the family $\{S_\alpha, \alpha \in T\}$ is mutually *orthogonal* if $(x_\alpha, x_\beta)=0$ whenever $\alpha, \beta \in T$, $\alpha \neq \beta$, $x_\alpha \in S_\alpha$, $x_\beta \in S_\beta$. In this case we write $\bigoplus_\alpha S_\alpha$ for $\bigvee_\alpha S_\alpha$ and call it the *direct sum* of the S_α's. If $S_1 \subset S_2$, we write $S_2 \ominus S_1$ for $S_2 \cap S_1^{\perp}$ and call it the *orthogonal complement* of S_1 *in* S_2.

For any set $A \subset \mathcal{H}$ we call the smallest subspace containing, A, *the span of A* and denote it by span A.

Proposition. 41.12. For any Hilbert space $\mathcal{H}$, let $L(\mathcal{H})$ denote the class of all subspaces of $\mathcal{H}$. Then the set theoretic inclusion $\subset$ makes $L(\mathcal{H})$ a partially ordered set. If we write 0 to denote the subspace consisting of 0 element, then $0 \subset S \subset \mathcal{H}$ for all $S \in L(\mathcal{H})$. Further the following properties hold for any three elements $S, S_1, S_2 \in L(\mathcal{H})$.

(i) $S_1 \vee S_2 \supset S_i$, $i=1, 2$; if $S \supset S_i$, $i=1, 2$ then $S \supset S_1 \vee S_2$;

(ii) $S_1 \wedge S_2 \subset S_i$, $i=1, 2$; if $S \subset S_i$, $i=1, 2$ then $S \subset S_1 \wedge S_2$;

(iii) $(S^{\perp})^{\perp} = S$

(iv) $(S_1 \vee S_2)^{\perp} = S_1^{\perp} \wedge S_2^{\perp}$; $(S_1 \wedge S_2)^{\perp} = S_1^{\perp} \vee S_2^{\perp}$; $0^{\perp} = \mathcal{H}$;

(v) if $S_1 \subset S_2$, then $S_1^{\perp} \supset S_2^{\perp}$.

Proof. Properties (i), (ii) and the fact that inclusion is a partial ordering such that $0 \subset S \subset \mathcal{H}$ for all $S \in L(\mathcal{H})$ are obvious. Property (iii) follows immediately from property (iv) of Proposition 41.8. Property (v) is an immediate consequence of the definition of orthogonal complement. Thus we have to prove only (iv). Let $x \in (S_1 \vee S_2)^{\perp}$. Then $x \in S_1^{\perp}$ and $S_2^{\perp}$. Hence $(S_1 \vee S_2)^{\perp} \subset S_1^{\perp} \wedge S_2^{\perp}$. If $x \in S_1^{\perp} \wedge S_2^{\perp}$ then $(x, y_1+y_2)=0$ for $y_i \in S_i$, $i=1, 2$. Since any element of $S_1 \vee S_2$ can be approximated by a sequence $y_{n1}+y_{n2}$, where $y_{ni} \in S_i$, $i=1, 2$; $n=1, 2, \ldots$ it follows from Proposition 41.6 that $x \in (S_1 \vee S_2)^{\perp}$. This proves the first half of (iv). The second half follows from (iii).

Remark. 41.13. The collection $L(\mathcal{H})$ with the partial ordering '$\subset$' and three operations $\vee$, $\wedge$ and $\perp$ have all the features of set theory except one. In fact $\subset$ behaves likes set theoretic inclusion, $\vee$ like the union $\cup$, $\wedge$ like the intersection $\cap$ and $\perp$ like set theoretic complement. However, $\wedge$ does not distribute with $\vee$, whereas $\cap$ does distribute with $\cup$. Whereas the boolean algebra of subsets of a set forms the domain of definition of a probability distribution, the collection $L(\mathcal{H})$ of subspaces of a Hilbert space $\mathcal{H}$ is the domain of definition of what is called a 'state' in quantum mechanics. Indeed, a *state* μ is a map from $L(\mathcal{H})$ into the unit interval [0, 1] such that (i) $\mu(\mathcal{H})=1$, $\mu(0)=0$; (ii) $\mu(S_1 \vee S_2)=\mu(S_1)+\mu(S_2)$ if $S_1 \perp S_2$. The state μ is said to be *countably additive* if $\mu(\vee_i S_i)=\Sigma_i \mu(S_i)$ whenever the sequence $\{S_i\}$ is mutually orthogonal. The study of such countably additive states constitutes what is known as 'non commutative probability theory'. The reader may refer to [15], [18], [22].

As an application of the projection theorem we shall now identify the dual of $\mathcal{H}$.

Proposition. 41.14. (Riesz representation theorem for Hilbert space). Let $\mathcal{H}$ be a Hilbert space and let λ be a linear functional on $\mathcal{H}$. Then there exists a unique element x_λ in $\mathcal{H}$ such that

$$\lambda(x)=(x, x_\lambda) \text{ for all } x. \tag{41.1}$$

Proof. Let $S_\lambda=\{x : \lambda(x)=0\}$. Then S_λ is a subspace of $\mathcal{H}$. Suppose $\lambda\neq 0$. Then S_λ is not the whole space. Hence by projection theorem $S_\lambda^\perp$ is a non-zero subspace. Choose an $x_0 \in S_\lambda^\perp$ such that $\| x_0 \|=1$. Since $\lambda(x_0)\neq 0$, the element $x-\dfrac{\lambda(x)}{\lambda(x_0)} x_0$ is well-defined and belongs to S_λ. Hence

$$\left(x-\frac{\lambda(x)}{\lambda(x_0)} x_0, x_0\right)=0.$$

Thus

$$\lambda(x)=\lambda(x_0)(x, x_0).$$

Put $x_\lambda=\overline{\lambda(x_0)}\, x_0$. Then Eq. (41.1) holds. Uniqueness of x_λ is obvious. This completes the proof.

Corollary. 41.15. Let A be an operator on $\mathcal{H}$. Then there exists a unique operator A^* on $\mathcal{H}$ such that $(Ax, y) = (x, A^* y)$ for all x, y. If A, B are any two operators the following properties are fulfilled:

(i) $(aA + bB)^* = \bar{a}A^* + \bar{b}B^*$ for all scalars a, b;

(ii) $(AB)^* = B^* A^*$;

(iii) $A^{**} = A$;

(iv) $\| A^* \| = \| A \|$.

(The operator A^* is called the *adjoint* of A. A is self adjoint if and only if $A^* = A$).

Proof. Consider the linear functional λ defined by

$$\lambda(x) = (Ax, y)$$

for any fixed y. By the preceding proposition there exists a unique element $A^* y \in \mathcal{H}$ such that

$$(Ax, y) = (x, A^* y) \text{ for all } x.$$

We have

$$\begin{aligned}(x, A^*(ay_1 + by_2)) &= (Ax, ay_1 + by_2) \\ &= \bar{a}(Ax, y_1) + \bar{b}(Ax, y_2) \\ &= \bar{a}(x, A^* y_1) + \bar{b}(x, A^* y_2) \\ &= (x, \bar{a}A^* y_1 + \bar{b}A^* y_2).\end{aligned}$$

Since this holds for all x, we have

$$A^*(ay_1 + by_2) = aA^* y_1 + bA^* y_2.$$

By Schwarz inequality

$$|(x, A^* y)| = |(Ax, y)| \leqslant \| Ax \| \, \| y \|.$$

If $A^* y \neq 0$, put $x = \dfrac{A^* y}{\| A^* y \|}$ in the above inequality. Then

$$\| A^* y \| \leqslant \| A \| \, \| y \|.$$

This shows that $\| A^* \| \leqslant \| A \|$. Thus A^* is an operator. Properties (i), (ii) and (iii) are immediate consequences of the identity $(Ax, y) = (x, A^* y)$ for all $x, y \in \mathcal{H}$ and all operators A. Now (iv) follows from (iii). The proof is complete.

Definition. 41.16. We shall say that an operator A on $\mathcal{H}$ is *invertible* if there exists an operator B such that $AB = BA = I$ where I is the identity operator defined by $Ix = x$ for all x. In such a case we shall write $B = A^{-1}$ and call it the *inverse* of A.

Exercise. 41.17. If A and B are invertible operators on $\mathcal{H}$, so is AB and $(AB)^{-1} = B^{-1} A^{-1}$. Further A^* is invertible and $A^{*-1} = (A^{-1})^*$.

We shall now prove some elementary algebraic properties of projections.

Proposition. 41.18. Let S_1, S_2 be two subspaces of a Hilbert space $\mathcal{H}$. Then $P^{S_1} + P^{S_2}$ is a projection if and only if S_1 and S_2 are orthogonal. In such a case $P^{S_1} + P^{S_2} = P^{S_1 \oplus S_2}$.

$P^{S_1} P^{S_2}$ is a projection if and only if $P^{S_1} P^{S_2} = P^{S_2} P^{S_1}$. In such a case $P^{S_1} P^{S_2} = P^{S_1 \wedge S_2}$.

Proof. Let $P^{S_1} + P^{S_2}$ be a projection. By Proposition 41.10, the square of $P^{S_1} + P^{S_2}$ is itself. Hence

$$P^{S_1} P^{S_2} + P^{S_2} P^{S_1} = 0. \tag{41.2}$$

If $x \in S_1 \wedge S_2$, then

$$2x = P^{S_1} P^{S_2} x + P^{S_2} P^{S_1} x = 0.$$

Thus $S_1 \wedge S_2 = 0$. By property (iv) of Proposition 41.12, $S_1^{\perp} \vee S_2^{\perp} = \mathcal{H}$. If $x \in S_1^{\perp}$ or $S_2^{\perp}$, Eq. (41.2) implies that

$$P^{S_1} P^{S_2} x = P^{S_2} P^{S_1} x = 0.$$

Hence $P^{S_1} P^{S_2} = P^{S_2} P^{S_1} = 0$. It is now clear that

$$P^{S_1} + P^{S_2} = P^{S_1 \oplus S_2}.$$

Conversely, if S_1 and S_2 are mutually orthogonal subspaces the above equation holds. This proves the first part.

To prove the second part suppose that $P^{S_1} P^{S_2}$ is a projection. The self adjointness of $P^{S_1} P^{S_2}$ and property (ii) of Corollary 41.15 imply that

$$P^{S_1} P^{S_2} = (P^{S_1} P^{S_2})^* = P^{S_2} P^{S_1}.$$

Conversely, if $P^{S_1} P^{S_2} = P^{S_2} P^{S_1}$, then

$$(P^{S_1} P^{S_2})^2 = P^{S_1} P^{S_2}; \; (P^{S_1} P^{S_2})^* = P^{S_1} P^{S_2}.$$

Proposition 41.10 implies that $P^{S_1} P^{S_2}$ is a projection. In such a case

$$\begin{aligned} P^{S_1} P^{S_2} x &= x \text{ if } x \in S_1 \wedge S_2 \\ &= 0 \text{ if } x \in S_1^{\perp} \text{ or } S_2^{\perp}. \end{aligned}$$

Hence $P^{S_1} P^{S_2} = P^{S_1 \wedge S_2}$. This completes the proof.

Now we shall prove a continuity property of the projections P^S.

Proposition. 41.19. Let $S_1 \subset S_2 \subset \ldots$ be an increasing sequence of subspaces in $\mathcal{H}$ and let $S_\infty = \bigvee_i S_i$. Then

$$\lim_{n \to \infty} P^{S_n} x = P^{S_\infty} x \text{ for all } x.$$

If $S_1 \supset S_2 \supset \ldots$ is a decreasing sequence of subspaces in $\mathcal{H}$ and $S^\infty = \bigwedge_i S_i$, then

$$\lim_{n \to \infty} P^{S_n} x = P^{S^\infty} x.$$

Proof. Let $\{S_n\}$ be increasing and let $m<n$. Then by the projection theorem

$$0\leqslant \| P^{S_n} x - P^{S_m} x \|^2 = \| P^{S_n} x \|^2 + \| P^{S_m} x \|^2 - 2\mathrm{Re}(P^{S_n} x, P^{S_m} x)$$
$$= \| P^{S_n} x \|^2 - \| P^{S_m} x \|^2.$$

This shows that $\| P^{S_n} x \|^2$ is monotonic increasing and bounded and therefore converges to a limit. Thus $\lim_{m,\, n\to\infty} \| P^{S_n} x - P^{S_m} x \| = 0$. By the completeness of $\mathcal{H}$ the limit $\lim_{n\to\infty} P^{S_n} x = Px$, say, exists. Since P^{S_n} is an operator for every n and $\| P^{S_n} x \| \leqslant \| x \|$, it follows that P is an operator satisfying $\| Px \| \leqslant \| x \|$ for all x. Since $(P^{S_n}x, P^{S_n}y)=(x, P^{S_n}y)=(P^{S_n} x, y)$ for all n and all x, y, we obtain by taking limits,

$(Px, Py)=(x, Py)=(Px, y)$ for all x, y.

Hence $P^*=P=P^2$ and P is a projection. If $x\in S_k$ for some k, then $x\in S_n$ for $n\geqslant k$ and

$$Px = \lim_{n\to\infty} P^{S_n} x = x.$$

Since P is an operator $Px=x$ for all $x\in \vee_n S_n=S_\infty$. If $x\in S_n^\perp$ for every n, then $P^{S_n} x = 0$ for all n and hence $Px=0$. This shows that $P=P^{S_\infty}$.

The second part follows from the first if we observe that for a decreasing sequence $\{S_n\}$ of subspaces, the sequences $\{S_n^\perp\}$ increases and $S^\infty = \wedge_n S_n = (\vee_n S_n^\perp)^\perp$. This completes the proof.

Example. 41.20. Let S be the subspace in $\mathcal{H}$ spanned by a finite number of elements $x_1, x_2, \ldots, x_n$ in $\mathcal{H}$. Suppose these are *linearly independent*, i.e., $\sum_i c_i x_i=0$ for some scalars $c_1, c_2, \ldots, c_n$ if and only if $c_i=0$ for all i. Let $x\in\mathcal{H}$ be arbitrary and

$$\Sigma=((\sigma_{ij})),\ \sigma_{ij}=(x_i, x_j),$$
$$\xi_i=(x_i, x),\ 1\leqslant i, j\leqslant n.$$

The linear independence of the elements x_j, $1\leqslant j\leqslant n$ and the properties of the inner product imply that Σ is a non-singular hermitian matrix of order n. We shall now explicitly evaluate $P^S x$ in terms of the matrix Σ and the column vector ξ with coordinates ξ_i. Since $P^S x\in S$ it can be expressed as

$$P^S x = \sum_{i=1}^{n} a_i x_i,$$

where a_i's are scalars. By projection theorem

$$(x - P^S x, x_i) = 0 \text{ for all } i.$$

Thus

$$\bar{\xi}_i = \sum_j a_j \sigma_{ji} \text{ for all } i.$$

In matrix notation

$$\Sigma' \mathbf{a} = \bar{\boldsymbol{\xi}},$$

where $\bar{\boldsymbol{\xi}}$ denotes the column vector with coordinates $\bar{\xi}_i$ and Σ' is the transpose of Σ. Thus

$$\mathbf{a} = \Sigma'^{-1} \bar{\boldsymbol{\xi}}.$$

We can express $P^S x$ as

$$P^S x = \boldsymbol{\xi}^* \Sigma^{-1} \mathbf{x}, \tag{41.3}$$

where $\mathbf{x}$ is the 'column vector' whose i-th element is x_i, $\boldsymbol{\xi}^*$ is the row vector with coordinates $\bar{\xi}_i$. Thus the problem of finding the best approximator reduces to the problem of inverting the matrix Σ. The square of the error in approximating x by $P^S x$ is given by

$$\begin{aligned} \| x - P^S x \|^2 &= \| x \|^2 - \| P^S x \|^2 \\ &= \| x \|^2 - \boldsymbol{\xi}^* \Sigma^{-1} \boldsymbol{\xi}. \end{aligned}$$

Remark. 41.21. Let $(X, \mathscr{B}, \mu)$ be a probability space and let $\mathcal{H} = L_2(\mu)$. Let $g, f_1, f_2, \ldots, f_n$ be real valued square integrable random variables on $(X, \mathscr{B}, \mu)$. Suppose that it is possible to observe the random variables $f_1, f_2, \ldots, f_n$ but not g. Then we can 'predict' g by a linear combination of the constant random variable 1 and the random variables f_j, $j = 1, 2, \ldots, n$ by projecting g on the span S of $\{1, f_1, f_2, \ldots, f_n\}$. The discussion in the preceding example shows that

$$P^S g = Eg + \boldsymbol{\xi}' \Sigma^{-1} (\mathbf{f} - E\mathbf{f}) \tag{41.4}$$

where Σ is the covariance matrix of the random variables f_j, $j=1, 2, \ldots, n$, and

$$\mathbf{f} = \begin{pmatrix} f_1 \\ f_2 \\ \cdot \\ \cdot \\ \cdot \\ f_m \end{pmatrix}, \quad \mathbf{Ef} = \begin{pmatrix} \mathbf{E}f_1 \\ \mathbf{E}f_2 \\ \cdot \\ \cdot \\ \cdot \\ \mathbf{E}f_m \end{pmatrix}, \quad \boldsymbol{\xi} = \begin{pmatrix} \xi_1 \\ \xi_2 \\ \cdot \\ \cdot \\ \cdot \\ \xi_n \end{pmatrix},$$

and $\xi_i = \mathbf{cov}\,(g, f_i)$, $i = 1, 2, \ldots, n$. Formula (41.4) is known as the *linear regression* of g on $f_1, f_2, ..., f_n$. The *mean square error* of prediction is defined as

$$E(g - P^S g)^2 = \mathbf{V}(g) - \boldsymbol{\xi}' \, \Sigma^{-1} \, \boldsymbol{\xi}, \tag{41.5}$$

where $\mathbf{V}(g)$ is the variance of g.

Example. 41.22. Let $(X, \mathscr{B}, \mu)$ be a σ-finite measure space. For any $A \in \mathscr{B}$, let $P_\mu(A)$ be the operator on $L_2(\mu)$ defined by

$$P_\mu(A) f = \chi_A f, \; f \in L_2(\mu).$$

Then $P_\mu(A)$ is the projection on the subspace

$$S = \{f : f(x) = 0 \text{ a.e. on } A'\},$$

where A' is the complement of A. Further

(i) $P_\mu(A) \; P_\mu(B) = P_\mu(A \cap B)$ for all $A, B \in \mathscr{B}$;

(ii) If A_n is a sequence of pair wise disjoint sets in $\mathscr{B}$, then $P_\mu(\bigcup_i A_i) = \sum_{i=1}^{\infty} P_\mu(A_i)$, where the infinite series on the right hand side converges strongly.

Definition. 41.23. Let $(X, \mathscr{B})$ be a borel space and let $\mathscr{H}$ be a Hilbert space. Let $A \to P(A)$ be a map from $\mathscr{B}$ into the space of all projection operators on $\mathscr{H}$ such that

(i) $P(A) \; P(B) = P(A \cap B)$ for all $A, B \in \mathscr{B}$;

(ii) $P(\bigcup_i A_i) = \sum_{i=1}^{\infty} P(A_i)$ if $A_1, A_2, ...$ is a sequence of disjoint elements from $\mathscr{B}$ and the infinite series is understood to converge in the strong sense. Then P is called a *projection valued measure* or *spectral measure*. In particular, in the preceding example P_μ is a spectral measure and it is called the canonical spectral measure in $L_2(\mu)$.

It is possible to develop a theory of integration with respect to such spectral measures. However, it is outside the scope of this book. The reader may refer to [8].

§42. Orthonormal Sequences

Definition. 42.1. A Hilbert space $\mathscr{H}$ is said to be *separable* if it is separable as a metric space under the metric induced by its norm.

A sequence $\{x_n\}$, $n = 1, 2, ...$ of elements in $\mathscr{H}$ is said to be *orthonormal* if $(x_m, x_n) = \delta_{mn}$ for all m and n, where $\delta_{mn} = 1$ if $m = n$ and 0 otherwise. An orthonormal sequence $\{x_n\}$ is said to be *complete* if the span of $\{x_n\}$ is $\mathscr{H}$.

Proposition. 42.2. Let $\{x_n\}$ be an arbitrary sequence in a Hilbert space $\mathcal{H}$ and let S_n be the subspace spanned by the first n elements $x_1, x_2, \ldots, x_n$ for every n. Then there exists an orthonormal sequence $\{y_n\}$ such that $S_n \subset \tilde{S}_n$ for every n, where $\tilde{S}_n$ is the span of $y_1, y_2, \ldots, y_n$ for all n. If the sequence $\{x_n\}$ spans $\mathcal{H}$ so does the sequence $\{y_n\}$. In particular, every separable Hilbert space $\mathcal{H}$ has a complete orthonormal sequence.

Proof. To prove the proposition we may assume without loss of generality that $x_1 \neq 0$. Put $z_1 = \dfrac{x_1}{\| x_1 \|}$. If $S_n \ominus S_{n-1}$ is a non-zero subspace choose an element $z_n \in S_n \ominus S_{n-1}$ such that $\| z_n \| = 1$. Otherwise put $z_n = 0$. Since, by the projection theorem we have

$$S_n = S_{n-1} \oplus (S_n \ominus S_{n-1}),$$

it follows by induction that the span of $z_1, z_2, \ldots, z_n$ is S_n for all n. Now drop the zero elements from the sequence $\{z_n\}$ and call the new sequence obtained thereby as $\{y_n\}$. Since any $z_n \in S_n \ominus S_{n-1}$, the sequence $\{z_n\}$ is orthogonal. Since $\| z_n \| = 1$ or 0, it follows that $\{y_n\}$ is an orthonormal sequence. It is clear that S_n is contained in the span of $y_1, y_2, \ldots, y_n$. If $\bigvee_n S_n = \mathcal{H}$, we have $\bigvee_n \tilde{S}_n = \mathcal{H}$.

Now suppose that $\mathcal{H}$ is separable. Then choose a dense sequence $\{x_n\}$ of elements in $\mathcal{H}$. In particular, the sequence $\{x_n\}$ spans $\mathcal{H}$. Then the preceding discussion shows that the sequence $\{y_n\}$ also spans $\mathcal{H}$. This completes the proof.

Remark. 42.3. Let $\mathcal{H}$ be a separable Hilbert space. If a complete orthonormal sequence terminates at a finite stage then $\mathcal{H}$ is called a *finite dimensional Hilbert space.* In this case all complete orthonormal sequences have the same number of elements. This follows from the linear independence of any finite set of orthonormal elements. Then $\mathcal{H}$ is a finite dimensional vector space and any complete orthonormal sequence has k elements, where k is the dimension of $\mathcal{H}$.

Thus all complete orthonormal sequences in a separable Hilbert space $\mathcal{H}$ have the same number of elements. If this number is infinite we say that $\mathcal{H}$ is *infinite dimensional.*

Proposition. 42.4. (Parseval's identity.) Let $\mathcal{H}$ be a separable Hilbert space and let $\{x_n\}$ be a complete orthonormal sequence. Then

(i) $\lim_{n\to\infty} \sum_1^n (x, x_i)\, x_i = x$ for all x;

(ii) $\sum_1^\infty (x, x_i)\, \overline{(y, x_i)} = (x, y)$ for all x, y, where the infinite series converges absolutely.

(If the complete orthonormal sequence terminates at the nth stage then we can drop the limit and replace ∞ by n.)

Proof. Let S_n be the span of $x_1, x_2, \ldots, x_n$. Since these are orthonormal it follows, in particular, from Example 41.21 (See also Eq. (41.3)) that

$$P^{S_n} x = \sum_1^n (x, x_i)\, x_i. \tag{42.1}$$

Since S_n increases to the whole space $\mathcal{H}$, Proposition 41.19 implies that $P^{S_n} x \to x$ as $n \to \infty$. This proves (i). Further

$$(x, y) = \lim_{n\to\infty} (P^{S_n} x, P^{S_n} y)$$

and Eq. (42.1) implies that

$$(x, y) = \lim_{n\to\infty} \sum_{i=1}^n (x, x_i)\, \overline{(y, x_i)}.$$

This proves the equation in (ii). To prove absolute convergence we note that

$$\| x \|^2 = \lim_n \| P^{S_n} x \|^2 = \sum_1^\infty |(x, x_i)|^2 < \infty \text{ for all } x.$$

Now Schwarz inequality implies

$$\Big(\sum_1^\infty |(x, x_i)\, \overline{(y, x_i)}|\Big)^2 \leqslant \sum_1^\infty |(x, x_i)|^2 . \sum_1^\infty |(y, x_i)|^2 < \infty.$$

The proof is complete.

Remark. 42.5. If $\{x_n\}$ is a complete orthonormal sequence we can expand any element as an infinite series $\sum_1^\infty (x, x_i)\, x_i$, where the convergence is in $\mathcal{H}$. In such a case if $x = \sum_1^\infty a_i x_i$, it follows from projection theorem that $a_i = (x, x_i)$ for all i. Thus the expansion in terms of the x_i's is unique. In view of this fact a complete orthonormal sequence is also called an *orthonormal basis*.

Exercise. 42.6. Let $\mathcal{H}$ be a separable Hilbert space. Then an orthonormal sequence $\{x_n\}$ is complete if and only if the following holds: any element $x = 0$ whenever $(x, x_n) = 0$ for every n.

Example. 42.7. Let $L_2([0, 1])$ denote the Hilbert space of all complex borel functions which are square integrable with respect to the Lebesgue measure in [0, 1]. Let

$$g_n = \exp 2\pi\, inx, \; n = 0, \pm 1, \pm 2, \ldots \tag{42.2}$$

Then

$$(g_m, g_n) = \int_0^1 \exp 2\pi i\,(m-n)\,x\,dx = \delta_{mn}, \text{ for all } m, n.$$

Thus the (bilateral) sequence $\{g_n\}$ is orthonormal. Soon we shall establish that this is a complete orthonormal sequence.

To any Lebesgue integrable function ϕ on [0, 1], let

$$a_n = \int_0^1 \phi(x)\, \overline{g_n(x)}\, dx, \; n = 0, \pm 1, \pm 2, \ldots$$

$\{a_n\}$ is called the sequence of *Fourier coefficients* of ϕ. The function

$$s_N(\phi, x) = \sum_{-N}^{+N} a_n e^{inx}$$

is called the N th *partial sum* of the *Fourier series*

$$\sum_{-\infty}^{+\infty} a_n e^{inx}.$$

It is a classical theorem of Fejer that

$$\lim_{n\to\infty} \sup_x \left| \frac{s_1(\phi, x) + s_2(\phi, x) + \ldots + s_n(\phi, x)}{n} - \phi(x) \right| = 0$$

$$\text{if } \phi \in C^t[0, 1] \tag{42.3}$$

where $C^t[0, 1]$ is the space of all continuous functions ϕ on [0, 1] satisfying the condition $\phi(0) = \phi(1)$. (Here the suffix t has been put to indicate the torus!) For a proof of this theorem the reader may refer to [23].

Suppose $f \in L_2([0, 1])$ and $(f, g_n) = 0$ for all n. Then Fejer's theorem implies that $(f, \phi) = 0$ for every $\phi \in C^t([0, 1])$. Since such functions are dense in $L_2([0, 1])$ (see Exercise 39.10) it follows that $(f, f) = 0$. Hence $f = 0$. This shows that $\{g_n\}$ is a complete orthonormal sequence.

Remark. 42.8. Since $\{g_n\}$, $n = 0, \pm 1, \pm 2, \ldots$ defined by Eq. (42.2) is a complete orthonormal sequence it follows from Parseval's identities that for any $f \in L_2([0, 1])$ the Fourier series $\sum_{n=-\infty}^{+\infty} a_n g_n$ where $a_n = (f, g_n)$ for all n, converges in $L_2([0, 1])$ to f. If $\{a_n\}$, $\{b_n\}$ are the Fourier coefficients of two functions f_1, f_2 in $L_2([0, 1])$ then

$$(f_1, f_2) = \int_0^1 f_1 \bar{f}_2 \, dx = \sum_{-\infty}^{+\infty} a_n \bar{b}_n \tag{42.4}$$

where the right hand side converges absolutely.

As an illustration of the power of the Parseval identity (42.4), we shall give a solution of the *isoperimetric problem.* Let $C = \{(x(t), y(t)), t \in [0, 1]\}$ be a closed curve in the plane R^2 so that $x(0) = x(1)$, $y(0) = y(1)$. Let us suppose that $\frac{dx}{dt}$ and $\frac{dy}{dt}$ belong to $L_2([0, 1])$.

Consider a particle travelling from $(x(0), y(0))$ back to the same point in unit time along the curve C with uniform speed. Then this speed is equal to the perimeter of C, which is l, say. We can express this by

$$\left(\frac{dx}{dt}\right)^2 + \left(\frac{dy}{dt}\right)^2 = l^2. \tag{42.5}$$

The area enclosed by the curve C is given by the classical formula

$$\int y \, dx = \int_0^1 y \frac{dx}{dt} dt = F, \text{ say.}$$

Equation (42.5) implies that

$$\int_0^1 \left[\left(\frac{dx}{dt}\right)^2 + \left(\frac{dy}{dt}\right)^2\right] dt = l^2.$$

Now we ask the following question: if the perimeter l is fixed and the curve C is varied which curve maximises the area enclosed, namely F? In other words we wish to find

$$\sup \left\{ \int_0^1 y \frac{dx}{dt} : x(t) \text{ and } y(t) \text{ vary over functions on } [0, 1] \right.$$

$$\text{such that } x(0) = x(1), y(0) = y(1) \text{ and}$$

$$\left. \int_0^1 \left[\left(\frac{dx}{dt}\right)^2 + \left(\frac{dy}{dt}\right)^2\right] dt = l^2 \right\},$$

and see where it is attained.

Let $x(t)$ and $y(t)$ have Fourier series:

$$x(t) = \sum_{-\infty}^{+\infty} a_n e^{2\pi int},$$

$$y(t) = \sum_{-\infty}^{+\infty} b_n e^{2\pi int}.$$

A simple integration by parts gives the Fourier coefficients of $\frac{dx}{dt}$ and $\frac{dy}{dt}$. Indeed,

$$\frac{dx}{dt} = 2\pi i \sum_{-\infty}^{+\infty} na_n e^{2\pi int},$$

$$\frac{dy}{dt} = 2\pi i \sum_{-\infty}^{+\infty} nb_n e^{2\pi int}$$

Now Parseval identity gives

$$F = \left(y, \frac{dx}{dt}\right) = -2\pi i \sum_{-\infty}^{+\infty} nb_n \bar{a}_n$$

$$l^2 = \left(\frac{dx}{dt}, \frac{dx}{dt}\right) + \left(\frac{dy}{dt}, \frac{dy}{dt}\right) = 4\pi^2 \sum_{-\infty}^{+\infty} n^2(\,|\,a_n\,|^2 + |\,b_n\,|^2).$$

Let

$$a_n = \alpha_n + i\beta_n,\ b_n = \gamma_n + i\delta_n,$$

where $\alpha_n, \beta_n, \gamma_n, \delta_n$ are real. Since $x(t), y(t)$ are real valued functions,

$$a_{-n} = \bar{a}_n,\ b_{-n} = \bar{b}_n;$$

$$\frac{l_2}{4\pi^2} = 2 \sum_{n=1}^{+\infty} n^2\left(\alpha_n^2 + \beta_n^2 + \gamma_n^2 + \delta_n^2\right);$$

$$\frac{F}{2\pi} = 2 \sum_{n=1}^{\infty} n\,(\alpha_n\,\delta_n - \beta_n\,\gamma_n).$$

Thus

$$l^2 - 4\pi F = 8\pi^2 \sum_{n=1}^{\infty} \left\{ (n\alpha_n - \delta_n)^2 + (n\beta_n + \gamma_n)^2 + (n^2 - 1)\left(\gamma_n^2 + \delta_n^2\right)\right\}$$

$$\geqslant 0.$$

Further $l^2 = 4\pi F$ if and only if

$$n\alpha_n - \delta_n = 0,\ n\beta_n + \gamma_n = 0 \text{ for all } n \geqslant 1,$$
$$\gamma_n = \delta_n = 0 \text{ for all } n \geqslant 2.$$

This implies that $a_n = b_n = 0$ for all $|n| \geqslant 2$. If $n = 1$, then

$$\alpha_1 = \delta_1,\ \beta_1 = -\gamma_1$$
$$a_1 = \alpha_1 + i\beta_1,\ b_1 = -\beta_1 + i\alpha_1.$$

Hence

$$x(t) = a_0 + a_1 e^{2\pi it} + \bar{a}_1 e^{-2\pi it},$$
$$y(t) = b_0 + b_1 e^{2\pi it} + \bar{b}_1 e^{-2\pi it},$$

for all $t \in [0, 1]$. Thus $(x(t) - a_0)^2 + (y(t) - b_0)^2 = 4\left(\alpha_1^2 + \beta_1^2\right)$ for all t.

In other words $l^2 = 4\pi F$ if and only if the curve C is a circle. Thus the closed curve which maximises the area enclosed for a given perimeter l is a circle of radius $l/2\pi$.

Example. 42.9. Let μ be a probability measure on R with the property that all the moments

$$m_n = \int x^n \, d\mu,\ n = 0, 1, 2, \ldots$$

are finite. Then the functions $\{x^n\}$, $n = 0, 1, 2\ldots$ belong to $L_2(\mu)$. By proposition 42.2 we can construct an orthonormal sequence of polynomials $\{p_n\}$, where p_n is of degree n for every n. (The sequence terminates at the n-th stage if and only if there exists a finite set A of $n+1$ points such that $\mu(A) = 1$.) Such a sequence of polynomials is uniquely determined if we put the condition that the leading coefficient of p_n (i.e., the coefficient of x^n) is strictly positive for each n. p_n is called the *orthogonal polynomial* of degree n.

Now the following question naturally arises: when is the sequence $\{p_n\}$ of orthogonal polynomials a complete orthonormal sequence in $L_2(\mu)$? There are distributions for which the sequence $\{p_n\}$ is not complete.

We shall soon prove the completeness of $\{p_n\}$ in the case when there exists an $\alpha \neq 0$ such that $\int e^{\alpha |x|} \, d\mu < \infty$.

Exercise. 42.10. Let $(X_i, \mathscr{B}_i, \mu_i)$, $i=1, 2$ be two σ-finite measure spaces. Let $\{f_n\}$ and $\{g_n\}$, $n=1, 2, 3\ldots$ be complete orthonormal sequences of functions in $L_2(\mu_1)$ and $L_2(\mu_2)$ respectively. Let h_{mn} be defined on $X_1 \times X_2$ by

$$h_{mn}(x_1, x_2) = f_m(x_1)\, g_n(x_2),\ m=1, 2, \ldots;\ n=1, 2\ldots$$

Then the family $\{h_{mn}\}$ is a colmplete orthonormal sequence (when enumerated suitably) in $L_2(\mu_1 \times \mu_2)$.

Exercise. 42.11. Let $(X_i, \mathscr{B}_i, \mu_i)$, $i=1, 2,\ldots$ be a sequence of probability spaces. Let $\mu=\mu_1\times\mu_2\times\ldots$ be the infinite product measure on $(X_1, X_2, \times\ldots, \mathscr{B}_1\times\mathscr{B}_2\times\ldots)$ whose finite dimensional projections are $\mu_1\times\mu_2\times\ldots\times\mu_n$ (for varying n). Let $\{f_{in}\}$, $n=0, 1, 2, \ldots$ be a complete orthonormal sequence of functions in $L_2(\mu_i)$ such that $f_{io}\equiv 1$ for every i. For any finite sequence $(n_1, n_2, \ldots, n_k)$ of non-negative integers let $g_{n_1, n_2, \ldots, n_k}$ be the function on the infinite product space $X_1\times X_2\times\ldots$, defined by

$$g_{n_1 n_2 \ldots n_k}(x_1, x_2,\ldots) = f_{1n_1}(x_1) f_{2n_2}(x_2)\ldots f_{kn_k}(x_k).$$

Then the family $\{g_{n_1 n_2 \ldots n_k}\}$ is a complete orthonormal sequence in $L_2(\mu)$.

Exercise. 42.12. Let $X_i, \mathscr{B}_i)$, $i=1, 2$ be borel spaces and let μ be a σ-finite measure on $(X_1, \mathscr{B}_1)$. Let $T: X_1\to X_2$ be a borel isomorphism such that μT^{-1} is also a σ-finite measure. A sequence $\{f_n\}$ in $L_2(\mu)$ is a complete orthonormal sequence if and only if $\{f_n\circ T^{-1}\}$ is a complete orthonormal sequence in $L_2(\mu T^{-1})$.

Exercise. 42.13. For every binary number of the form $i2^{-n}$, where i is a positive odd integer $< 2^n$, let

$$\begin{aligned} h_{i2^{-n}}(x) &= 2^{(n-1)/2} \text{ if } \frac{i-1}{2^n} \leqslant x < \frac{i}{2^n} \\ &= -2^{(n-1)/2} \text{ if } \frac{i}{2^n} \leqslant x < \frac{i+1}{2^n} \\ &= 0 \text{ otherwise,} \end{aligned}$$

for all $x \in [0, 1]$. Let $h_0(x)=1$ for all x. Then the family $\{h_0; h_{i2^{-n}}, 0<i<2^n, i \text{ odd integer}, n=1, 2, 3,\ldots\}$ is a complete orthonormal sequence in $L_2[0, 1]$. (The functions $h_{i2^{-n}}$ are known as Haar functions.)

§43. Completeness of Orthogonal Polynomials

In view of the importance of orthogonal polynomials in non-linear regression problems of prediction in statistics we shall take the trouble of proving the completeness of the sequence of orthogonal polynomials in $L_2(\mu)$ when μ is a distribution on R satisfying a regularity condition. This condition is fulfilled by most of the distributions which occur commonly in statistics.

Proposition. 43.1. Let μ be a probability measure on R such that for some $a>0$.

$$\int e^{a|x|} \, d\mu(x) < \infty. \tag{43.1}$$

Then all the moments of μ exist and the sequence of orthogonal polynomials is complete in $L_2(\mu)$.

Proof. Since

$$\frac{a^n}{n!}|x|^n \leqslant e^{a|x|},$$

it follows that all the moments

$$\int x^n \, d\mu, \; n=0, 1,2,\ldots$$

exist. Let $\{p_n\}$ be the sequence of orthogonal polynomials from the sequence $\{x^n\}$, $n=0, 1, 2, \ldots$. Let $f \in L_2(\mu)$ be such that

$$\int f p_n \, d\mu = 0, \text{ for all } n = 0, 1, 2,\ldots .$$

Since the subspace spanned by $p_0, p_1,\ldots,p_k$ in $L_2(\mu)$ is also the subspace spanned by $1, x, x^2,\ldots,x^k$ for every k, we have

$$\int f x^n \, d\mu = 0 \text{ for all } n = 0, 1,\ldots. \tag{43.2}$$

Since $f \in L_2(\mu)$ and (43.1) implies that the function $e^{\frac{a}{2}|x|} \in L_2(\mu)$ we have from Schwarz inequality

$$\int |f| \, e^{\frac{a}{2}|x|} \, d\mu(x) < \infty. \tag{43.3}$$

Without loss of generality we shall consider the real Hilbert space $L_2(\mu)$. In the complex plane consider the domain

$$D = \left\{ z : -\frac{a}{2} < \operatorname{Re} z < \frac{a}{2} \right\}.$$

Since $|e^{zx}| \leqslant e^{\frac{a}{2}|x|}$ for all $z \in D$, it follows from (43.3) that

$$\int |f| \, |e^{zx}| \, d\mu(x) < \infty \text{ for all } z \in D. \tag{43.4}$$

Let f^+, f^- be the positive and negative parts of f and let

$$\mu_1(E) = \int_E f^+ \, d\mu,$$

$$\mu_2(E) = \int_E f^- \, d\mu$$

for all borel sets $E \subset R$. By (43.3), μ_1 and μ_2 are totally finite measures on R. Let

$$\phi_i(z) = \int e^{zx} \, d\mu_i(x), \; i = 1, 2; \; z \in D.$$

Since

$$\left| \sum_{r=0}^{n} \frac{t^r x^r}{r!} \right| \leqslant e^{\frac{a}{2}|x|}$$

for all t in $\left(-\frac{a}{2}, \frac{a}{2}\right)$, we have from Lebesgue's dominated convergence theorem, Eqs. (43.2) and (43.3)

$$\int e^{tx} f(x)\, d\mu = \lim_{n\to\infty} \sum_{0}^{n} \frac{t^r}{r!} \int x^r f(x)\, d\mu(x) = 0.$$

In other words

$$\phi_1(t) = \phi_2(t) \text{ for } -\frac{a}{2} < t < \frac{a}{2}. \tag{43.5}$$

Now we shall prove that $\phi_i(z)$ is analytic in the domain D. First of all we observe that inequality (43.3) implies that

$$\int |f|\, |x|\, e^{\theta |x|}\, d\mu(x) < \infty \text{ for } 0 \leqslant \theta < \frac{a}{2}. \tag{43.6}$$

If Re $z = \xi$, we have

$$\left| \frac{e^{(z+h)x} - e^{zx}}{h} \right| \leqslant |x|\, e^{(|\xi| + |h|)\,|x|}.$$

If $z \in D$, it follows that for all sufficiently small $|h|$, $|Re z| + |h| < \frac{a}{2}$.

Hence by inequality (43.6) and Lebesgue's dominated convergence theorem,

$$\lim_{h\to 0} \frac{\phi_i(z+h) - \phi_i(z)}{h} = \int x\, e^{zx}\, d\mu_i(x),\ z \in D$$

exists. This proves the analyticity of ϕ_i. Now Eq. (43.5) implies that $\phi_1(z) = \phi_2(z)$ for all $z \in D$. In particular,

$$\int e^{itx}\, d\mu_1(x) = \int e^{itx}\, d\mu_2(x) \text{ for all } t \in R.$$

It follows from Proposition 53.9 (on the one to one correspondence between Fourier transforms and measures) that $\mu_1 = \mu_2$. In other words

$$\int_E f\, d\mu = 0 \text{ for all borel sets } E \subset R.$$

Thus $f = 0$ a.e. (μ). By Exercise 42.6, the sequence $\{p_n\}$ is a complete orthonormal sequence. This completes the proof.

Remark. 43.2. Before proceeding to the construction of specific examples of orthogonal polynomials we shall illustrate how orthogonal polynomials can be put to use in prediction problems. Suppose f, g are two real valued random variables with distribution P in the plane R^2 and $\mathbf{E}g^2 < \infty$ and $\mathbf{E}\,|f|^n < \infty$ for all $n = 0, 1, 2, \ldots$. Let $\mu = Pf^{-1}$ be the distribution of the random variable f on the real line, and let $\{p_n\}$, $n = 0, 1, 2, \ldots$ be the sequence of orthogonal polynomials in $L_2(\mu)$.

We now predict g on the basis of $\{1, f, f^2, \ldots, f^n\}$ by projecting g on the span of $\{1, f, f^2, \ldots, f^n\}$ in $L_2(P)$. The span of $\{1, f, f^2, \ldots, f^n\}$ is also the span of $p_0(f)=1, p_1(f), \ldots p_n(f)$. Hence the linear regression of g on $1, f, f^2, \ldots, f^n$ is simply $\sum_{j=0}^{n} [\mathbf{E} g p_j(f)]\, p_j(f)$. This is known as the n-th degree *polynomial regression* of g on f.

Example. 43.3. Let μ be the standard normal distribution in R defined by

$$\mu(E) = \frac{1}{\sqrt{2\pi}} \int_E e^{-\frac{1}{2}x^2}\, dx, \quad E \subset R.$$

Let $\{H_n(x)\}$, $n=0, 1, 2, \ldots$ be the sequence of polynomials defined by the identity

$$\rho_{\mathcal{N}}(t, x) = e^{-\frac{1}{2}(t^2 - 2tx)} = \sum_{n=0}^{\infty} \frac{t^n}{n!} H_n(x), \tag{43.7}$$

where t is a real parameter. A simple calculation shows that

$$\int \rho_{\mathcal{N}}(t, x)\, \rho_{\mathcal{N}}(s, x)\, d\mu(x) = e^{ts} \text{ for all } t, s \in R.$$

Identifying the coefficients of $t^k s^l$ on both sides for different pairs of positive integers k, l we obtain

$$\int \frac{H_k(x)}{\sqrt{k!}} \frac{H_l(x)}{\sqrt{l!}}\, d\mu(x) = \delta_{kl}, \quad k=0, 1, 2, \ldots \; l=0, 1, 2, \ldots$$

Since H_n is a polynomial of degree n it follows from Proposition 43.1 that the sequence $\left\{\frac{H_n(x)}{\sqrt{n!}}\right\}$, $n=0, 1, 2, \ldots$ is the complete orthonormal sequence of orthogonal polynomials for the standard normal distribution. H_n is known as the n-th degree *Hermite polynomial*.

Equation (43.7) implies that

$$e^{-\frac{1}{2}(x-t)^2} = \sum_{n=0}^{\infty} \frac{t^n}{n!} H_n(x)\, e^{-\frac{1}{2}x^2}.$$

Expanding the left hand side by Taylor's theorem we get

$$e^{-\frac{1}{2}(x-t)^2} = \sum_{n=0}^{\infty} (-1)^n \frac{t^n}{n!} \frac{d^n}{dx^n} (e^{-\frac{1}{2}x^2}).$$

Identifying the coefficients of t^n in both the expansions we get

$$H_n(x) = (-1)^n e^{\frac{1}{2}x^2} \frac{d^n}{dx^n} (e^{-\frac{1}{2}x^2}). \tag{43.8}$$

As an application of this orthonormal sequence we shall prove an interesting property of the normal distribution.

Proposition. 43.4. Let f, g be random variables with a bivariate normal distribution with mean vector zero and covariance matrix $\begin{pmatrix} 1 & \rho \\ \rho & 1 \end{pmatrix}$. Let $\phi(f)$ and $\psi(g)$ be two random variables which are functions of f alone and g alone respectively.
Suppose

$$\begin{aligned} \mathbf{E}\,\phi(f) &= \mathbf{E}\,\psi(g) = 0, \\ \mathbf{E}\,\phi(f)^2 &= \mathbf{E}\,\psi(g)^2 = 1. \end{aligned} \tag{43.9}$$

Then

$$|\,\mathbf{cov}\,(\phi(f), \psi(g))\,| \leqslant |\rho|. \tag{43.10}$$

Equality is attained if and only if $\phi(f) = \pm f$ and $\psi(g) = \pm g$.

Proof. Consider the functions $\phi(x)$ and $\psi(x)$ on the real line. From the conditions stated it follows that ϕ and ψ are square integrable with respect to the standard normal distribution on R. Hence by Example 43.3 we can expand ϕ and ψ in the orthonormal basis of Hermite polynomials. i.e.,

$$\phi(x) = \sum_0^\infty c_n\, p_n\,(x),$$

$$\psi(x) = \sum_0^\infty d_n\, p_n\,(x),$$

where

$$p_n(x) = \frac{H_n(x)}{\sqrt{n!}},\; n = 0, 1, 2, \ldots$$

Then Eq. (43.9) implies

$$\sum_0^\infty c_n^2 = 1,\; \sum_0^\infty d_n^2 = 1,\; c_0 = d_0 = 0. \tag{43.11}$$

Further

$$\mathbf{cov}\,(\phi\,(f), \psi\,(g)) = \sum_{k,\,l=1}^{\infty} \frac{c_k\, d_l}{2\pi\sqrt{1-\rho^2}} \int p_k\,(x)\; p_l\,(y) \times$$

$$\exp\left\{-\frac{1}{2(1-\rho^2)}\,(x^2+y^2-2\rho\,xy)\right\}\,dx\,dy.$$

Suppose $k \neq l$. More precisely, let $k < l$. Then the k, l-th term of the above double series is equal to

$$\frac{c_k\, d_l}{2\pi} \int p_k\,(\xi\sqrt{1-\rho^2} + \rho\eta)\; p_l\,(\eta)\; e^{-\frac{1}{2}(\xi^2+\eta^2)} d\xi\, d\eta.$$

By Fubini's theorem we can first integrate with respect to η and then with respect to ξ. Since every polynomial of degree k in η can be rewritten as a linear combination of p_j, $1 \leqslant j \leqslant k$, it follows that the above integral is 0.

If $k=l$, the same integral becomes

$$\frac{c_k d_k}{2\pi} \int p_k (\xi \sqrt{1-\rho^2} + \rho\eta) \, p_k (\eta) \, e^{-\frac{1}{2}(\xi^2+\eta^2)} \, d\xi \, d\eta.$$

But

$$p_k (\xi \sqrt{1-\rho^2} + \rho\eta) = \rho^k p_k (\eta) + q (\xi, \eta),$$

where $q(\xi, \eta)$ is a polynomial in η (for fixed ξ) of degree less than or equal to $k-1$. Hence the above integral is $c_k \, d_k \, \rho^k$. Thus

$$\mathbf{cov} \, (\phi (f), \psi (g)) = \rho \, (\sum_1^\infty c_k \, d_k \, \rho^{k-1}).$$

This together with Eq. (43.11) implies inequality (43.10). Equality is attained if and only if $c_k = d_k = 0$ for all $k > 1$ and $c_1 = \pm 1$, $d_1 = \pm 1$. Since $p_1 (x) = x$ the proof is complete.

Exercise. 43.5. Let f, g be real valued random variables where f has normal distribution with mean m and variance σ^2. Then the n-th degree polynomial regression of g on f is equal to

$$\sum_{j=0}^{n} \frac{1}{j!} \left\{ \mathbf{E} \left[g \, H_j \left(\frac{f-m}{\sigma} \right) \right] \right\} H_j \left(\frac{f-m}{\sigma} \right),$$ where H_j is the jth degree Hermite polynomial.

Example. 43.6. Let μ be the gamma distribution in $(0, \infty)$ defined by

$$\mu (E) = \frac{1}{\Gamma(a)} \int_E x^{a-1} e^{-x} \, dx,$$

where $a > 0$. Let $\{L_n(x)\}$ be a sequence of polynomials for $n=0$, 1 2, ... defined by the identity

$$\rho_G (t, x) = \frac{1}{(1-t)^a} e^{-\frac{tx}{1-t}}$$

$$= \sum_{n=0}^{\infty} \frac{t^n}{n!} L_n (x), \qquad (43.12)$$

where t is a parameter in $(-1, 1)$. A fairly simple calculation shows that

$$\int \rho_G(t, x)\rho_G(s, x)\, d\mu(x) = \frac{1}{(1-ts)^a}$$

$$= \sum_0^\infty \frac{a(a+1)\dots(a+n-1)}{n!} t^n s^n. \tag{43.13}$$

As before identification of coefficients of $t^k s^l$ on both sides shows that

$$\int p_k(x)\, p_l(x) d\mu(x) = \delta_{kl},\ k=0, 1, 2, \dots;\ l=0, 1, 2, \dots$$

where

$$p_k(x) = \left\{\frac{1}{k!\, a(a+1)\dots(a+k-1)}\right\}^{\frac{1}{2}} L_k(x),\ k=0, 1, 2, \dots \tag{43.14}$$

It is not too difficult to show that

$$p_k(x) = c_k \sum_{r=0}^{k} \binom{k}{r} (-1)^r \frac{x^r}{\Gamma(r+a)},$$

where c_k is a normalising constant. The polynomial L_n is known as the nth degree *Laguerre Polynomial*.

Since $\int e^{tx}\, d\mu < \infty$ for all $t \in (-1, 1)$ it follows from Proposition 43.1 that the sequence $\{p_k\}$ defined by Eq. (43.14) is a complete orthonormal sequence in $L_2(\mu)$.

Exercise. 43.7. Let μ be the Poisson distribution with parameter λ so that

$$\mu(\{x\}) = e^{-\lambda} \frac{\lambda^x}{x!},\ x = 0, 1, 2, \dots$$

Let

$$\rho_P(t, x) = e^{-t}(1 + \lambda^{-1} t)^x \tag{43.15}$$

where t is a real parameter. Then

$$\int \rho_P(t, x)\, \rho_P(s, x)\, d\mu(x) = \exp\left\{\frac{ts}{\lambda}\right\} \tag{43.16}$$

$$\rho_P(t, x) = \sum_0^\infty \frac{t^n}{n!} P_n(x), \tag{43.17}$$

$$p_n(x) = \sum_0^n \binom{n}{r} (-1)^{n-r} \lambda^{-r} x(x-1)\dots(x-r+1). \tag{43.18}$$

If

$$p_n(x) = \frac{\lambda^{n/2}}{\sqrt{n\,!}} P_n(x),\ n = 0, 1, 2 \ldots$$

then $\{p_n\}$ is a complete orthonormal sequence in $L_2(\mu)$. (P_n is known as the n-th degree Poisson Charlier polynomial).

Exercise. 43.8. Let μ be the binomial distribution defined on the set of integers 0, 1, 2, ..., n by

$$\mu(\{x\}) = \binom{n}{x} p^x q^{n-x},\ x = 0, 1, 2, \ldots, n.$$

Let

$$\rho_B(t, x) = (1+tq)^x (1-tp)^{n-x}. \tag{43.19}$$

Then

$$\int \rho_B(t, x)\, \rho_B(s, x)\, d\mu(x) = (1+pqts)^n. \tag{43.20}$$

Further

$$\rho_B(t, x) = \sum_{k=0}^{\infty} \frac{t^k}{k\,!} \left\{ \sum_{j=0}^{k} (-1)^j \binom{k}{j} p^j q^{k-j} x(x-1) \right.$$
$$\left. \ldots(x-k+j-1)\,(n-x)\,(n-x-1)\ldots(n-x-j+1) \right\}$$
$$= \sum_{0}^{\infty} \frac{t^k}{k\,!} K_k(x). \tag{43.21}$$

The sequence $\{p_k\}$ defined by

$$p_k(x) = \frac{(pq)^{-k/2} \binom{n}{k}^{-\frac{1}{2}}}{k\,!} K_k(x),\ k = 0, 1, \ldots, n$$

is an orthonormal basis in $L_2(\mu)$. (The polynomial K_k is known as the *Krawtchouk Polynomial* of degree k.)

Exercise. 43.9. Let ρ_N, ρ_P, ρ_B be defined by Eqs. (43.7), (43.15) and (43.19) respectively. Then

(i) $\lim\limits_{n\to\infty} \rho_B\left(\dfrac{t}{\sqrt{npq}}, np + x\sqrt{npq}\right) = \rho_N(t, x)$;

(ii) $\lim\limits_{\substack{n\to\infty \\ np\to\lambda}} \rho_B(t\lambda^{-1}, x) = \rho_P(t, x)$.

(Property (i) may be compared with the limit theorem of Laplace-DeMoivre in Proposition 6.1. Property (ii) may be compared with Proposition 5.1.)

Remark. 43.10. In the examples and exercises above we have constructed the so called *generating function* $\rho(t, x)$ with two properties:

(i) $\rho(t, x) = \sum_{n=0}^{\infty} \frac{t^n}{n!} q_n(x)$;

(ii) $\int \rho(t, x) \rho(s, x) \, d\mu(x) = \phi(ts)$,

where q_n is a polynomial of degree n and ϕ is a function of a real parameter. Then the q_n's become a complete orthonormal sequence after a suitable normalisation. However, given an arbitrary probability distribution μ on R there is no systematic procedure for directly constructing such a generating function $\rho(t, x)$. The interested reader may look into the book of G. Szego [21] for a wealth of material on this subject.

§44. Conditional Expectation

Let $(X, \mathscr{B}, \mu)$ be a probability space and let f be a real valued random variable which is observed in an experiment. Suppose we wish to predict another real valued random variable g on the basis of f. In the preceding section, in Remark 43.2, we saw how g could be predicted as a polynomial in f by using the theory of orthogonal polynomials. However, while adopting such a procedure we are not making full use of the observed variable f.

For any borel set $E \subset R$ we can say whether or not the event $f^{-1}(E)$ has occurred as soon as we have observed f. All events of the form $f^{-1}(E)$ as E varies in the borel σ-algebra $\mathscr{B}_R$ constitute a sub σ-algebra $\mathscr{B}_0 \subset \mathscr{B}$. We also write $f^{-1}(\mathscr{B}_R)$ for $\mathscr{B}_0$. We shall say that a borel function on $(X, \mathscr{B}_0)$ is a $\mathscr{B}_0$ *measurable* function on X. We note that if ϕ is any borel function on the real line then $\phi(f)$ is measurable with respect to the σ-algebra $\mathscr{B}_0$. We shall now prove the converse.

Proposition. 44.1. Let ψ be any real valued random variable on $(X, \mathscr{B})$ which is $\mathscr{B}_0$ measurable, where $\mathscr{B}_0 = f^{-1}(\mathscr{B}_R)$ and f is a real valued random variable on $(X, \mathscr{B})$. Then there exists a borel function ϕ on the real line such that $\psi = \phi \circ f$.

Proof. Let ψ be a simple random variable which is B_0 measurable. Then ψ is of the form

$$\psi(x) = \sum_{i=1}^{k} a_i \chi_{f^{-1}(E_i)}(x).$$

where k is a positive integer, a_i's are real and E_i's are borel subsets of R. Hence

$$\psi(x) = \left(\sum_{i=1}^{k} a_i \chi_{E_i}\right)(f(x)).$$

This shows that the proposition holds when ψ is simple. If ψ is a general random variable which is $\mathscr{B}_0$ measurable construct a sequence ψ_n of simple $\mathscr{B}_0$ measurable random variables such that $\psi_n(x) \to \psi(x)$ for every x. Let $\psi_n(x) = s_n(f(x))$, where s_n are simple borel functions on the real line. Let

$$\begin{aligned} \phi(t) &= \overline{\lim_{n\to\infty}}\, s_n(t) \text{ if } \overline{\lim_{n\to\infty}}\, s_n(t) < \infty, \\ &= 0 \text{ otherwise.} \end{aligned}$$

Then ϕ is a borel function on R and $\psi(x) = \phi(f(x))$ for all x. This completes the proof.

The above proposition shows that in order to predict a random variable g in $L_2(\mu)$ on the basis of an observed random variable f we can project g on the subspace of all random variables which are functions of f or, equivalently, we can project on the subspace of $\mathscr{B}_0$ measurable random variables in $L_2(\mu)$.

Now we can forget about f and straightaway consider the subspace of all square integrable functions which are measurable with respect to any sub σ-algebra $\mathscr{B}_0 \subset \mathscr{B}$. If S_0 is this subspace we write for any $g \in L_2(\mu)$,

$$\mathbf{E}(g \mid \mathscr{B}_0) = P^{S_0} g, \tag{44.1}$$

where P^{S_0} is the orthogonal projection onto S_0. We call the expression (44.1) the *conditional expectation of g given the sub σ-algebra $\mathscr{B}_0$*. We can use this to predict g on the basis of the subspace S_0.

We shall now deduce a few elementary properties of conditional expectation from the properties of orthogonal projection. To this end we shall consider a fixed probability space $(X, \mathscr{B}, \mu)$ and the real Hilbert space $L_2(\mu)$.

Proposition. 44.2. Let $\mathscr{B}_0 \subset \mathscr{B}$ be any sub σ-algebra and let $g \in L_2(\mu)$. Then

$$\int_B g\, d\mu = \int_B \mathbf{E}(g \mid \mathscr{B}_0) d\mu \text{ for any } B \in \mathscr{B}_0. \tag{44.2}$$

Proof. If χ_B is the characteristic function of B then it is $\mathscr{B}_0$ measurable. Let S_0 be the subspace of $\mathscr{B}_0$ measurable functions in $L_2(\mu)$. Then

$$\int_B g\, d\mu = (g, \chi_B) = (g, P^{S_0} \chi_B) = (P^{S_0} g, \chi_B) = \int_B \mathbf{E}(g \mid \mathscr{B}_0)\, d\mu.$$

This completes the proof.

Proposition. 44.3. Let $\mathscr{B}_0$, g be as in the preceding proposition. If h is a $\mathscr{B}_0$ measurable function such that

$$\int_B h \, d\mu = \int_B g \, d\mu \text{ for all } B \in \mathscr{B}_0,$$

then $h = \mathbf{E}(g \mid \mathscr{B}_0)$.

Proof. From Proposition 44.2 we have

$$\int_B [h - E(g \mid \mathscr{B}_0)] \, d\mu = 0, \text{ for all } B \in \mathscr{B}_0.$$

Since the integrand is a random variable on $(X, \mathscr{B}_0, \mu)$ the required result follows.

Remark. 44.4. So far we have assumed that $g \in L_2(\mu)$. The left handside of Eq. (44.2) is meaningful even when $g \in L_1(\mu)$. Since μ is a probability measure $L_1(\mu) \supset L_2(\mu)$. It is only natural to ask the question whether we can define a function $\mathbf{E}(g \mid \mathscr{B}_0)$ satisfying Eq. (44.2). We shall now prove that this is indeed so.

Proposition. 44.5. Let $(X, \mathscr{B}, \mu)$ be a probability space and let $\mathscr{B}_0 \subset \mathscr{B}$ be a sub σ-algebra. For any $g \in L_1(\mu)$ there exists a unique $\mathscr{B}_0$ measurable function $h = \mathbf{E}(g \mid \mathscr{B}_0)$ such that $h \in L_1(\mu)$ and

$$\int_B g \, d\mu = \int_B h \, d\mu \text{ for all } B \in \mathscr{B}_0. \tag{44.3}$$

Such a h is unique.

Proof. For any positive integer n, let

$$\begin{aligned} g_n(x) &= g(x) \text{ if } |g(x)| \leqslant n \\ &= 0 \text{ otherwise.} \end{aligned}$$

Then by Lebesgue's dominated convergence theorem we have

$$\lim_{n \to \infty} \int |g_n(x) - g(x)| \, d\mu(x) = 0. \tag{44.4}$$

Since g_n is bounded, $g_n \in L_2(\mu)$. We define

$$h_n = \mathbf{E}(g_n \mid \mathscr{B}_0)$$

by Eq. (44.1). By Proposition 44.2 we have

$$\int_B g_n \, d\mu = \int_B h_n \, d\mu \text{ for all } B \in \mathscr{B}_0, \; n = 1, 2, \ldots. \tag{44.5}$$

Then

$$\sup_{B \in \mathscr{B}_0} \left| \int_B h_m \, d\mu - \int_B h_n \, d\mu \right| \leqslant \int |g_m - g_n| \, d\mu.$$

If $B_1 = \{x : (h_m(x) - h_n(x)) \geqslant 0\}$, $B_2 = \{x : (h_m(x) - h_n(x)) \leqslant 0\}$, then $B_1, B_2 \in \mathscr{B}_0$ and the above inequality implies

$$\int_{B_i} |h_m - h_n| \, d\mu \leqslant \int |g_m - g_n| \, d\mu, \; i = 1, 2.$$

Adding over $i=1, 2$ we have

$$\int |h_m - h_n| \, d\mu \leqslant 2 \int |g_m - g_n| \, d\mu.$$

From Eq. (44.4) and Riesz-Fischer theorem we now conclude that h_n converges to a limit h in $L_1(\mu)$. Now Eqs. (44.4) and (44.5) imply (44.3). Since h is a limit of h_n, it follows that h is $\mathscr{B}_0$ measurable. The uniqueness of h follows exactly as in the proof of Proposition 44.3.

Remark. 44.6. The unique $\mathscr{B}_0$ measurable function h which satisfies Eq. (44.3) is called the *conditional expectation* of g given $\mathscr{B}_0$ and denoted by $\mathbf{E}(g|\mathscr{B}_0)$. When $\mathscr{B}_0$ is the whole σ-algebra $\mathscr{B}$, $\mathbf{E}(g|\mathscr{B})=g$. When $\mathscr{B}_0$ is the trivial σ-algebra consisting of the empty set $\emptyset$ and the whole space X, $\mathbf{E}(g \mid \mathscr{B}_0)=\mathbf{E}g$, a constant.

If f is a real valued random variable on X and $\mathscr{B}_0=f^{-1}(\mathscr{B}_R)$ then by Proposition 44.1, there exists a borel function ϕ on the real line such that $\mathbf{E}(g \mid \mathscr{B}_0)=\phi(f)$. The function $\phi(t)$, $t \in R$ is defined a.e. μf^{-1} uniquely. It is called the *regression* of g on f. One writes

$$\phi(t)=\mathbf{E}(g|f=t).$$

The random variable $\phi(f)$ may be used to predict g on the basis of f.

Proposition. 44.7. Conditional expectation operation satisfies the following properties:

(i) $\mathbf{E}(ag_1+bg_2 \mid \mathscr{B}_0)=a\mathbf{E}(g_1 \mid \mathscr{B}_0)+b\mathbf{E}(g_2 \mid \mathscr{B}_0)$ for all $g_1, g_2 \in L_1(\mu)$ and constants a, b;

(ii) if $g(x) \geqslant 0$ a.e. (μ) and $g \in L_1(\mu)$, then $\mathbf{E}(g \mid \mathscr{B}_0) \geqslant 0$ a.e. (μ);

(iii) if $g \in L_1(\mu)$, h is $\mathscr{B}_0$ measurable and $gh \in L_1(\mu)$ then $\mathbf{E}(gh \mid \mathscr{B}_0)=h\,\mathbf{E}(g|\mathscr{B}_0)$;

(iv) $\mathbf{E}(\mathbf{E}(g|\mathscr{B}_0))=\mathbf{E}g$ for all $g \in L_1(\mu)$;

(v) if $\mathscr{B}_1 \subset \mathscr{B}_2 \subset \mathscr{B}$ are sub σ-algebras then $\mathbf{E}(\mathbf{E}(g \mid \mathscr{B}_2) \mid \mathscr{B}_1) = \mathbf{E}(g \mid \mathscr{B}_1)$ for all $g \in L_1(\mu)$.

Proof. Property (i) follows immediately from the linearity of integrals and the uniqueness of conditional expectation. To prove (ii) we note that for $B \in \mathscr{B}_0$,

$$\int_B \mathbf{E}(g \mid \mathscr{B}_0) \, d\mu = \int_B g \, d\mu \geqslant 0 \text{ if } g \geqslant 0 \text{ a.e.}$$

Since $\mathbf{E}(g \mid \mathscr{B}_0)$ is $\mathscr{B}_0$ measurable it follows that $\mathbf{E}(g|\mathscr{B}_0) \geqslant 0$ a.e. To prove (iii) we note that $|g|$ and $|g|\,|h| \in L_1(\mu)$ and hence it is enough to prove the result when both g and h are non-negative functions. For any two $B, C \in \mathscr{B}_0$, we have

$$\int_B g \chi_C \, d\mu = \int_{BC} g \, d\mu = \int_{BC} \mathbf{E}(g \mid \mathscr{B}_0) \, d\mu$$
$$= \int_B \chi_C \mathbf{E}(g \mid \mathscr{B}_0) \, d\mu.$$

Hence for any non-negative $\mathscr{B}_0$ measurable simple function s

$$\int_B g s \, d\mu = \int_B s \mathbf{E}(g \mid \mathscr{B}_0) \, d\mu \text{ for all } B \in \mathscr{B}_0.$$

If h is any non-negative $\mathscr{B}_0$ measurable function we can approximate it by an increasing sequence of $\mathscr{B}_0$ measurable simple functions s_n. An application of monotone convergence theorem implies that

$$\int_B g h \, d\mu = \int_B h \mathbf{E}(g \mid \mathscr{B}_0) \, d\mu \text{ for all } B \in \mathscr{B}_0.$$

Since the integrand on the right hand side is $\mathscr{B}_0$ measurable and conditional expectation is uniquely defined (iii) is proved. Property (iv) is obtained by putting $B = X$ in Eq. (44.3). To prove (v) we note that for all $B \in \mathscr{B}_1$,

$$\int_B g \, d\mu = \int_B \mathbf{E}(g \mid \mathscr{B}_2) \, d\mu$$
$$= \int_B \mathbf{E}(\mathbf{E}(g \mid \mathscr{B}_2) \mid \mathscr{B}_1) \, d\mu.$$

Since the integrand in the last integral is $\mathscr{B}_1$ measurable uniqueness of conditional expectation implies (v). This completes the proof.

Corollary. 44.8. Let $(X, \mathscr{B}, \mu)$ be a probability space and let $\mathscr{B}_0 \subset \mathscr{B}$ be a sub σ-algebra. Then the map

$$g \to \mathbf{E}(g \mid \mathscr{B}_0), \; g \in L_1(\mu)$$

is an operator P from the B-space $L_1(\mu)$ onto the B-space $L_1(X, \mathscr{B}_0, \mu)$ such that (a) $\|P\| = 1$; (b) $P^2 = P$; (c) if $g \geqslant 0$, $g \in L_1(\mu)$ then $Pg \geqslant 0$, where $L_1(X, \mathscr{B}_0, \mu)$ is the subspace of all $\mathscr{B}_0$ measurable functions in $L_1(\mu)$.

Proof. If $g \in L_1(\mu)$, then $|g| + g \geqslant 0$, $|g| - g \geqslant 0$ and

$$-\mathbf{E}(|g| \mid \mathscr{B}_0) \leqslant \mathbf{E}(g \mid \mathscr{B}_0) \leqslant \mathbf{E}(|g| \mid \mathscr{B}_0).$$

Hence

$$|\mathbf{E}(g \mid \mathscr{B}_0)| \leqslant \mathbf{E}(|g| \mid \mathscr{B}_0) \text{ for all } g \in L_1(\mu). \tag{44.6}$$

Integrating both sides and using property (iv) of the proposition we get

$$\|\mathbf{E}(g \mid \mathscr{B}_0)\|_1 \leqslant \|g\|_1.$$

Property (i) of Proposition 44.7 implies that P is an operator of norm $\leqslant 1$. Putting g equal to the constant function 1 we have $P1=1$. Hence $\|P\|=1$. Putting $\mathscr{B}_1=\mathscr{B}_2=\mathscr{B}_0$ in property (v) of the proposition we have $P^2=P$. Now (c) is a restatement of property (ii) of the proposition. P is onto because it is the identity operator on the subspace $L_1(X, \mathscr{B}_0, \mu)$. The proof is complete.

Corollary. 44.9. If $g_n \to g$ in $L_1(\mu)$ as $n \to \infty$, then $\mathbf{E}(g_n \mid \mathscr{B}_0) \to \mathbf{E}(g \mid \mathscr{B}_0)$ in $L_1(\mu)$ as $n \to \infty$.

Proof. This is a restatement of the preceding corollary because conditional expectation is an operator on $L_1(\mu)$.

Proposition. 44.10. (Jensen's inequality for conditional expectation.) Let $(X, \mathscr{B}, \mu)$ be a probability space and let $\mathscr{B}_0 \subset \mathscr{B}$ be a sub σ-algebra. Let U be an open convex set in R^n and let $\mathbf{g}: X \to U$ be a vector valued random variable which is represented as a column vector of random variables g_i in $L_1(\mu)$. Suppose

$$\mathbf{E}(\mathbf{g} \mid \mathscr{B}_0) = \begin{pmatrix} \mathbf{E}(g_1 \mid \mathscr{B}_0) \\ \mathbf{E}(g_2 \mid \mathscr{B}_0) \\ \cdot \\ \cdot \\ \cdot \\ \mathbf{E}(g_n \mid \mathscr{B}_0) \end{pmatrix} \in U \text{ a.e. } (\mu).$$

If ϕ is a real valued twice differentiable function on U such that the matrix $\left(\left(\dfrac{\partial^2 \phi}{\partial t_i\, \partial t_j}\right)\right)$, $1 \leqslant i, j \leqslant n$ is positive semi-definite and continuous at every $\mathbf{t} \in U$ and $\phi(\mathbf{g}) \in L_1(\mu)$ then

$$\mathbf{E}(\phi(\mathbf{g}) \mid \mathscr{B}_0) \geqslant \phi(\mathbf{E}(\mathbf{g} \mid \mathscr{B}_0)) \text{ a.e. } (\mu). \tag{44.7}$$

Proof. We denote the points of R^n by column vectors and indicate transpose of a matrix or a vector by a prime "'".

Let

$$\nabla\phi = \begin{bmatrix} \dfrac{\partial\phi}{\partial t_1} \\ \dfrac{\partial\phi}{\partial t_2} \\ \cdot \\ \cdot \\ \cdot \\ \dfrac{\partial\phi}{\partial t_n} \end{bmatrix}, \quad \mathcal{J}(\mathbf{t}) = \left(\left(\frac{\partial^2 \phi}{\partial t_i\, \partial t_j}\right)\right), \quad 1 \leqslant i, j \leqslant n.$$

For any $\mathbf{t}, \boldsymbol{\xi} \in U$ we have by Taylor's theorem

$$\phi(\mathbf{t}) = \phi(\boldsymbol{\xi}) + (\mathbf{t}-\boldsymbol{\xi})' (\nabla\phi)(\boldsymbol{\xi}) + \tfrac{1}{2}(\mathbf{t}-\boldsymbol{\xi})' \mathcal{J}(\boldsymbol{\xi}+\theta(\mathbf{t}-\boldsymbol{\xi}))(\mathbf{t}-\boldsymbol{\xi}),$$

where θ is a scalar in (0, 1) depending on $\mathbf{t}$ and $\boldsymbol{\xi}$. The positive semi-definiteness of $\mathcal{J}$ implies that

$$\phi(\mathbf{t}) \geqslant \phi(\boldsymbol{\xi}) + (\mathbf{t}-\boldsymbol{\xi})' (\nabla\phi)(\boldsymbol{\xi}).$$

Putting $\mathbf{t}=\mathbf{g}$ and $\boldsymbol{\xi}=\mathbf{E}(\mathbf{g} \mid \mathscr{B}_0)$ we have

$$\phi(\mathbf{g}) \geqslant \phi(\mathbf{E}(\mathbf{g} \mid \mathscr{B}_0)) + (\mathbf{g}-\mathbf{E}(\mathbf{g} \mid \mathscr{B}_0))' (\nabla\phi)(\mathbf{E}(\mathbf{g} \mid \mathscr{B}_0)) \text{ a.e. } (\mu). \tag{44.8}$$

Let

$$E_k = \left\{ x : |\phi(\mathbf{E}(\mathbf{g} \mid \mathscr{B}_0))(x)| \leqslant k, \left|\frac{\partial\phi}{\partial t_i}(\mathbf{E}(\mathbf{g} \mid \mathscr{B}_0))(x)\right| \leqslant k \right.$$
$$\left. \text{for all } i=1, 2, \ldots, n \right\}.$$

Then $E_k \in \mathscr{B}_0$. Multiply both sides of inequality (44.8) by χ_{E_k} and take conditional expectation given $\mathscr{B}_0$. (Such a multiplication is done in order to ensure that the random variables for which we take conditional expectation are in $L_1(\mu)$.) Then we have from Proposition 44.7

$$\chi_{E_k} \mathbf{E}(\phi(\mathbf{g}) \mid \mathscr{B}_0) \geqslant \chi_{E_k} \phi(\mathbf{E}(\mathbf{g} \mid \mathscr{B}_0)) \text{ a.e. } (\mu).$$

Letting $k \to \infty$ we get inequality (44.7). This completes the proof.

We shall now prove another version of Jensen's inequality for conditional expectation when ϕ is a 'smooth' convex function on an interval of the real line.

Proposition. 44.11. Let $(X, \mathscr{B}, \mu)$ be a probability space and let $\mathscr{B}_0 \subset \mathscr{B}$ be a sub σ-algebra. Let g be a real valued random variable on X, taking values in an open interval U. If ϕ is a real continuous function on U such that ϕ' is continuous, ϕ'' exists and remains non-negative in U and $g \in L_1(\mu)$, $\phi(g) \in L_1(\mu)$, then $\mathbf{E}(g \mid \mathscr{B}_0) \in U$ a.e. (μ) and

$$\mathbf{E}(\phi(g) \mid \mathscr{B}_0) \geqslant \phi(\mathbf{E}(g \mid \mathscr{B}_0)) \text{ a.e. } (\mu). \tag{44.9}$$

Proof. Suppose $g > c$ a.e. (μ), where c is a constant. Then for any $B \in \mathscr{B}_0$ such that $\mu(B) > 0$, we have

$$\int_B \mathbf{E}(g \mid \mathscr{B}_0)\, d\mu = \int_B g\, d\mu > c\mu(B).$$

Thus

$$\int_B [\mathbf{E}(g \mid \mathscr{B}_0) - c]\, d\mu > 0,$$

for all B such that $\mu(B) > 0$ and $B \in \mathscr{B}_0$. Hence

$$\mathbf{E}(g \mid \mathscr{B}_0) > c \text{ a.e. } (\mu).$$

Similarly if $g<c$ a.e. μ, then $\mathbf{E}(g \mid \mathscr{B}_0)<c$ a.e. (μ). Thus $\mathbf{E}(g \mid \mathscr{B}_0) \in U$ a.e. (μ) whenever $g \in U$ a.e. (μ).

Now we use the special form of Taylor's theorem as in [10 (p. 286)]. Then for any $t,\ \xi \in U$, we obtain

$$\phi(t) = \phi(\xi) + (t-\xi)\ \phi'(\xi) + \tfrac{1}{2}(t-\xi)^2\ \phi''(\xi+\theta\ (t-\xi)),$$

where $0<\theta<1$. Hence

$$\phi(t) \geqslant \phi(\xi)+(t-\xi)\ \phi'(\xi).$$

Putting $t=g$ and $\xi=\mathbf{E}(g \mid \mathscr{B}_0)$ and proceeding exactly as in the proof of the preceding proposition we get inequality (44.9). This completes the proof.

Proposition. 44.12. Let $(X, \mathscr{B}, \mu)$ be a probability space and let $\mathscr{B}_0 \subset \mathscr{B}$ be a sub σ-algebra. If $g \in L_p(\mu)$ for some $p>1$, then for any $1 \leqslant p_1 < p$,

$$\{\mathbf{E}(\,|g|^{p_1} \mid \mathscr{B}_0)\}^{1/p_1} \leqslant \{\mathbf{E}(\,|g|^{p} \mid \mathscr{B}_0)\}^{1/p} \text{ a.e. } (\mu).$$

In particular, $\mathbf{E}(g \mid \mathscr{B}_0) \in L_p(\mu)$ and

$$\| \mathbf{E}(g \mid \mathscr{B}_0) \|_p \leqslant \| g \|_p \text{ for all } g \in L_p(\mu).$$

Thus conditional expectation is an operator of norm unity from the B-space $L_p(\mu)$ onto the subspace of $\mathscr{B}_0$ measurable functions in $L_p(\mu)$. If $g_n \rightarrow g$ in $L_p(\mu)$ as $n \rightarrow \infty$ then $\mathbf{E}(g_n \mid \mathscr{B}_0) \rightarrow \mathbf{E}(g \mid \mathscr{B}_0)$ in $L_p(\mu)$. As $n \rightarrow \infty$.

Proof. Let $g > 0$ a.e. (μ) and let $g \in L_p(\mu)$. Let $h=g^{p_1}$ and let $\phi(t)=t^{p/p_1}$ for $t>0$. Then

$$\phi''(t) = \frac{p}{p_1}\left(\frac{p}{p_1}-1\right) t^{\frac{p}{p_1}-2} > 0 \text{ for all } t > 0.$$

By Proposition 34.13, $h \in L_1(\mu)$. Further $\phi(h)=g^p \in L_1(\mu)$. Hence by Proposition 44.11,

$$\mathbf{E}(g^p \mid \mathscr{B}_0) \geqslant [\mathbf{E}(g^{p_1} \mid \mathscr{B}_0)]^{p/p_1}. \tag{44.10}$$

Raising both sides to the $1/p$ th power we get the required inequality. If $g \geqslant 0$ a.e. (μ), then for any $\varepsilon>0$, inequality (44.10) holds with g replaced by $g+\varepsilon$. Letting $\varepsilon \rightarrow 0$, we get the required result. This completes the proof.

Exercise. 44.13. (Conditional Holder's inequality). Let $0< \alpha \leqslant 1$, $0< \beta \leqslant 1$, $\alpha + \beta \leqslant 1$. If $f, g \in L_1(\mu)$ then

$$\mathbf{E}(\,|f|^\alpha\, |g|^\beta \mid \mathscr{B}_0) \leqslant [\mathbf{E}(\,|f| \| \mathscr{B}_0)]^\alpha\ [\mathbf{E}(\,|g| \| \mathscr{B}_0)]^\beta.$$

Putting $\alpha=1/p$, $\beta=1/q$, $|f|=\phi^p$, $|g|=\psi^q$, where $\phi \in L_p(\mu)$, $\psi \in L_q(\mu)$, we have

$$\mathbf{E}(\phi\psi \mid \mathscr{B}_0) \leqslant [\mathbf{E}(\,|\phi|^p \mid \mathscr{B}_0)]^{1/p}\ [\mathbf{E}(\,|\psi|^q \mid \mathscr{B}_0)]^{1/q}.$$

(Hint: The function $\phi(x, y) = -x^{\alpha} y^{\beta}$ satisfies the conditions of Proposition 44.10 in the set $U = \{(x, y) : x > 0, y > 0\}$, whenever $0 < \alpha \leq 1$, $0 < \beta \leq 1$, $\alpha + \beta < 1$).

Remark. 44.14. So far we have analysed the properties of $\mathbf{E}(g | \mathscr{B}_0)$ when the sub σ-algebra $\mathscr{B}_0$ is fixed and the random variable g varies. We shall now analyse the properties of conditional expectation when the sub σ-algebra $\mathscr{B}_0$ varies. To this end we fix our notations. If $\{\mathscr{B}_\alpha, \alpha \in T\}$ is a family of σ-algebras we write $\bigvee_\alpha \mathscr{B}_\alpha$ for the smallest α-algebra containing $\bigcup_\alpha \mathscr{B}_\alpha$. We write $\bigwedge_\alpha \mathscr{B}_\alpha$ for the σ-algebra $\bigcap_\alpha \mathscr{B}_\alpha$.

Proposition. 44.15. (Doob's inequality) Let $\mathscr{B}_1 \subset \mathscr{B}_2 \subset \dots$ be an increasing sequence of sub σ-algebras of $\mathscr{B}$ in the probability space $(X, \mathscr{B}, \mu)$. Let $g \in L_1(\mu)$ and let $g_i = \mathbf{E}(g \mid \mathscr{B}_i)$, $i = 1, 2, \dots$ Then for any $\varepsilon > 0$.

$$\mu\{x : \sup_{1 \leq i \leq n} |g_i(x)| > \varepsilon\} \leq \frac{\int |g| \, d\mu}{\varepsilon}. \tag{44.11}$$

Proof. Without loss of generality we may assume that g is a non-negative random variable. Let

$$E = \{x : \sup_{1 \leq i \leq n} g_i(x) > \varepsilon\},$$

$$E_j = \{x : g_1(x) \leq \varepsilon, g_2(x) \leq \varepsilon, \dots, g_{j-1}(x) \leq \varepsilon, g_j(x) > \varepsilon\},$$

for $j = 1, 2, \dots, n$. Then E_j are disjoint and $E = \bigcup_{j=1}^{n} E_j$. Since $E_j \in \mathscr{B}_j$ for all for all $j = 1, 2, \dots, n$ we have

$$\int g \, d\mu \geq \int_E g \, d\mu = \sum_j \int_{E_j} g \, d\mu = \sum_j \int_{E_j} g_j \, d\mu \geq \varepsilon \sum_j \mu(E_j) = \varepsilon \, \mu(E).$$

This is same as inequality (44.11) and hence the proof is complete.

Proposition. 44.16. Let $(X, \mathscr{B}, \mu)$ be a probability space and let $\mathscr{B}_1 \subset \mathscr{B}_2 \subset \dots$ be an increasing sequence of sub σ-algebras of $\mathscr{B}$. Let $\mathscr{B}_\infty = \bigvee_n \mathscr{B}_n$. For any $g \in L_1(\mu)$.

$$\lim_{n \to \infty} \mathbf{E}(g \mid \mathscr{B}_n) = \mathbf{E}(g \mid \mathscr{B}_\infty) \text{ a.e. } (\mu).$$

Further

$$\lim_{n \to \infty} \mathbf{E}(g \mid \mathscr{B}_n) = \mathbf{E}(g \mid \mathscr{B}_\infty) \text{ in } L_1(\mu).$$

Proof. By property (v) of Proposition 44.7,

$$\mathbf{E}(g \mid \mathscr{B}_n) = \mathbf{E}(\mathbf{E}(g \mid \mathscr{B}_\infty) \mid \mathscr{B}_n) \text{ for all } n.$$

Hence we may assume without loss of generality that $\mathscr{B}_\infty$ is the entire σ-algebra $\mathscr{B}$ and prove that $\mathbf{E}(g \mid \mathscr{B}_n)$ converges to g a.e. (μ). Let S_n be the subspace of $\mathscr{B}_n$ measurable functions in $L_1(\mu)$. Then $S_1 \subset S_2 \subset \ldots$ and $\bigcup_n S_n$ is dense in $L_1(\mu)$. Let $S = \bigvee_n S_n$. Let $\varepsilon > 0$ be arbitrary. Then there exists an $h \in S$ such that $\mathbf{E}|g-h| < \varepsilon^2$. We have

$$\begin{aligned}\tilde{g} &= \overline{\lim_{n\to\infty}} \,|\mathbf{E}(g \mid \mathscr{B}_n) - g| \\ &\leqslant \overline{\lim_{n\to\infty}} \,|\mathbf{E}(g-h \mid \mathscr{B}_n)| + \overline{\lim_{n\to\infty}} \,|\mathbf{E}(h \mid \mathscr{B}_n) - h| + |g-h|.\end{aligned}$$

Since $h \in S_n$ for all large n, $\mathbf{E}(h \mid \mathscr{B}_n) = h$ for all large n. Thus the second term on the right hand side of the above inequality is zero. By Doob's inequality

$$\begin{aligned}\mu\{x : \overline{\lim_{n\to\infty}} \,|\mathbf{E}(g-h \mid \mathscr{B}_n)| > \varepsilon\} &\leqslant \mu\{x : \sup_n |\mathbf{E}(g-h \mid \mathscr{B}_n)| > \varepsilon\} \\ &\leqslant \frac{\mathbf{E}|g-h|}{\varepsilon} < \varepsilon.\end{aligned}$$

By Chebyshev's inequality

$$\mu\{x : |g-h| > \varepsilon\} \leqslant \frac{\mathbf{E}|g-h|}{\varepsilon} < \varepsilon.$$

Thus

$$\begin{aligned}\mu\{x : \tilde{g}(x) > 2\varepsilon\} \leqslant \mu\{x : &\overline{\lim_{n\to\infty}} \,\mathbf{E}(g-h \mid \mathscr{B}_n) > \varepsilon\} + \\ &\mu\{x : |g-h| > \varepsilon\} \leqslant 2\varepsilon.\end{aligned}$$

Since ε is arbitrary, we have $\tilde{g}(x) = 0$ a.e. $x(\mu)$. This completes the proof of the first part. To prove the second part we observe that

$$\begin{aligned}\mathbf{E}|\mathbf{E}(g \mid \mathscr{B}_n) - g| &\leqslant 2\,\mathbf{E}|g-h| + \mathbf{E}|\mathbf{E}(h \mid \mathscr{B}_n) - h| \\ &\leqslant 2\varepsilon^2 \text{ for all large } n.\end{aligned}$$

This completes the proof.

Exercise. 44.17. Let $\mathscr{B}_1 \supset \mathscr{B}_2 \supset \ldots$ be a decreasing sequence of sub σ-algebras of $\mathscr{B}$ and let $\mathscr{B}^\infty = \bigwedge_n \mathscr{B}_n$. For any $g \in L_1(\mu)$,

$$\lim_{n\to\infty} \mathbf{E}(g \mid \mathscr{B}_n) = \mathbf{E}(g \mid \mathscr{B}^\infty) \text{ a.e. } (\mu),$$

$$\lim_{n\to\infty} \mathbf{E}(g \mid \mathscr{B}_n) = \mathbf{E}(g \mid \mathscr{B}^\infty) \text{ in } L_1(\mu).$$

(Hint: Use Doob's inequality for every finite increasing sequence $\mathscr{B}_n \subset \mathscr{B}_{n-1} \subset \cdots \subset \mathscr{B}_1$.)

We shall now analyse the convergence of $\mathbf{E}(g \mid \mathscr{B}_n)$ in $L_p(\mu)$ for $p > 1$. To this end we need an elementary lemma.

Proposition. 44.18. Let f be a non-negative random variable on the probability space $(X, \mathscr{B}, \mu)$. Let $p > 1$ and

$$\int_0^\infty t^{p-1} \mu\{x : f(x) > t\} \, dt < \infty. \tag{44.12}$$

Then $f \in L_p(\mu)$.

Proof. For any positive integer n, we have

$$\begin{aligned} &\int_{n-1}^{n} t^{p-1} \mu\{x : f(x) > t\} \, dt \\ &\geqslant (n-1)^{p-1} \mu\{x : f(x) > n\} \\ &\geqslant (n-1)^{p-1} \mu\{x : [f(x)] > n\}, \end{aligned} \tag{44.13}$$

where $[f(x)]$ denotes the integral part of $f(x)$. Let

$$p_n = \mu\{x : [f(x)] = n\}.$$

From inequalities (44.12) and (44.13) we have

$$\begin{aligned} \infty &> \sum_{n=1}^{\infty} \int_{n-1}^{n} t^{p-1} \mu\{x : f(x) > t\} \, dt \\ &\geqslant \sum_{n=1}^{\infty} (n-1)^{p-1} \mu\{x : [f(x)] > n\} \\ &= \sum_{n=1}^{\infty} (n-1)^{p-1} (p_{n+1} + p_{n+2} + \ldots) \\ &= \sum_{n=2}^{\infty} p_{n+1} (1^{p-1} + 2^{p-1} + \ldots + (n-1)^{p-1}) \\ &\geqslant \sum_{n=2}^{\infty} p_{n+1} \sum_{k=1}^{n-1} \int_{k-1}^{k} x^{p-1} \, dx \\ &= \sum_{n=2}^{\infty} p_{n+1} \frac{(n-1)^p}{p}. \end{aligned}$$

Hence $\sum_{n=0}^{\infty} n^p p_n < \infty$, or equivalently, $[f(x)] \in L_p(\mu)$. Since fractional part of f is bounded it follows that $f \in L_p(\mu)$. This completes the proof.

Proposition. 44.19. (Dominated L_p convergence theorem for conditional expectation). Let $(X, \mathscr{B}, \mu)$ be a probability space and let $\mathscr{B}_1 \subset \mathscr{B}_2 \subset \ldots$ be an increasing sequence of sub σ-algebras of $\mathscr{B}$. Suppose that $\mathscr{B}_\infty = \bigvee_n \mathscr{B}_n$, $g \in L_p(\mu)$ where $p > 1$ and

$$g_n = \mathbf{E}(g \mid \mathscr{B}_n), g^* = \sup_{n \geqslant 1} |g_n|.$$

Then $g^* \in L_p(\mu)$ and g_n converges to $\mathbf{E}(g \mid \mathscr{B}_\infty)$ in $L_p(\mu)$ as $n \to \infty$.

Proof. To prove the proposition we may assume without loss of generality that g is a non-negative random variable. Let h be the random variable defined by

$$\begin{aligned} h(x) &= g(x) \text{ if } g(x) > \frac{t}{2}, \\ &= 0 \text{ otherwise.} \end{aligned}$$

Then

$$g(x) \leqslant h(x) + \tfrac{1}{2} t \text{ for all } x.$$

If $h_n = \mathbf{E}(h \mid \mathscr{B}_n)$ and $h^* = \sup_n \mathbf{E}(h \mid \mathscr{B}_n)$ then $g_n \leqslant h_n + \frac{1}{2} t$ for all n and hence $g^* \leqslant h^* + \frac{t}{2}$. Now Doob's inequality implies

$$\mu\{x : g^*(x) > t\} \leqslant \mu\left\{x : h^*(x) > \frac{t}{2}\right\}$$

$$\leqslant \frac{2}{t} \int h \, d\mu = \frac{2}{t} \int\limits_{\left\{x : g(x) > \frac{t}{2}\right\}} g \, d\mu.$$

Hence by Fubini's theorem we have

$$\int_0^\infty t^{p-1} \mu\{x : g^*(x) > t\} \, dt$$

$$\leqslant \int_0^\infty t^{p-1} \left[\frac{2}{t} \int\limits_{\{x : g(x) > \frac{1}{2} t\}} g \, d\mu \right] dt$$

$$= 2 \int_0^\infty t^{p-2} \left[\int_R \chi_{(\frac{1}{2} t, \infty)}(\xi) \, \xi \, \mu g^{-1}(d\xi) \right] dt$$

$$= 2 \int_R \int_0^\infty t^{p-2} \chi_{[0, 2\xi)}(t) \, \xi \, dt \, \mu g^{-1}(d\xi)$$

$$= 2 \int_R \xi \int_0^{2\xi} t^{p-2} \, dt \, \mu g^{-1}(d\xi)$$

$$= \frac{2^p}{p-1} \int \xi^p \, d\mu g^{-1}(\xi) = \frac{2^p}{p-1} \int g^p \, d\mu < \infty.$$

By Proposition 44.18, $g^* \in L_p(\mu)$. Since g_n converges to $\mathbf{E}(g \mid \mathscr{B}_\infty)$ a.e. (μ) and g_n is dominated by the L_p-integrable function g^*, it follows from the Lebesgue's dominated convergence theorem that g_n converges to g in the space $L_p(\mu)$. This completes the proof.

Exercise. 44.20. Proposition 44.19 holds when $\mathscr{B}_1 \supset \mathscr{B}_2 \supset \ldots$, and $\mathscr{B}_\infty$ is replaced by $\mathscr{B}^\infty = \bigwedge_n \mathscr{B}_n$.

Exercise. 44.21. Let $\mathscr{B}_1$, $\mathscr{B}_2$ be two sub σ-algebras of $\mathscr{B}$ in the probability space $(X, \mathscr{B}, \mu)$. Let $\mathbf{E}_1$, $\mathbf{E}_2$ be the conditional expectation operators given $\mathscr{B}_1$, $\mathscr{B}_2$ respectively in the B-space $L_1(\mu)$. Then $(\mathbf{E}_1 \mathbf{E}_2)^n$ converges strongly to the conditional expectation operator given $\mathscr{B}_1 \wedge \mathscr{B}_2$. (Hint: Use Exercise 41.20 and the fact that $L_2(\mu)$ is dense in $L_1(\mu)$.)

§45. Conditional Probability

Let $(X, \mathscr{B}, \mu)$ be a fixed probability space and let $\mathscr{B}_0 \subset \mathscr{B}$ be a fixed sub σ-algebra. For any, $A \in \mathscr{B}$, let

$$P(x, A) = \mathbf{E}(\chi_A \mid \mathscr{B}_0)(x).$$

Then $P(x, A)$ is a function on $X \times \mathscr{B}$ with the following properties:

(i) $0 \leqslant P(x, A) \leqslant 1$;

(ii) $P(x, X) = 1$;

(iii) for fixed A, $P(x, A)$ is a $\mathscr{B}_0$ measurable function of x;

(iv) for $B \in \mathscr{B}_0$, $A \in \mathscr{B}$,

$$\int_B P(x, A)\, d\mu(x) = \mu(A \cap B);$$

(v) if $A_1, A_2, \ldots$ is a sequence of disjoint sets belonging to $\mathscr{B}$, then there exists a set $\mathscr{N} \in \mathscr{B}_0$ such that $\mu(\mathscr{N}) = 0$ and

$$\sum_{n=1}^{\infty} P(x, A_i) = P(x, \bigcup_i A_i) \text{ if } x \notin \mathscr{N}.$$

Indeed, the first four properties follow immediately from the definition of conditional expectation and Proposition 44.7. To see (v) we note that $\Sigma \chi_{A_i}$ converges in $L_1(\mu)$ to $\chi_{\bigcup_i A_i}$ and apply Corollary 44.9.

The function $P(x, A)$ is called a *version of the conditional probability of A given* $\mathscr{B}_0$. Sometimes it is denoted by $P(A|\mathscr{B}_0)$. Now we ask the following natural question: does there exist a transition probability (see Definition 35.4) $\tilde{P}(x, A)$ such that the first four properties mentioned above are fulfilled? We shall first examine this question in the real line.

Proposition. 45.1. Let μ be a probability measure on $(R, \mathscr{B}_R)$ and let $\mathscr{B}_0 \subset \mathscr{B}_R$ be a sub σ-algebra. Then there exists a transition probability $\tilde{P}(x, A)$ on $R \times \mathscr{B}_R$ such that

(a) $\tilde{P}(x, A)$ is $\mathscr{B}_0$ measurable in x for fixed A;

(b) $\int_B \tilde{P}(x, A)\, d\mu(x) = \mu(A \cap B)$ for all $A \in \mathscr{B}$, $B \in \mathscr{B}_0$.

Proof. Let $P(x, A)$ be a version of the conditional probability of A given $\mathscr{B}_0$. For any rational number r, let

$$F(x, r) = P(x, (-\infty, r]).$$

From the basic properties of conditional expectation we have

(i) $F(x, r) \leqslant F(x, s)$ a.e. $x(\mu)$ if $r \leqslant s$ and r, s are rational;

(ii) $\lim_{r \to -\infty} F(x, r) = 0$ a.e. $x(\mu)$;

(iii) $\lim_{r \to +\infty} F(x, r) = 1$ a.e. $x(\mu)$.

The countability of the set of all rationals implies the existence of a set $\mathcal{N} \in \mathscr{B}_0$ such that $\mu(\mathcal{N}) = 0$ and for *all* $x \notin \mathcal{N}$, properties (i), (ii) and (iii) hold good. For $t \in R$, let

$$\begin{aligned} \tilde{F}(x, t) &= \lim_{r \to t+0} F(x, r) \text{ for } x \notin \mathcal{N}, \\ &= F(t) \quad \text{for } x \in \mathcal{N}, \end{aligned}$$

where $F(t)$ is a fixed probability distribution function in t. Then $\tilde{F}(x, t)$ is a probability distribution function in t for each fixed x and a $\mathscr{B}_0$ measurable function in x for each fixed t. Indeed, the right continuity of $\tilde{F}(x, t)$ in t follows from the monotonicity of $F(x, r)$ in the rational variable r when $x \notin \mathcal{N}$. Since

$$\chi_{(-\infty, t]}(x) = \lim_{r \to t+0} \chi_{(-\infty, r]}(x)$$

in $L_1(\mu)$, it follows from Corollary 44.9 that for any t,

$$\tilde{F}(x, t) = \mathbf{E}(\chi_{(-\infty, t]} \mid \mathscr{B}_0)(x) \text{ a.e. } x(\mu).$$

Let $\tilde{P}(x, .)$ be the unique probability measure whose distribution function is $\tilde{F}(x, .)$ for each x. Let

$$\mathscr{L} = \Big\{A : \tilde{P}(x, A) \text{ is a } \mathscr{B}_0 \text{ measurable function of } x, \int_B \tilde{P}(x, A)\, d\mu(x) = \mu(A \cap B) \text{ for all } B \in \mathscr{B}_0\Big\}.$$

(See Remark 13.10). From the discussion above it follows that every interval of the form $(-\infty, t] \in \mathscr{L}$. Since $\tilde{P}(x, .)$ and μ are measures, finite disjoint unions of intervals of the form $(a, b]$ lie in $\mathscr{L}$. Further $\mathscr{L}$ is a monotone class. Hence $\mathscr{L} = \mathscr{B}$. Thus $\tilde{P}$ satisfies properties (a) and (b) of the proposition. This completes the proof.

Corollary. 45.2. If $\tilde{P}$ is the transition probability of the above proposition then for any $g \in L_1(\mu)$,

$$\mathbf{E}(g \mid \mathscr{B}_0)(x) = \int g(y) \tilde{P}(x, dy) \text{ a.e. } (\mu).$$

Proof. If A is a borel set and $g = \chi_A$, this is just a restatement of Proposition 45.1. Hence the same holds for simple functions. Since any non-negative borel function is a limit of an increasing sequence of simple functions an application of monotone convergence theorem yields the required result when g is non-negative. For an arbitrary g we decompose g as $g^+ - g^-$ and complete the proof.

Remark. 45.3. If μ is a probability measure on $(R^k, \mathscr{B}_{R^k})$ and $\mathscr{B}_0$ is a sub σ-algebra of $\mathscr{B}_{R^k}$, Proposition 45.1 can be proved in the same manner after taking note of Remark 19.7. Instead of rational numbers one has to use rational vectors.

Remark. 45.4. The transition probability $\tilde{P}(x, A)$ of Proposition 45.1 is called a *version of the conditional probability distribution given* $\mathscr{B}_0$.

§46. Regular Conditional Probability Distributions

Consider the probability space $(R, \mathscr{B}_R, \mu)$ and a real valued random variable f on this space. We shall now explain the concept of conditional probability distribution given that f takes the value ξ.

Proposition. 46.1. Let f be a real valued random variable on the probability space $(R, \mathscr{B}_R, \mu)$ where $\mathscr{B}_R$ is the borel σ-algebra of the real line R. Then there exists a function $p(\xi, A)$ on $R \times \mathscr{B}_R$ such that the following properties hold:

(i) $p(\xi, A)$ is a transition probability on $R \times \mathscr{B}_R$

(ii) $p(\xi, f^{-1}(\{\xi\})) = 1$ a.e. ξ (μf^{-1})

(iii) for any $g \in L_1(\mu)$,

$$\int \left[\int g(x)\, p(\xi, dx) \right] d\mu f^{-1}(\xi) = \int g(x)\, d\mu(x).$$

Proof. Let $\mathscr{B}_0 = f^{-1}(\mathscr{B}_R)$. Consider a version $\tilde{P}(x, A)$ of the conditional probability distribution given $\mathscr{B}_0$. Since $\tilde{P}(x, A)$ is $\mathscr{B}_0$

measurable in x for any fixed A, by Proposition 44.1, there exists a function $Q(\xi, A)$ such that

$$\tilde{P}(x, A) = Q(f(x), A) \text{ for all } x \in R, A \in \mathscr{B}_R. \tag{46.1}$$

By Corollary 24.23 there exists a borel set $B_0 \subset R$ such that $f(B_0)$ is also a borel set and

$$\mu(B_0) = 1, \ (\mu f^{-1}) \ (f(B_0)) = 1. \tag{46.2}$$

Define

$$\begin{aligned} p_1(\xi, A) &= Q(\xi, A) \text{ if } \xi \in f(B_0), \\ &= \lambda(A) \text{ if } \xi \notin f(B_0), \end{aligned} \tag{46.3}$$

where λ is any arbitrary but fixed probability measure on $\mathscr{B}_R$. From Eq. (46.1) it is clear that p_1 is a transition probability on $R \times \mathscr{B}_R$. If E, F are any two borel subsets of the real line we have from property (b) of Proposition 45.1, and Eqs, (46.1) to (46.3)

$$\begin{aligned} \mu f^{-1} (E \cap F) &= \mu (f^{-1} (E) \cap f^{-1} (F)) \\ &= \int_{f^{-1}(F)} \tilde{P}(x, f^{-1} (E)) \, d\mu(x) \\ &= \int_{f^{-1}(F)} p_1(f(x), f^{-1} (E)) \, d\mu(x) \\ &= \int_F p_1(\xi, f^{-1} (E)) \, d\mu f^{-1} (\xi). \end{aligned} \tag{46.4}$$

On the other hand

$$\mu f^{-1}(E \cap F) = \int_F \chi_E (\xi) \, d\mu f^{-1} (\xi). \tag{46.5}$$

Comparing Eqs. (46.4) and (46.5) for varying F, we have

$$p_1(\xi, f^{-1} (E)) = \chi_E (\xi) \text{ a.e. } \xi \ (\mu f^{-1})$$

for each $E \in \mathscr{B}_R$. Now choose the boolean algebra $\mathscr{F}$ generated by left open right closed intervals with rational end points. $\mathscr{F}$ is a countable family. Choose a borel set $N \subset R$ such that $\mu f^{-1}(N) = 0$ and

$$p_1(\xi, f^{-1} (E)) = \chi_E (\xi) \text{ for all } E \in \mathscr{F}, \ \xi \notin N.$$

Since both sides of the above equation are probability measures in the variable E it follows that

$$p_1(\xi, f^{-1} (E)) = \chi_E (\xi) \text{ for all } E \in \mathscr{B}_R, \ \xi \notin N.$$

In particular,

$$p_1 (\xi, f^{-1} (\{\xi\})) = 1 \text{ for all } \xi \notin N.$$

Now define

$$\begin{aligned} p(\xi, A) &= p_1(\xi, A) \text{ if } \xi \notin N, A \in N, \\ &= \lambda(A) \text{ if } \xi \in N, A \in \mathscr{B}. \end{aligned}$$

Then

$$p(\xi, f^{-1}(\{\xi\}))=1 \text{ for all } \xi \notin N$$

and $p(\xi, A)$ is a transition probability. Finally, for any $g \in L_1(\mu)$, we have by Corollary 45.2,

$$\begin{aligned}\int [\int g(x)\, p(\xi, dx)]\, d\mu f^{-1}(\xi) &= \int [\int g(x)\, p_1(\xi, dx)]\, d\mu f^{-1}(\xi)\\ &= \int [\int g(x)\, Q(\xi, dx)]\, d\mu\, f^{-1}(\xi)\\ &= \int [\int g(x)\, Q(f(y), dx)]\, d\mu(y)\\ &= \int [\int g(x) \widetilde{P}(y, dx)]\, d\mu(y)\\ &= \mathbf{E}(\mathbf{E}(g \mid \mathscr{B}_0)) = \mathbf{E}g.\end{aligned}$$

This completes the proof.

Corollary. 46.2. Let B_0 be a borel subset of the real line R and let μ be a probability measure on $\mathscr{B}_R \cap B_0$. If f is a borel map from B_0 into R then there exists a transition probability $p(\xi, A)$ on $R \times (\mathscr{B}_R \cap B_0)$ such that

(i) $p(\xi, f^{-1}(\{\xi\}))=1$ a.e. ξ (μf^{-1});

(ii) for any $g \in L_1(\mu)$,

$$\int [\int g(x)\, p(\xi, dx)]\, d\mu f^{-1}(\xi) = \int g(x) d\mu(x).$$

Proof. Define the measure μ_1 and function f_1 on R by

$$\begin{aligned}\mu_1(A) &= \mu(A \cap B_0),\ A \in \mathscr{B}_R,\\ f_1(x) &= f(x) \text{ if } x \in B_0,\\ &= a \text{ if } x \notin B_0,\end{aligned}$$

where a is any point in R such that $\mu f^{-1}(\{a\})=0$. Now apply Proposition 46.1 and construct a transition probability $p_1(\xi\ A)$ satisfying properties (i), (ii) and (iii) of the same proposition. Then

$$\int p_1(\xi, B_0)\, d\mu_1 f_1^{-1}(\xi) = \int \chi_{B_0}(x)\, d\mu_1(x) = \mu_1(B_0)=1.$$

Hence

$$p_1(\xi, B_0)=1 \text{ a.e. } \xi.$$

In other words there exists a borel set $N \subset R$ such that $\mu_1 f_1^{-1}(N)=0$ and

$$p_1(\xi, B_0)=1 \text{ for all } \xi \notin N.$$

Define

$$\begin{aligned}p(\xi, A) &= p_1(\xi, A) \text{ for } A \in \mathscr{B}_R \cap \mathscr{B}_0,\ \xi \notin N \cup \{a\}\\ &= \lambda(A) \text{ otherwise,}\end{aligned}$$

where λ is an arbitrary but fixed probability measure on $\mathscr{B}_R \cap B_0$. It is clear that $\mu_1 f_1^{-1} = \mu f^{-1}$. If $\xi \notin N \cup \{a\}$,

$$p(\xi, f^{-1}(\{\xi\}))=p_1(\xi, f_1^{-1}(\{\xi\}))=1 \text{ a.e. } \xi(\mu f^{-1}).$$

Thus p satisfies property (i). If $g \in L_1(\mu)$ then

$$\begin{aligned}&\int \Big[\int g(x)\, p\,(\xi, dx) \Big]\, d\mu f^{-1}(\xi)\\ &= \int \Big[\int g(x)\, p(\xi, dx) \Big]\, d\mu_1 f_1^{-1}(\xi)\\ &= \int \Big[\int_{B_0} g(x)\, p_1(\xi, dx) \Big]\, d\mu_1 f_1^{-1}\,(\xi) = \int g\, d\mu.\end{aligned}$$

This completes the proof.

Now we can prove Proposition 46.1 for all standard borel spaces.

Proposition. 46.3. Let $(X, \mathscr{B})$ and $(Y, \mathscr{C})$ be standard borel spaces and let $f : X \to Y$ be a borel map. If μ is a probability measure on $(X, \mathscr{B})$, then there exists a transition probability $p(y, A)$ on $Y \times \mathscr{B}$ such that

(i) $p(y, f^{-1}(\{y\})) = 1$ a.e. y (μf^{-1});

(ii) for any $g \in L_1(\mu)$,

$$\int \Big[\int g(x)\, p\,(y, dx) \Big]\, d\mu f^{-1}(y) = \int g(x)\, d\mu(x).$$

Proof. By the definition of standard borel space we may assume X and Y to be borel subsets of complete and separable metric spaces. Let $\nu = \mu f^{-1}$. By the isomorphism theorem (Proposition 26.6) and Remark 26.8 there exists a probability measure ν' on the borel σ-algebra $\mathscr{B}_R$ of R such that the probability spaces $(Y, \mathscr{C}, \nu)$ and $(R, \mathscr{B}_R, \nu')$ are isomorphic. Let the isomorphism be τ'. By definition 26.1 we can choose a set $Y_1 \subset Y$ such that

$$Y_1 \in \mathscr{C};\ \nu(Y_1) = 1;\ \tau'\,(Y_1) \in \mathscr{B}_R, \tag{46.6}$$

and τ' is a borel isomorphism between Y_1 and $\tau'\,(Y_1)$. Let $X_1 = f^{-1}(Y_1)$. Then $X_1 \in \mathscr{B}$ and $\mu(X_1) = 1$. By the same argument as above there exists a probability measure μ' on $\mathscr{B}_R$ such that the probability spaces $(X_1, \mathscr{B} \cap X_1, \mu)$ and $(R, \mathscr{B}_R, \mu')$ are isomorphic. Let the isomorphism be τ. Choose $X_2 \subset X_1$ such that

$$X_2 \in \mathscr{B};\ \mu(X_2) = 1;\ \tau(X_2) \in \mathscr{B}_R \tag{46.7}$$

and τ is a borel isomorphism between X_2 and $\tau(X_2)$. By Corollary 24.23 choose $X_3 \subset X_2$ such that

$$X_3 \in \mathscr{B};\ f\,(X_3) \in \mathscr{C};\ \mu(X_3) = 1.$$

Let $Y_3 = f\,(X_3)$. Then we have the diagram

$$\begin{array}{ccc} X_3 & \xrightarrow{\quad f \quad} & Y_3 \\ \Big\downarrow \tau & & \Big\downarrow \tau' \\ \tau(X_3) & \xrightarrow[\phi = \tau' f \tau^{-1}]{} & \tau'(Y_3) \end{array}$$

where $Y_3 = f(X_3) \subset f(X_2) \subset f(X_1) = Y_1$. In the diagram all the sets are borel, f is onto, τ, τ' are borel isomorphisms and

$$\tau(X_3) \in \mathscr{B}_R;\ \tau'(Y_3) \in \mathscr{B}_R;$$
$$\mu(X_3) = \nu(Y_3) = \mu'(\tau(X_3)) = \nu'(\tau'(Y_3)) = 1;$$
$$\mu\tau^{-1} = \mu',\ \nu\tau'^{-1} = \nu'.$$

Since the diagram commutes and by Corollary 46.2 the Proposition holds for the probability measure μ' and the map ϕ it follows that the same holds for the measure μ and the map f. This completes the proof.

Remark. 46.4. We can understand the meaning of Proposition 46.3 in terms of the following illustration.

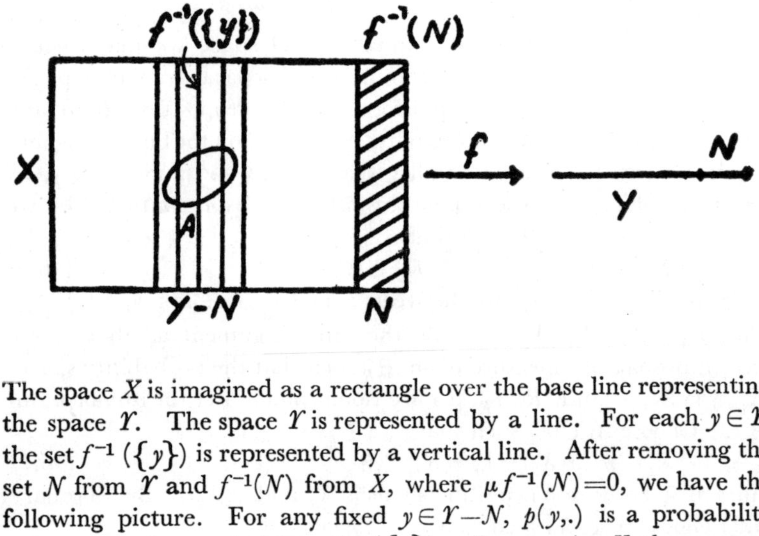

The space X is imagined as a rectangle over the base line representing the space Y. The space Y is represented by a line. For each $y \in Y$, the set $f^{-1}(\{y\})$ is represented by a vertical line. After removing the set N from Y and $f^{-1}(N)$ from X, where $\mu f^{-1}(N) = 0$, we have the following picture. For any fixed $y \in Y - N$, $p(y,.)$ is a probability measure on the vertical line $f^{-1}(\{y\})$. For any $A \subset X$, the measure $\mu(A)$ of A is obtained by the formula

$$\mu(A) = \int_{Y-N} p(y, A \cap f^{-1}(\{y\}))\, d\mu f^{-1}(y).$$

In other words the measure μ has been expressed as a '*continuous*' sum of the measures $p(y,.)$ on the spaces $f^{-1}(\{y\})$.

The distribution $p(y,.)$ on $f^{-1}(\{y\})$ is called the *regular conditional probability distribution* given $f = y$. It is, of course, defined almost everywhere $y(\mu f^{-1})$. Sometimes one writes $p(y, A) = P(A \mid f = y)$. For almost all $y(\mu f^{-1})$ it is a probability measure on the *fibre* $f^{-1}(\{y\})$.

Proposition. 46.5. Let $(X, \mathscr{B})$ be a standard borel space and let $\mathscr{B}_0 \subset \mathscr{B}$ be a sub σ-algebra. Let μ be a probability measure on $\mathscr{B}$. Then there exists a transition probability $P(x, A)$ on $X \times \mathscr{B}$ such that

(i) for fixed $A \in \mathscr{B}$, $P(x, A)$ is $\mathscr{B}_0$ measurable;

(ii) $\int_B P(x, A)\, d\mu(x) = \mu(A \cap B)$ for all $A \in \mathscr{B}$, $B \in \mathscr{B}_0$;

(iii) for any $g \in L_1(\mu)$,

$$\mathbf{E}(g \mid \mathscr{B}_0)(x) = \int g(\xi)\, P(x, d\xi) \text{ a.e. } x\ (\mu).$$

Proof. Consider the Hilbert space $L_2(\mu)$ and the subspace S of all $\mathscr{B}_0$ measurable functions. Since $L_2(\mu)$ is separable we can choose a dense sequence $\phi_1, \phi_2, \ldots$ in S. Let $Y = R^\infty$ be the countable product of copies of the real line R. Consider the map $f : X \to Y$, defined by

$$f(x) = (\phi_1(x), \phi_2(x), \ldots).$$

Then f is a borel map from the standard borel space X into the standard borel space Y. Let $\mathscr{B}_1 = f^{-1}(\mathscr{B}_{R^\infty})$. Since $\phi_1, \phi_2, \ldots$ are $\mathscr{B}_0$ measurable it follows that $\mathscr{B}_1 \subset \mathscr{B}_0$. Since the span of all ϕ_j's is S it follows that the completion of $\mathscr{B}_1$ and $\mathscr{B}_0$ under μ are same. Now put

$$P(x, A) = p(f(x), A)$$

where $p(y, A)$ is a regular conditional probability distribution given $f = y$. For fixed A, $P(x, A)$ is $\mathscr{B}_1$ measurable and hence $\mathscr{B}_0$ measurable. If $B \in \mathscr{B}_0$, then there exists $B_1 \in \mathscr{B}_1$ such that $\mu(B \,\Delta\, B_1) = 0$ and

$$\begin{aligned} \int_B P(x, A)\, d\mu(x) &= \int_{B_1} P(x, A)\, d\mu(x) \\ &= \int_{B_1} p(f(x), A)\, d\mu(x). \end{aligned}$$

Let $C \subset Y$ be a borel set such that $B_1 = f^{-1}(C)$. Since $p(y, f^{-1}(\{y\})) = 1$ a.e. $y(\mu f^{-1})$, we have

$$\begin{aligned} \int_B P(x, A)\, d\mu(x) &= \int_C p(y, A)\, d\mu f^{-1}(y) \\ &= \int p(y, A \cap f^{-1}(C))\, d\mu f^{-1}(y) \\ &= \mu(A \cap f^{-1}(C)) = \mu(A \cap B_1) \\ &= \mu(A \cap B). \end{aligned}$$

This proves property (ii). Property (iii) is same as property (ii) if $g = \chi_A$, $A \in \mathscr{B}$. By taking linear combinations of χ_A's and going to limits we complete the proof.

Proposition. 46.6. (Jensen's inequality for conditional expectation.) Let $(X, \mathscr{B}, \mu)$ be a standard probability space and let $\mathscr{B}_0 \subset \mathscr{B}$ be a sub σ-algebra. Let $E \subset R^k$ be a convex set of the form $\bigcup_i K_i$

where $\{K_n\}$ is an increasing sequence of compact convex sets. Let ϕ be a real valued continuous convex function on E. Let $f_1, f_2, \ldots, f_k$ be real valued random variables on $(X, \mathcal{B}, \mu)$ such that the map

$$x \to \mathbf{f}(x) = \begin{pmatrix} f_1(x) \\ f_2(x) \\ \cdot \\ \cdot \\ \cdot \\ f_k(x) \end{pmatrix}$$

takes values in E. If $\mathbf{E}\,|\phi(\mathbf{f})| < \infty$ and $\mathbf{E}\,\|\mathbf{f}\| < \infty$, where $\|\mathbf{f}\|$ is the function $(\sum_i f_i(x)^2)^{\frac{1}{2}}$, then $\mathbf{E}(\mathbf{f} \mid \mathcal{B}_0)$ takes values in E and

$$\phi(\mathbf{E}(\mathbf{f} \mid \mathcal{B}_0)) \leqslant \mathbf{E}(\phi(\mathbf{f}) \mid \mathcal{B}_0) \text{ a.e. } (\mu).$$

Proof. This follows from Remark 34.29 and the fact that

$$\mathbf{E}(\mathbf{f} \mid \mathcal{B}_0)(x) = \int \mathbf{f}(\xi)\, P(x, d\xi), \text{ a.e. } x(\mu),$$

where $P(x, A)$ is a transition probability satisfying conditions (i)-(ii) of Proposition 46.5.

§47. Radon-Nikodym Theorem and Lebesgue Decomposition

Let $(X, \mathcal{B})$ be a borel space and let λ, μ be two totally finite measures on $\mathcal{B}$. We shall now decompose X into three disjoint parts with certain special properties relative to λ and μ. We state it in the form of a proposition.

Proposition. 47.1. Let λ, μ be two totally finite measures in $(X, \mathcal{B})$. Then there exist three disjoint sets $X_1, X_2, X_3 \in \mathcal{B}$ such that

(i) $X = \bigcup_{i=1}^{3} X_i$;

(ii) $\lambda(X_3) = \mu(X_1) = 0$;

(iii) there exists a strictly positive borel function g on X_2 such that for any $E \subset X_2$, $E \in \mathcal{B}$,

$$\lambda(E) = \int_E g\, d\mu,$$

$$\mu(E) = \int_E g^{-1}\, d\lambda.$$

In particular, for $E \subset X_2$, $E \in \mathcal{B}$, $\lambda(E)$ vanishes if and only if $\mu(E)$ vanishes.

Proof. Let $\lambda+\mu=\nu$. Consider the real Hilbert space $L_2(\nu)$. Since ν is a totally finite measure any $f \in L_2(\nu)$ also belongs to $L_1(\nu)$ and hence to $L_1(\lambda)$. Let

$$\Lambda(f) = \int f\, d\lambda, f \in L_2(\nu).$$

By Schwarz's inequality,

$$\begin{aligned} |\Lambda(f)| &\leqslant \left(\int |f|^2\, d\lambda\right)^{\frac{1}{2}} \lambda(X)^{\frac{1}{2}} \\ &\leqslant \lambda(X)^{\frac{1}{2}} \left(\int |f|^2\, d\nu\right)^{\frac{1}{2}}. \end{aligned}$$

Hence Λ is a bounded linear functional on the Hilbert space $L_2(\nu)$. By Riesz's theorem (Proposition 41.14) there exists a borel function $f_0 \in L_2(\nu)$ such that

$$\Lambda(f) = \int f\, d\lambda = \int f f_0\, d\nu \text{ for all } f \in L_2(\nu). \tag{47.1}$$

Putting $f = \chi_E$, $E \in \mathscr{B}$,

$$\lambda(E) = \int_E f_0\, d\nu \geqslant 0 \text{ for all } E \in \mathscr{B}. \tag{47.2}$$

Hence

$$f_0 \geqslant 0 \text{ a.e. } (\nu).$$

Further

$$\nu(E) \geqslant \lambda(E) = \int_E f_0\, d\nu,$$

or equivalently,

$$\int_E (1 - f_0)\, d\nu \geqslant 0 \text{ for all } E \in \mathscr{B}.$$

Hence

$$1 - f_0 \geqslant 0 \text{ a.e. } (\nu).$$

Thus we can choose f_0 such that $0 \leqslant f_0(x) \leqslant 1$ for all x and Eq. (47.1) is fulfilled. Let

$$\begin{aligned} X_1 &= \{x : f_0(x) = 1\}, \\ X_2 &= \{x : 0 < f_0(x) < 1\}, \\ X_2 &= \{x : f_0(x) = 0\}. \end{aligned}$$

Then X_1, X_2, X_3 are three disjoint sets such that condition (i) is fulfilled. Further by Eq. (47.2)

$$\lambda(X_3) = \int_{X_3} f_0\, d\nu = 0$$

$$\begin{aligned} \mu(X_1) &= \nu(X_1) - \lambda(X_1) \\ &= \int_{X_1} [1 - f_0(x)]\, d\nu(x) = 0. \end{aligned}$$

Hence condition (ii) is also fulfilled. Now let $E \subset X_2$, $E \in \mathscr{B}$. Then Eq. (47.2) implies that

$$\int_E (1 - f_0)\, d\lambda = \int_E f_0\, d\mu.$$

Hence for any non-negative borel function f

$$\int_{X_2} f(1 - f_0)\, d\lambda = \int_{X_2} f f_0\, d\mu. \tag{47.3}$$

Put $f = \dfrac{\chi_E}{1 - f_0}$. Then

$$\lambda(E) = \int_E \frac{f_0}{1 - f_0}\, d\mu, \; E \subset X_2.$$

Put

$$f = \frac{\chi_E}{f_0}$$

in Eq. (47.3). Then

$$\mu(E) = \int_E \frac{1 - f_0}{f_0}\, d\mu, \; E \subset X_2.$$

If we define $g = f_0 / 1 - f_0$ on X_2, then g is a strictly positive borel function on X_2 and condition (iii) is fulfilled. This completes the proof.

Exercise. 47.2. Proposition 47.1 holds in the case when λ and μ are σ-finite measures. (Hint: Decompose X into a sequence $\{A_i\}$ of disjoint sets such that $\mu(A_i) + \lambda(A_i) < \infty$ for each i and use Proposition 47.1 in each piece A_i.)

Corollary. 47.3. (Radon-Nikodym theorem) Let λ, μ be two σ-finite measures on $(X, \mathscr{B})$. Suppose $\lambda(E) = 0$ whenever $\mu(E) = 0$ for $E \in \mathscr{B}$. Then there exists a non-negative borel function f on X such that

$$\lambda(E) = \int_E f\, d\mu \text{ for all } E \in \mathscr{B}. \tag{47.4}$$

Conversely, if λ is a σ-finite measure defined by the above equation where f is a non-negative borel function and μ is a σ-finite measure then $\lambda(E) = 0$ whenever $\mu(E) = 0$ for all $E \in \mathscr{B}$.

If f_1 is another function such that Eq. (47.4) holds with f replaced by f_1, then $f = f_1$ a.e. (μ).

Proof. It is enough to prove in the case when λ and μ are totally finite. Using Proposition 47.1 divide X into three parts X_1, X_2, X_3 satisfying conditions (i) to (iii). Then $\lambda(X_3) = \mu(X_1) = 0$. The condition of the corollary implies that $\lambda(X_1) = 0$. Thus

$$\lambda(E) = \lambda(E \cap X_2) = \int_E g\, \chi_{x2}\, d\mu \text{ for all } E \in \mathscr{B}.$$

Putting $f = g\chi_{x2}$ the proof of the first part is complete. The second and third parts trivially follow from the basic properties of integrals.

Remark. 47.4. If λ, μ are σ-finite measures on $(X, \mathscr{B})$ such that $\lambda(E)=0$ whenever $\mu(E)=0$ for all $E \in \mathscr{B}$, then λ is said to be ***absolutely continuous*** with respect to μ or *dominated* by μ. The function f defined by Corollary 47.3 and satisfying Eq. (47.4) is called the *Radon-Nikodym derivative* of λ with respect to μ. We write

$$f=\frac{d\lambda}{d\mu}$$

in analogy with the derivative in classical differential calculus. The derivative $\frac{d\lambda}{d\mu}$ is defined a.e. (μ). If λ is absolutely continuous with respect to μ, we write $\lambda \ll \mu$. If $\lambda \ll \mu$ and $\mu \ll \lambda$ we write $\lambda \equiv \mu$ and say that λ is *equivalent* to μ. Then $\ll$ is a partial order in the space of all σ-finite measures on $(X, \mathscr{B})$ and $\equiv$ is an equivalence relation.

Corollary. 47.5. Let λ, μ be σ-finite measures on $(X, \mathscr{B})$ and let $\lambda \ll \mu$. If $g \in L_1(\lambda)$, then $g \dfrac{d\lambda}{d\mu} \in L_1(\mu)$ and

$$\int g\, d\lambda = \int g \frac{d\lambda}{d\mu}\, d\mu.$$

Proof. The result holds when g is of the form χ_E. The corollary now follows from the linearity of integrals and the fact that non-negative borel functions are limits of increasing sequences of non-negative simple functions.

Exercise. 47.6. If $\lambda \ll \mu \ll \nu$ for three σ-finite measures on $(X, \mathscr{B})$, then

$$\frac{d\lambda}{d\nu}=\frac{d\lambda}{d\mu}\frac{d\mu}{d\nu} \text{ a.e. } (\mu).$$

If $\lambda \equiv \mu$, then

$$\frac{d\lambda}{d\mu}=\left(\frac{d\mu}{d\lambda}\right)^{-1} \text{ a.e. } (\mu).$$

Exercise. 47.7. (Lebesgue decomposition theorem) Given two σ-finite measures λ, μ on $(X, \mathscr{B})$ there exist two σ-finite measures λ_1 and λ_2 such that

(i) $\lambda=\lambda_1+\lambda_2$,

(ii) $\lambda_1 \ll \mu$,

(iii) there exists a set $A \in \mathscr{B}$ with the property $\mu(A)=0$, $\lambda_2(A')=0$, where A' is the complement of A. Such a decomposition is unique.

Remark. 47.8. Two measures λ and μ on a borel space $(X, \mathscr{B})$ are said to be *singular* or *orthogonal* with respect to each other if there exists a set $A \in \mathscr{B}$ such that $\lambda(A')=0$ and $\mu(A)=0$. Lebesgue's decomposition theorem can now be restated as follows: Given any two σ-finite measures λ, μ on $(X, \mathscr{B})$ we can decompose λ uniquely into a sum of two measures λ_1 and λ_2 such that $\lambda_1 \ll \mu$ and λ_2 orthogonal to μ. λ_1 is called the *absolutely continuous part* of λ with respect to μ. λ_2 is called the *singular part* of λ with respect to μ.

Exercise. 47.9. Let λ be any σ-finite measure on $(R^k, \mathscr{B}_{R^k})$. Then λ can be uniquely decomposed as $\lambda=\lambda_1+\lambda_2+\lambda_3$ where λ_1 is absolutely continuous with respect to the Lebesgue measure, λ_2 is a non-atomic measure singular with respect to Lebesgue measure and λ_3 is a purely atomic measure, i.e., there exists a countable subset A of R^k such that $\lambda_3(A')=0$. (Regarding the existence of nonatomic measures singular with respect to Lebesgue measure see Remark 25.4.)

§48. Elementary Properties of Radon-Nikodym Derivatives

The following proposition is extremely useful in the computation of Radon-Nikodym derivatives in many statistical problems.

Proposition. 48.1. Let λ, μ be two probability measures on $(X, \mathscr{B})$ such that $\lambda \ll \mu$. Let $\mathscr{B}_1 \subset \mathscr{B}_2 \subset \ldots$ be a sequence of sub σ-algebras such that $\mathscr{B} = \bigvee_n \mathscr{B}_n$. Let f_n be the Radon-Nikodym derivative of λ with respect to μ in $(X, \mathscr{B}_n)$. Then

$$\lim_{n\to\infty} f_n = \frac{d\lambda}{d\mu} \text{ a.e. } (\mu),$$

$$\lim_{n\to\infty} f_n = \frac{d\lambda}{d\mu} \text{ in } L_1\,(\mu).$$

Proof. We have

$$\lambda(E) = \int_E f\,d\mu \text{ for all } E \in \mathscr{B},$$

where $f=\frac{d\lambda}{d\mu}$. If $E \in \mathscr{B}_n$, then

$$\lambda(E) = \int_E f\,d\mu = \int_E \mathbf{E}_\mu\,(f \mid \mathscr{B}_n)\,d\mu,$$

where $\mathbf{E}_\mu\,(f \mid \mathscr{B}_n)$ is the conditional expectation of f given $\mathscr{B}_n$ with respect to the probability measure μ. These conditional expectations are well defined since $f \in L_1(\mu)$. By the uniqueness of Radon-Nikodym derivative it follows that

$$f_n = \mathbf{E}_\mu (f | \mathscr{B}_n),$$

since both sides are $\mathscr{B}_n$ measurable. An application of Proposition 44.16 completes the proof.

Exercise. 48.2. Let R^∞ be the countable product of copies of the real line with the product topology. Let λ, μ be two probability distributions on the borel σ-algebra of R^∞. Let π_n be the projection map

$$\pi_n(\mathbf{x}) = (x_1, x_2, \ldots, x_n),$$

where $\mathbf{x} \in R^\infty$ and $\mathbf{x} = (x_1, x_2, \ldots)$. If $\lambda \overset{-1}{\pi_n}$ and $\mu \overset{-1}{\pi_n}$ have density functions (with respect to the Lebesgue measure) f_n and g_n in R^n, then the sequence $\{\rho_n\}$ defined by

$$\rho_n(\mathbf{x}) = \frac{f_n(x_1, x_2, \ldots, x_n)}{g_n(x_1, x_2, \ldots, x_n)}, \quad \mathbf{x} \in R^\infty$$

converges a.e. $\mathbf{x}$ (μ) to a function ρ. Further

$$\rho(\mathbf{x}) = \frac{d\lambda_1}{d\mu}(\mathbf{x}) \text{ a.e. } \mathbf{x}\ (\mu),$$

where λ_1 is the absolutely continuous part of λ with respect to μ. (Hint: If $\mathscr{B}_n = \overset{-1}{\pi_n}(\mathscr{B}_{R^n})$, then $\{\mathscr{B}_n\}$ is an increasing sequence of sub σ-algebras such that $\bigvee_n \mathscr{B}_n = \mathscr{B}_{R^\infty}$.)

Remark. 48.3. Let $(X, \mathscr{B})$ be a borel space. By a finite or countable partition $\mathscr{P}$ of X we mean a finite or infinite sequence $\{A_j\}$ of disjoint sets in $\mathscr{B}$ such that $X = \bigcup_j A_j$. Each A_j is called an element of $\mathscr{P}$. If $\mathscr{P}_1$ and $\mathscr{P}_2$ are two partitions we say that $\mathscr{P}_2$ is finer than $\mathscr{P}_1$ and write $\mathscr{P}_1 < \mathscr{P}_2$ if every element of $\mathscr{P}_1$ is the union of some elements in $\mathscr{P}_2$.

Suppose $\mathscr{P}_1 < \mathscr{P}_2 < \ldots < \mathscr{P}_n < \ldots$ is an increasing sequence of finite or countable partitions such that the σ-algebras $\mathscr{B}_n$ generated by all the elements of $\mathscr{P}_n$ have the property $\bigvee_n \mathscr{B}_n = \mathscr{B}$. If λ and μ are two probability measures on $\mathscr{B}$, let

$$\begin{aligned} \rho_n(x) &= \frac{\lambda(A)}{\mu(A)} \text{ if } x \in A,\ A \in \mathscr{P}_n \text{ and } \mu(A) \neq 0, \\ &= 0 \text{ otherwise,} \end{aligned}$$

for each n. Let λ_1 be the absolutely continuous part of λ with respect to μ. Then ρ_n is the Radon-Nikodym derivative of λ_1 with respect to μ in the borel space $(X, \mathscr{B}_n)$. Proposition 48.1 implies that

$$\lim_{n\to\infty} \rho_n(x) = \frac{d\lambda_1}{d\mu}(x) \text{ a.e. } x(\mu)$$

and in $L_1(\mu)$.

Our next proposition concerns the usefulness of the Radon-Nikodym derivative as a *statistic* for discriminating between two distributions λ and μ which are equivalent. To state the proposition we introduce some notation. Let $(X, \mathscr{B}, \mu)$ be a probability space and let g be a real valued random variable on X. For any $f \in L_1(\mu)$, we note that $\mathbf{E}(f \mid g^{-1}(\mathscr{B}_R))$ is a function of the form $\rho(g)$, where ρ is a borel function on R. (See Remark 44.6). We write

$$\rho(g) = \mathbf{E}_\mu(f \mid g),$$

where the suffix μ indicates the measure μ with respect to which conditional expectation is taken. The function ρ is uniquely determined almost everywhere μg^{-1}. Now we are ready to state our next proposition.

Proposition. 48.4. Let λ, μ be two equivalent probability measures on the borel space $(X, \mathscr{B})$. Let $g(x) = \dfrac{d\lambda}{d\mu}(x)$. Then

$$\mathbf{E}_\lambda(f \mid g) = \mathbf{E}_\mu(f \mid g) \text{ a.e. } \lambda,$$

for every bounded real valued random variable f on $(X, \mathscr{B})$.

Proof. For any borel set $A \subset R$, we have

$$\int_{g^{-1}(A)} f\,d\lambda = \int_{g^{-1}(A)} \mathbf{E}_\lambda(f \mid g)\,d\lambda. \tag{48.1}$$

On the other hand

$$\begin{aligned}\int_{g^{-1}(A)} f\,d\lambda &= \int_{g^{-1}(A)} f\frac{d\lambda}{d\mu}\,d\mu \\ &= \int_{g^{-1}(A)} \mathbf{E}_\mu(fg \mid g)\,d\mu \\ &= \int_{g^{-1}(A)} g\,\mathbf{E}_\mu(f \mid g)\,d\mu \\ &= \int_{g^{-1}(A)} \mathbf{E}_\mu(f \mid g)\,d\lambda. \end{aligned} \tag{48.2}$$

Comparing Eqs. (48.1) and (48.2) we have

$$\mathbf{E}_\lambda(f \mid g) = \mathbf{E}_\mu(f \mid g) \text{ a.e. } (\lambda).$$

This completes the proof.

Corollary. 48.5. If λ and μ are equivalent probability measures on $(X, \mathscr{B})$ then the conditional probability distributions given $\frac{d\lambda}{d\mu}$ with respect to λ and μ agree almost everywhere. (In statistical language one says that $\frac{d\lambda}{d\mu}$ is a *sufficient statistic* for the family of distributions containing just λ and μ).

Example. 48.6. As an illustration of the above ideas we shall study a simple example. Let $X=R^\infty$ and $\mathscr{B}$ be its borel σ-algebra. Let μ be the probability measure on $\mathscr{B}$ according to which the coordinate maps $x_1, x_2, \ldots$ of $\mathbf{x}\in X$ are independent and identically distributed random variables with standard normal distribution. Let λ be the measure on $\mathscr{B}$ according to which $x_1, x_2, \ldots$ are independently normally distributed with variance unity and means $m_1, m_2, \ldots$. If π_n is the projection map from X onto R^n, described in Exercise 48.2, then

$$\frac{d\lambda\pi_n^{-1}}{d\mu\pi_n^{-1}} = \exp\left[\sum_{i=1}^n m_i x_i - \sum_{i=1}^n \frac{m_i^2}{2}\right].$$

This converges a.e. (μ) as $n\to\infty$ if and only if $\sum_i m_i^2$ converges. In such a case the limit is positive a.e. (μ). Thus $\lambda\equiv\mu$ if and only if $\sum_i m_i^2$ is convergent. In such a case

$$\frac{d\lambda}{d\mu}(\mathbf{x}) = \exp\left[\sum_{i=1}^\infty m_i x_i - \tfrac{1}{2}\sum_{i=1}^\infty m_i^2\right].$$

The linear functional $h(\mathbf{x}) = \sum_{i=1}^\infty m_i x_i$ is a sufficient statistic for discriminating between λ and μ, i.e., $\mathbf{E}_\lambda(f\,|\,h) = \mathbf{E}_\mu(f\,|\,h)$ a.e. (λ) for any bounded borel function f on $(X, \mathscr{B})$.

Now we shall examine how the Radon-Nikodym derivative changes under transformations.

Proposition. 48.7. Let $(X, \mathscr{B})$, $(Y, \mathscr{C})$ be borel spaces and let T be a borel isomorphism from $(X, \mathscr{B})$ to $(Y, \mathscr{C})$. Let λ and μ be σ-finite measures on $\mathscr{B}$ such that λT^{-1} and μT^{-1} are σ-finite on $\mathscr{C}$. If $\lambda\ll\mu$, then $\lambda T^{-1}\ll\mu T^{-1}$ and

$$\frac{d\lambda T^{-1}}{d\mu T^{-1}}(y) = \frac{d\lambda}{d\mu}(T^{-1}y) \text{ a.e. } y(\mu T^{-1}).$$

Proof. We have for $E \in \mathscr{C}$,

$$\lambda T^{-1}(E) = \int_{T^{-1}(E)} \frac{d\lambda}{d\mu}(x)\, d\mu(x) = \int_E \frac{d\lambda}{d\mu}(T^{-1}y)\, d\mu T^{-1}(y).$$

Hence by the uniqueness of Radon-Nikodym derivatives the required result follows.

Exercise. 48.8. Let μ be the Lebesgue measure on $(R^n, \mathscr{B}_{R^n})$. Suppose T is a homeomorphism of R^n with continuous first order derivatives. Then $\mu T \equiv \mu$ and

$$\frac{d\mu T}{d\mu}(\mathbf{x}) = |\det \mathcal{J}(T, \mathbf{x})|,$$

where $\mathcal{J}$ is the Jacobian matrix of the transformation T. (Hint: use Proposition 37.4).

Exercise. 48.9. Let $(X, \mathscr{B}, \mu)$ be a σ-finite measure space and let G be the class of all one-one onto maps $T : X \to X$ such that

(a) T and T^{-1} are borel maps

(b) $\mu(TE)=0$ whenever $\mu(E)=0$ for all $E \in \mathscr{B}$.

Let $\alpha(T, x) = \dfrac{d\mu T}{d\mu}(x)$. Then for any $T_1, T_2, \in G$,

$$\alpha(T_1 T_2, x) = \alpha(T_1, T_2 x)\, \alpha(T_2, x) \text{ a.e. } x(\mu). \tag{48.3}$$

For any $f \in L_2(\mu)$, let

$$[U(T)f](x) = [\alpha(T, T^{-1}x)]^{-1/2} f(T^{-1}x)$$

Then $U(T) f \in L_2(\mu)$. $U(T)$ is an operator in the Hilbert space $L_2(\mu)$ satisfying the following conditions:

(i) $U(T_1)\, U(T_2) = U(T_1 T_2)$ for all $T_1, T_2 \in G$;

(ii) $U(T)^{-1} = U(T)^*$.

(Equation 48.3 is known as *cocycle identity*. The map $T \to U(T)$ which is a homomorphism from the group G into the group of all unitary operators on $L_2(\mu)$ is called a *unitary representation* of G. This is one of the most well-known methods of constructing unitary representations.)

§49. Law of Large Numbers and Ergodic Theorem

In Section 11 we considered averages of the type

$$S_n = \frac{s_1 + s_2 + \ldots + s_n}{n}$$

where $s_1, s_2, \ldots$ are independent and identically distributed random variables taking a finite number of values. We would like to study the behaviour of such averages in the case when the assumption of independence is dropped and the random variables s_n are not necessarily simple. To this end we shall reformulate the problem in terms of measure preserving transformations.

Let (Ω, S, P) be a probability space and let $f_1, f_2, \ldots$ be a sequence of independent and identically distributed random variables on Ω. With the help of such a sequence we shall construct a map $\mathbf{f}$ from Ω into the space R^∞ of sequences of real numbers by

$$\mathbf{f}(\omega)=(f_1(\omega), f_2(\omega), \ldots).$$

Denote a general point of R^∞ by $\mathbf{x}=(x_1, x_2, \ldots)$. Let $\mathscr{B}^\infty$ be the σ-algebra generated by the open sets of the product topology in R^∞ or equivalently the σ-algebra generated by the family of all finite dimensional borel cylinder sets. Then $\mathbf{f}$ is a borel map from (Ω, S) into $(R^\infty, \mathscr{B}^\infty)$. Let $\mu=P\mathbf{f}^{-1}$. Then $(R^\infty, \mathscr{B}^\infty, \mu)$ is a probability space. In R^∞ we consider the map T defined by

$$(T\mathbf{x})_n=x_{n+1},\ n=1, 2, \ldots,$$

i.e., the sequence $(x_1, x_2, \ldots)$ goes to the sequence $(x_2, x_3, \ldots)$ under T. For any borel cylinder of the form

$$E=\{\mathbf{x} : x_j \in E_j, j=1, 2, \ldots, k\} \subset R^\infty,$$

we have

$$\begin{aligned}\mu T^{-1}(E) &= \mu\{\mathbf{x}: x_2 \in E_1, x_3 \in E_2, \ldots x_{k+1} \in E_k\}\\ &= P\{\omega : f_2(\omega) \in E_1, f_3(\omega) \in E_2, \ldots, f_{k+1}(\omega) \in E_k\}\\ &= \prod_{j=1}^{k} P\{\omega : f_{j+1}(\omega) \in E_j\}\\ &= \prod_{j=1}^{k} P\{\omega : f_j(\omega) \in E_j\}\\ &= \mu(E),\end{aligned}$$

because $f_1, f_2, \ldots$ are independent and identically distributed. Thus μ and μT^{-1} are two measures which agree on all finite dimensional cylinders which constitute a boolean algebra generating $\mathscr{B}^\infty$. Hence $\mu=\mu T^{-1}$. In other words the *shift* T preserves the measure μ. It is now clear that the averages

$$\tilde{f}_n=\frac{f_1+f_2+\ldots+f_n}{n}$$

will converge a.e. (P) if and only if, in R^∞,

$$\mu\left\{\mathbf{x} : \frac{x_1+x_2+\ldots+x_n}{n} \text{ converges as } n\to\infty\right\}=1.$$

On R^∞, define the function g by

$$g(\mathbf{x}) = x_1 \text{ for all } \mathbf{x}. \tag{49.1}$$

Then

$$\frac{g(\mathbf{x}) + g(T\mathbf{x}) + \ldots + g(T^{n-1}\mathbf{x})}{n} = \frac{x_1 + x_2 + \ldots + x_n}{n}$$

for all n. Thus the study of laws of large numbers reduces to the study of averages of the form

$$\frac{g + gT + gT^2 + \ldots + gT^{n-1}}{n} \tag{49.2}$$

where T is a measure preserving transformation and g is a real valued random variable.

We shall say that a set $E \in \mathscr{B}^\infty$ is T-*invariant* if $E \subset T^{-1} E$. Since $\mu(E) = \mu T^{-1}(E)$ it follows that $\mu(T^{-1}(E) - E) = 0$. Thus, for any invariant set E,

$$\chi_E(\mathbf{x}) = \chi_E(T\mathbf{x}) \text{ a.e. } (\mu).$$

Hence by induction

$$\chi_E(\mathbf{x}) = \chi_E(T^k \mathbf{x}) \text{ a.e. } (\mu),$$

for each k. In other words, for any fixed k, χ_E is a function of x_{k+1}, x_{k+2}, ... For g defined by Eq. (49.1), g, gT, gT^2, ... is a sequence of independent and identically distributed random variables on R^∞. Hence, for any T-invariant set E, it follows that χ_E is independent of all the random variables g, gT, ..., gT^k for fixed k. In other words, for any finite dimensional cylinder $C \subset R^\infty$, E and C are independent, i.e.

$$\mu(E \cap C) = \mu(E)\,\mu(C).$$

Both sides are measures (in the variable C) agreeing on all finite dimensional cylinders. Hence

$$\mu(E \cap F) = \mu(E)\,\mu(F) \text{ for all } F \in \mathscr{B}^\infty.$$

In particular,

$$\mu(E) = \mu(E \cap E) = \mu(E)^2.$$

Thus for every T invariant borel set E,

$$\mu(E) = 0 \text{ or } 1.$$

In view of these properties that we have come across about the sequence of independent and identically distributed random variables we introduce the following definition.

Definition. 49.1. Let $(X, \mathscr{B}, \mu)$ be a proability space. A borel map $T : X \to X$ is said to be μ-*measure preserving* if $\mu(E) = \mu T^{-1}(E)$ for all $E \in \mathscr{B}$. A set $E \in \mathscr{B}$ is T-invariant if $\mu(E \,\Delta\, T^{-1} E) = 0$. The measure μ is said to be *ergodic* if every T-invariant set E has probability 0 or 1. In such a case we also say that T is ergodic with respect to μ. The σ-algebra of all T-invariant sets is called the *invariant σ-algebra.*

The discussion preceding this definition can now be summarised in the form of a proposition.

Proposition. 49.2 Let $(X, \mathcal{B}, \lambda)$ be a probability space and let $(X^\infty, \mathcal{B}^\infty, \lambda^\infty)$ be its countable product. For any $\mathbf{x} \in X^\infty$, $\mathbf{x}=(x_1, x_2, \ldots)$, $x_i \in X$ for all i, let $T\mathbf{x}$ be the sequence $(x_2, x_3, \ldots)$. The map $T : X^\infty \to X^\infty$ is λ^∞-measure preserving and is ergodic.

We shall now study the convergence of averages of the form (49.2) in the general case of a measure preserving transformation. The proofs we present are due to Garsia.

Proposition. 49.3. (Maximal inequality). Let $(X, \mathcal{B}, \mu)$ be a probability space and let $T : X \to X$ be a borel map such that $\mu T^{-1} = \mu$. Let $f \in L_1(\mu)$,

$$f_n(x) = f(x) + f(Tx) + \ldots + f(T^{n-1}x),$$
$$f_0(x) = 0,$$
$$E_N = \{x \colon \max(f_1(x), f_2(x), \ldots, f_N(x)) \geqslant 0\}.$$

Then

$$\int_{E_N} f\, d\mu \geqslant 0.$$

Proof. Let

$$F(x) = \max(f_0(x), f_1(x), \ldots, f_N(x)).$$

Then

$$F(Tx) \geqslant f_j(Tx), j=0, 1, 2, \ldots, N;$$
$$f(x) + F(Tx) \geqslant f(x) + f_j(Tx) = f_{j+1}(x), \quad j=0, 1, 2, \ldots, N.$$

Hence

$$f(x) + F(Tx) \geqslant \max(f_1(x), f_2(x), \ldots, f_{N+1}(x)).$$

Thus for any $x \in E_N$,

$$f(x) + F(Tx) \geqslant \max(f_0(x), f_1(x), \ldots, f_N(x)) = F(x).$$

Thus

$$\int_{E_N} f\, d\mu \geqslant \int_{E_N} F\, d\mu - \int_{E_N} FT\, d\mu.$$

Since $F(x) = 0$ on E'_N and $F(x) \geqslant 0$ for all x we have from the invariance of μ,

$$\int_{E_N} f\, d\mu \geqslant \int_X F\, d\mu - \int_X FT\, d\mu = 0$$

This completes the proof.

Proposition. 49.4. (Birkhoff's individual ergodic theorem.) Let $(X, \mathscr{B}, \mu)$ be a probability space and let $T: X \to X$ be a borel map such that $\mu T^{-1} = \mu$. For any $f \in L_1(\mu)$, the sequence

$$A_n f = \frac{f + fT + fT^2 + \ldots + fT^{n-1}}{n}$$

converges almost everywhere to $\mathbf{E}(f \mid \mathscr{I})$, where $\mathscr{I}$ is the invariant σ-algebra. Further $\mathbf{E} \mid A_n f - \mathbf{E}(f \mid \mathscr{I}) \mid \to 0$ as $n \to \infty$.

Proof. Let $a < b$ and let

$$X_{a,b} = \left\{ x : \varliminf_{n\to\infty} (A_n f)(x) < a < b < \varlimsup_{n\to\infty} (A_n f)(x) \right\}. \quad (49.3)$$

Then $X_{a,b}$ is an invariant borel set. We shall now prove that $\mu(X_{a,b}) = 0$. Suppose this is not true. Consider the probability space $(X_1, \mathscr{B}_1, \mu_1)$, where

$$X_1 = X_{a,b}, \mathscr{B}_1 = \mathscr{B} \cap X_1,$$

$$\mu_1(E) = \frac{\mu(E)}{\mu(X_1)}, \quad E \in \mathscr{B}_1.$$

Then T restricted to X_1 preserves the measure μ_1. Put

$$g(x) = f(x) - b,$$
$$h(x) = a - f(x). \quad (49.4)$$

From Eq. (49.3), we have

$$\varlimsup_{n\to\infty} (A_n g)(x) > 0;$$

$$\varlimsup_{n\to\infty} (A_n h)(x) > 0, \text{ for all } x \in X_1.$$

In particular,

$$\sup_{n \geqslant 1} \{g(x) + g(Tx) + \ldots + g(T^{n-1}x)\} > 0,$$

$$\sup_{n \geqslant 1} \{h(x) + h(Tx) + \ldots + h(T^{n-1}x)\} > 0.$$

Hence

$$X_1 = \bigcup_{N=1}^{\infty} E_N = \bigcup_{N=1}^{\infty} F_N$$

where

$$E_N = \left\{ x : \max_{1 \leqslant n \leqslant N} (g(x) + g(Tx) + \ldots + g(T^{n-1}x)) \geqslant 0 \right\},$$

$$F_N = \left\{ x : \max_{1 \leqslant n \leqslant N} (h(x) + h(Tx) + \ldots + h(T^{n-1}x)) \geqslant 0 \right\}.$$

By Proposition 49.3

$$\int_{E_N} g \, d\mu_1 \geqslant 0, \int_{F_N} h \, d\mu_1 \geqslant 0 \text{ for all } N.$$

Since $\{E_N\}$ and $\{F_N\}$ increase to X_1, we have

$$\int_{X_1} g\, d\mu_1 \geqslant 0, \quad \int_{X_1} h\, d\mu_1 \geqslant 0.$$

Adding the two and using Eq. (49.4) we see that $a-b \geqslant 0$. This contradicts the choice of a, b. Hence $\mu(X_{a,b})=0$. This implies

$$\mu\left(\bigcup_{a<b,\ a,\ b \text{ rational}} X_{a,b}\right) = 0.$$

Equivalently,

$$\mu\left\{x : \underline{\lim}_{n\to\infty} (A_n f)(x) = \overline{\lim}_{n\to\infty} (A_n f)(x)\right\} = 1.$$

Let $A f(x) = \underline{\lim}_{n\to\infty} (A_n f)(x)$. Then $(Af)(Tx)=(Af)(x)$. Thus Af is an invariant borel function and hence $\mathcal{I}$ measurable. If $f \geqslant 0$ then $A_n f \geqslant 0$ for all n and by Fatou's lemma

$$\int Af\, d\mu \leqslant \underline{\lim}_{n\to\infty} \int A_n f\, d\mu = \int f\, d\mu < \infty.$$

Thus $Af \in L_1(\mu)$. If f is not non-negative we conclude by splitting f into f^+ and f^-, that Af is still integrable. Thus $A_n f$ converges almost everywhere to an integrable function Af. If f is bounded then $|A_n f|$ is uniformly bounded by a constant and hence by Lebesgue-dominated convergence theorem we obtain

$$\lim_{n\to\infty} \mathbf{E}\,|A_n f - Af| = 0.$$

If B is an invariant measurable set we have

$$\int_B f\, d\mu = \int_B A_n f\, d\mu = \lim_{n\to\infty} \int_B A_n f\, d\mu = \int_B Af\, d\mu.$$

The $\mathcal{I}$ measurability of Af shows that $Af=\mathbf{E}(f|\mathcal{I})$. If f is not bounded then for any $\varepsilon > 0$, we choose a bounded function g such that $\mathbf{E}\,|f-g| < \varepsilon$. Then

$$\begin{aligned}\mathbf{E}\,|A_n f - \mathbf{E}(f|\mathcal{I})| &\leqslant \mathbf{E}\,|A_n(f-g)| + \mathbf{E}\,|A_n g - \mathbf{E}(g|\mathcal{I})| + \\ &\qquad \mathbf{E}\,|\mathbf{E}(g-f|\mathcal{I})| \\ &\leqslant 2\mathbf{E}\,|f-g| + \mathbf{E}\,|A_n g - \mathbf{E}(g|\mathcal{I})|.\end{aligned}$$

Letting $n\to\infty$, we have

$$\overline{\lim}_{n\to\infty} \mathbf{E}\,|A_n f - \mathbf{E}(f|\mathcal{I})| < \varepsilon$$

Since ε is arbitrary the proof is complete.

Corollary. 49.5. Under the conditions of Proposition 49.4, if μ is ergodic under T, then

$$\lim_{n\to\infty} \frac{f+fT+\ldots+fT^{n-1}}{n} = \mathbf{E}f \text{ a.e. } (\mu).$$

Proof. In this case every set in $\mathcal{I}$ has probability 0 or 1. Hence $\mathbf{E}(f|\mathcal{I})=\mathbf{E}f$. This completes the proof.

49.6. Corollary. Let $(X, \mathcal{B})$ be a borel space and let T be any borel automorphism of X. If μ, ν are two ergodic invariant probability measures then $\mu \perp \nu$ or $\mu = \nu$.

Proof. If $\mu \neq \nu$, then there exists a borel set E such that $\mu(E) \neq \nu(E)$. Let

$$A = \left\{x : \frac{\chi_E(x)+\chi_E(Tx)+\ldots+\chi_E(T^{n-1}x)}{n} \to \mu(E) \text{ as } n\to\infty\right\},$$

$$B = \left\{x : \frac{\chi_E(x)+\chi_E(Tx)+\ldots+\chi_E(T^{n-1}x)}{n} \to \nu(E) \text{ as } n\to\infty\right\}.$$

Then $A \cap B = \emptyset$ and $\mu(A) = \nu(B) = 1$. Hence $\mu \perp \nu$.

Corollary. 49.7. Let $(X, \mathcal{B}, \mu)$ be a standard probability space and let T be a borel map of $(X, \mathcal{B})$ into itself such that $\mu T^{-1} = \mu$. Then there exists a standard borel space $(Y, \mathcal{C})$ and a map $y \to \mu_y$ from Y into the space of probability measures on $(X, \mathcal{B})$ with the following properties:

(i) $\mu_y T^{-1} = \mu_y$ and μ_y is ergodic for all $y \in Y$;

(ii) for any $E \in \mathcal{B}$, the map $y \to \mu_y(E)$ is borel;

(iii) the map $y \to \mu_y$ is $1-1$;

(iv) there exists a probability measure ν on Y such that

$$\mu(E) = \int \mu_y(E)\, d\nu(y) \text{ for all } E \in \mathcal{B}.$$

Proof. Let $\mathcal{I}$ be the σ-algebra of all invariant sets in $\mathcal{B}$. Choose a countable dense set $\{\phi_n\}$, $n = 1, 2, \ldots$ in the subspace of all $\mathcal{I}$-measurable functions in $L_2(\mu)$ and consider the map γ from X into R^∞ (the countable product of real lines) defined by

$$\gamma(x) = (\phi_1(x), \phi_2(x), \ldots).$$

Let $\mathcal{B}^\infty$ be the borel σ-algebra of R^∞. Then $(R^\infty, \mathcal{B}^\infty)$ is standard. Let $\mathcal{I}_0=\gamma^{-1}(\mathcal{B}^\infty)$. Since $\gamma(x)=\gamma(Tx)$, we have $\mathcal{I}_0 \subset \mathcal{I}$. Let $p(\xi, A)$ be a regular conditional probability distribution for μ given $\gamma = \xi$ and let $\nu = \mu\gamma^{-1}$. Choose and fix a sequence $E_1, E_2, \ldots$ of sets in $\mathcal{B}$ which constitute a boolean algebra and generate $\mathcal{B}$.

For any $F \in \mathscr{B}^\infty$, consider the function $\chi_F(\gamma(x))\, \chi_{E_j}(x)$. The invariance of μ under T implies

$$\int \chi_F(\gamma(x))\, \chi_{E_j}(Tx)\, d\mu(x) = \int \chi_F(\gamma(x))\, \chi_{E_j}(x)\, d\mu(x).$$

By the properties of regular conditional probability distribution we have

$$p(\xi, \gamma^{-1}(\{\xi\})) = 1 \text{ a.e. } \xi\ (\nu),$$

$$\int_F \Big[\int \chi_{E_j}(Tx)\, p(\xi, dx) \Big]\, d\nu(\xi)$$

$$= \int_F \Big[\int \chi_{E_j}(x)\, p(\xi, dx) \Big]\, d\nu(\xi) \text{ for all } F \in \mathscr{B}^\infty,$$

$$j = 1, 2, \ldots.$$

Hence there exists a set $N_0 \subset R^\infty$ such that $\nu(N_0) = 0$ and

$$p(\xi, \gamma^{-1}(\{\xi\})) = 1;$$

$$p(\xi, T^{-1} E_j) = p(\xi, E_j) \text{ for all } j = 1, 2, \ldots, \xi \notin N_0.$$

By the choice of the E_j's we have

$$p(\xi, T^{-1} E) = p(\xi, E) \text{ for all } E \in \mathscr{B},\ \xi \notin N_0. \tag{49.5}$$

If $A \in \mathscr{I}_0$ it is clear that $p(\xi, A) = 0$ or 1 for each $\xi \notin N_0$ according as $\gamma^{-1}(\{\xi\}) \subset A$ or $\gamma^{-1}(\{\xi\}) \subset A'$. Hence every $\mathscr{I}_0$ measurable function is a constant almost everywhere with respect to the measure $p(\xi, .)$, for each $\xi \notin N_0$. For any borel function f on X, let

$$(Af)(x) = \lim_{n\to\infty} \frac{f(x) + f(Tx) + \ldots + f(T^{n-1}x)}{n}.$$

For each E_j consider the function

$$\psi_j = A\chi_{E_j},\ j = 1, 2, \ldots.$$

By the choice of the functions $\{\phi_n\}$ it is clear that there exists an $\mathscr{I}_0$ measurable function $\tilde{\psi}_j$ such that

$$\psi_j = \tilde{\psi}_j \text{ a.e. } (\mu) \text{ for each } j = 1, 2, \ldots.$$

Since

$$\mu(E) = \int p(\xi, E)\, d\nu\, (\xi),$$

it follows that

$$\psi_j(x) = \text{const. a.e. } p(\xi, .)$$

for all $\xi \notin N_j$ where $\nu(N_j) = 0$. Let $\xi \notin \bigcup_0^\infty N_j$. Then

$$A\chi_{E_j} = \text{const. a.e. } p(\xi, .),\ j = 1, 2, \ldots. \tag{49.6}$$

We write

$$Y = R^\infty - \bigcup_0^\infty N_j$$

$$\mathscr{C} = \mathscr{B}^\infty \cap Y,$$

$$\mu_y(E) = p(y, E) \text{ for all } E \in \mathscr{B},\ y \in Y.$$

We denote conditional expectation with respect to μ_y given the invariant σ-algebra $\mathcal{I}$ by $\mathbf{E}_y(\,|\,\mathcal{I})$. Equations (49.5), (49.6) and Birkhoff's ergodic theorem imply

$$A\,\chi_{E_j} = \mathbf{E}_y(\chi_{E_j}\,|\,\mathcal{I}) = \text{const a.e. } (\mu_y)$$

for all $j=1, 2, \ldots$ and $y \in \Upsilon$. Since the χ_{E_j}'s span the space $L_1(\mu_y)$ for any fixed y we have

$$\mathbf{E}_y(f\,|\,\mathcal{I}) = \text{constant a.e. } (\mu_y)$$

for every $f \in L_1(\mu_y)$. This can happen if and only if $\mu_y(E)=0$ or 1 for every $E \in \mathcal{I}$. In other words μ_y is ergodic for each y. Since μ_y is still the regular conditional probability distribution given γ, properties (ii), (iii), (iv) follow immediately and the proof is complete.

Example. 49.8. Consider the unit interval [0, 1] with the uniform distribution. Let T be the transformation defined by

$$Tx = x+a \pmod 1)$$

where a is an irrational number. It is left as an exercise for the reader to show that T is measure preserving. We claim that T is an ergodic automorphism. Indeed, let f be any bounded T-invariant borel function on [0, 1]. Then we expand f by Fourier series:

$$f(x) = \sum_{-\infty}^{+\infty} a_n\, e^{2\pi i n x},$$

(See Example 42.7), Then

$$f(x+a) = \sum_{-\infty}^{+\infty} a_n\, e^{2\pi i n a} \cdot e^{2\pi i n x}.$$

Since $f(x+a) = f(x)$ for all x, we have by the uniqueness of Fourier coefficients,

$$a_n\, e^{2\pi i n a} = a_n \text{ for all } n=0, \pm 1, \pm 2, \ldots.$$

Since a is irrational $e^{2\pi i n a}=1$ if and only if $n=0$. Hence $a_n=0$ for all $n \neq 0$. In other words f is a constant almost everywhere. In particular, if E is a T invariant set then χ_E is an invariant borel function and χ_E = constant almost everywhere. This can happen if and only if E has measure 0 or 1. Thus T is ergodic.

This fact can be used to do numerical integration of functions. Indeed by Birkhoff's ergodic theorem we have for integrable functions f on the unit interval,

$$\int_0^1 f(x)\,dx = \lim_{n\to\infty} \frac{f(x)+f(x+a)+f(x+2a)+\ldots+f(x+\overline{n-1}\,a)}{n}$$

almost everywhere, where $x+ja$ is taken as mod 1.

Exercise. 49.9. Let $a_1, a_2, \ldots, a_k$ be k rationally independent irrational numbers, i.e., for no set of integers (positive or negative) $n_1, n_2, \ldots, n_k$ the number $n_1 a_1 + n_2 a_2 + \ldots + n_k a_k$ is rational. Let I^k be the k-fold product of the unit interval with Lebesgue measure. Let T be the transformation defined by

$$T\mathbf{x} = \mathbf{x} + \boldsymbol{\alpha} \pmod 1, \; \mathbf{x} \in I^k,$$

where $\mathbf{x}$ denotes an arbitrary point of I^k and addition is coordinate wise. Then T is ergodic.

Remark. 49.10. Combining Proposition 49.2 and Birkhoff's ergodic theorem we get the strong law of large numbers: if $f_1, f_2, \ldots$ are independent and identically distributed random variables with distribution μ and finite mean m, we have

$$\lim_{n \to \infty} \frac{f_1 + f_2 + \ldots + f_n}{n} = m \text{ a.e.}$$

Once again we can apply this result for doing numerical integration. Suppose μ is a given distribution on the real line and there is a method of simulating random variables $\zeta_1, \zeta_2, \ldots$ which are independent and which have distribution μ. For any $\phi \in L_1(\mu)$, Birkhoff's theorem shows that

$$\lim_{n \to \infty} \frac{\phi(\zeta_1) + \phi(\zeta_2) + \ldots + \phi(\zeta_n)}{n} = \int \phi \, d\mu \text{ a.e.}$$

Thus $\dfrac{\phi(\zeta_1) + \phi(\zeta_2) + \ldots + \phi(\zeta_n)}{n}$ can be used as an approximation for $\int \phi \, d\mu$.

In particular, if f is a function on R which is integrable with respect to Lebesgue measure and $\zeta_1, \zeta_2, \ldots$ are independent and identically distributed with standard normal distribution then we put

$$\phi(x) = \sqrt{2\pi} \, f(x) e^{\frac{1}{2} x^2}$$

and observe that

$$\lim_{n \to \infty} \frac{\phi(\zeta_1) + \phi(\zeta_2) + \ldots + \phi(\zeta_n)}{n} = \int_{-\infty}^{+\infty} f(x) \, dx$$

almost everywhere.

To do numerical integration on the half line $(0, \infty)$ we can consider, for example, the exponential distribution μ defined by $\mu((-\infty, x]) = 0$ if $x \leqslant 0$, $= 1 - e^{-x}$ if $x > 0$. If $\zeta_1, \zeta_2, \ldots$ are independent and identically distributed with distribution μ and f is an integrable function in $(0, \infty)$ then we put

$$\phi(x) = f(x)\, e^x$$

and observe that

$$\lim_{n\to\infty} \frac{\phi(\zeta_1) + \phi(\zeta_2) + \dots + \phi(\zeta_n)}{n} = \int_0^\infty f(x)\, dx$$

almost everywhere.

Simulating a sequence of random variables with a given distribution μ and using it to calculate functionals (like integrals) is usually known as *Monte Carlo method*. (See Remark 25.2.)

Exercise. 49.11. Let $(X, \mathscr{B}, \mu)$ be a probability space and let T be a borel automorphism of $(X, \mathscr{B})$ preserving μ. Let $k(x)$ be a function on X defined by

$$k(x) = \inf \{ j : j \geqslant 1,\ T^j x \in A \},$$

where A is some fixed set in $\mathscr{B}$ with $\mu(A) > 0$. Then $k(.)$ is an integer-valued random variable. Define a transformation T_A on A by

$$\begin{aligned} T_A x &= T^{k(x)} x \text{ if } k(x) < \infty, \\ &= x \text{ if } k(x) = \infty. \end{aligned}$$

Let $A_j = \{x : k(x) = j\}$. Then $\mu(A_\infty) = 0$. If μ_A is the probability measure defined on $\mathscr{B} \cap A$ by

$$\mu_A(E) = \frac{\mu(E)}{\mu(A)},\quad E \in \mathscr{B} \cap A,$$

Then T_A preserves μ_A. Further T_A is ergodic for μ_A if T is ergodic for μ. (T_A is known as the *induced transformation* on A).

Exercise. 49.12. Let $(X, \mathscr{B}, \mu)$ be a probability space and let T be a borel automorphism of $(X, \mathscr{B})$ such that $\mu T \equiv \mu$. Let ρ be the measure on $(X \times R, \mathscr{B} \times \mathscr{B}_R)$ defined by $\rho = \mu \times \nu$, where

$$\nu(E) = \int_E e^{-t}\, dt.$$

Let $\psi(x) = \log \dfrac{d\mu T}{d\mu}(x)$ be chosen to be finite at all x and let

$$\tilde{T}(x, t) = (Tx, t + \psi(x)).$$

Then $\tilde{T}$ is a borel automorphism of $X \times R$ which preserves the measure ρ. If T_1 and T_2 are two borel automorphisms for which $\mu T_i \equiv \mu$, $i = 1, 2$ then $(T_1 T_2)^\sim = \tilde{T}_1 \tilde{T}_2$. (The extension $\tilde{T}$ constructed on the space $X \times R$ is known as *skew product*.)

§50. Dominated Ergodic Theorem

In Section 49 we showed that for any $f \in L_1(\mu)$ and any measure preserving transformation T, the average

$$A_n f = \frac{f + fT + \ldots + fT^{n-1}}{n}$$

converges a.e. (μ) as well as in $L_1(\mu)$. If $f \in L_p(\mu)$ for some $p > 1$, it is natural to ask whether $A_n f$ converges in $L_p(\mu)$. To answer this question we shall prove what is known as dominated ergodic theorem. It is due to N. Wiener. The proofs are very similar to that of Proposition 44.19.

Proposition. 50.1. Let $(X, \mathscr{B}, \mu)$ be a probability space and let T be a μ-measure preserving transformation. Let $f \in L_1(\mu)$ and let

$$f^*(x) = \sup_{n \geq 1} |(A_n f)(x)|, \; x \in X.$$

Then, for any $\varepsilon > 0$,

$$\mu\{x : f^*(x) > \varepsilon\} \leq \frac{1}{\varepsilon} \int |f| \, d\mu.$$

Proof. Let

$$E_\varepsilon = \{x : f^*(x) > \varepsilon\}.$$

Without loss of generality we may assume that f is non-negative. Then $A_n f \geq 0$ for all n. If $\mu(E_\varepsilon) \neq 0$ we consider the function $f - \varepsilon$ and apply Proposition 49.3 to the space E_ε with the probability measure

$$\mu_\varepsilon(A) = \frac{\mu(A)}{\mu(E_\varepsilon)}, \; A \in \mathscr{B}, \; A \subset E_\varepsilon.$$

Exactly as in the beginning of the proof of Proposition 49.4 we get

$$\int_{E_\varepsilon} (f - \varepsilon) \, d\mu \geq 0.$$

Then

$$\mu(E_\varepsilon) \leq \frac{1}{\varepsilon} \int_{E_\varepsilon} f \, d\mu \leq \frac{1}{\varepsilon} \int |f| \, d\mu.$$

This completes the proof.

Proposition. 50.2. Let $(X, \mathscr{B}, \mu)$ be a probability space and let T, f, f^* be as in the preceding proposition. Let $f \in L_p(\mu)$ for some $p > 1$. Then $f^* \in L_p(\mu)$.

Proof. We proceed exactly as in the proof of Proposition 44.19. Without loss of generality we may assume $f \geqslant 0$. Define

$$g(x) = f(x) \text{ if } f(x) > \frac{t}{2},$$
$$= 0 \text{ otherwise.}$$

Then

$$f(x) < g(x) + \tfrac{1}{2} t,$$
$$f^*(x) < g^*(x) + \tfrac{1}{2} t,$$

where

$$g^*(x) = \sup_{n \geqslant 1} (A_n g)(x).$$

Exactly as in the proof of Proposition 44.19 (replacing g and h by f and g respectively) we get

$$\int_0^\infty t^{p-1} \mu\{x : f^*(x) > t\}\, dt < \infty.$$

By Proposition 44.18, it follows that $f^* \in L_p(\mu)$. This completes the proof.

Corollary. 50.3. (Dominated L_p ergodic theorem.) Let $(X, \mathscr{B}, \mu)$ be a probability space and let T be a borel map of $(X, \mathscr{B})$ into itself such that $\mu T^{-1} = \mu$. For any $p > 1$ and any $f \in L_p(\mu)$

$$\lim_{n \to \infty} \mathbf{E} \left| A_n f - \mathbf{E}(f \mid \mathscr{I}) \right|^p = 0,$$

where $\mathscr{I}$ is the σ-algebra of T invariant sets and

$$A_n f = \frac{f + fT + \ldots + fT^{n-1}}{n}.$$

Further there exists a non-negative function $f^* \in L_p(\mu)$ such that

$$|A_n f| \leqslant f^* \text{ for all } n = 1, 2, \ldots$$

Proof. Define $f^* = \sup_{n \geqslant 1} |A_n f|$. By Proposition 50.2, $f^* \in L_p(\mu)$. By Birkhoff's ergodic theorem and Lebesgue's dominated convergence theorem we get the required result. This completes the proof.

Exercise. 50.4. Let $(X, \mathscr{B}, \mu)$ and T be as in the preceding corollary. Let $\log^+ t = \log t$ if $t > 1$ and 0 otherwise. Let f be a random variable on X such that $|f| \log^+ |f| \in L_1(\mu)$. Then the function $f^* = \sup_{n \geqslant 1} |A_n f|$ is integrable.

We conclude the section with the note that the results of the last two sections constitute the beginnings of a new subject called ergodic theory. The interested reader may look for further developments in [3], [5] and [9].

CHAPTER SEVEN

Weak Convergence of Probability Measures

§51. Criteria for Weak Convergence in the Space of Probability Measures

Throughout this chapter we shall concern ourselves with the study of probability measures on separable metric spaces only. As usual, for any such metric space X we shall write $\mathscr{B}_X$ for the borel σ-algebra of subsets of X. We shall denote by $C(X)$ the space of all bounded real valued continuous functions on X and $M_0(X)$ the space of all probability measures on $\mathscr{B}_X$.

In the very first chapter we have seen the importance of limit theorems in probability theory. Distributions like binomial, multinomial, hypergeometric, Bose-Einstein distributions which involve complicated expressions like factorials of large integers were approximated by Poisson and normal distributions which have simpler expressions. Thus, from the computational point of view, the role of limit theorems in probability theory is quite clear. We shall try to formalise this idea into a theory.

Suppose $\{\mu_n\}$ is a sequence of probability distributions on X. We can say that μ_n *converges* to a distribution μ for some *large class* $\mathscr{E}$ of sets in $\mathscr{B}_X$ if $\mu_n(E) \to \mu(E)$ as $n \to \infty$. Equivalently, we can write $\int \chi_E \, d\mu_n \to \int \chi_E \, d\mu$ as $n \to \infty$. Thus *convergence* can be interpreted as

$$\lim_{n \to \infty} \int f \, d\mu_n = \int f \, d\mu \tag{51.1}$$

for a large class $\mathscr{D}$ of functions on X. Let us write for any $x \in X$, δ_x, for the probability measure degenerate at x. If $x_n \to x$ as $n \to \infty$ in X it is natural to demand that δ_{x_n} converges to δ_x as $n \to \infty$. Since $\int f \, d\delta_x = f(x)$, Eq. (51.1) means that $f(x_n) \to f(x)$ as $n \to \infty$ for $f \in \mathscr{D}$. In other words we demand that the functions in $\mathscr{D}$ be continuous. In view of this observation we introduce the following definition.

Definition. 51.1. Let X be a separable metric space. A sequence $\{\mu_n\}$ in $M_0(X)$ is said to *converge weakly* to an element μ in $M_0(X)$ if

$$\lim_{n\to\infty} \int f\,d\mu_n = \int f\,d\mu \text{ for every } f\in C(X).$$

In such a case we write $\mu_n \Longrightarrow \mu$ as $n\to\infty$.

Proposition. 51.2. Let μ_n be a sequence in $M_0(X)$ and let $\mu\in M_0(X)$. Then the following conditions are equivalent:

(i) $\mu_n \Longrightarrow \mu$;

(ii) $\lim_{n\to\infty} \int g\,d\mu_n = \int g\,d\mu$ for every $g\in U(X)$ where $U(X)$ is the space of all bounded real valued uniformly continuous functions on X;

(iii) $\overline{\lim}_{n\to\infty} \mu_n(C) \leqslant \mu(C)$ for every closed set C;

(iv) $\underline{\lim}_{n\to\infty} \mu_n(G) \geqslant \mu(G)$ for every open set G;

(v) $\lim_{n\to\infty} \mu_n(A) = \mu(A)$ for every borel set A whose boundary has μ-measure 0.

Proof. Since $C(X)\supset U(X)$ it is clear that (i) → (ii). We shall now prove that (ii) → (iii). For any closed set C consider the function $d(x, C)$ defined in Proposition 19.11. Then $d(x, C)\in U(X)$. Let $G_n = \left\{x : d(x, C) < \frac{1}{n}\right\}$. Then C and G_n' are disjoint closed sets. If

$$f_n(x) = \frac{d(x, C)}{d(x, C) + d(x, G_n')},$$

then $\inf_{x\in C,\, y\in G_n'} d(x, y) \geqslant \frac{1}{n}$ and Proposition 19.11 implies that $f_n\in U(X)$ for all n. Further $0\leqslant f_n\leqslant 1$, $f_n(x)=1$ for $x\in G_n'$, $f_n(x)=0$ for $x\in C$. We have $G_1\supset G_2\supset\ldots$ and $\bigcap G_n = C$. Thus

$$\begin{aligned}\overline{\lim}_{n\to\infty} \mu_n(C) &\leqslant \lim_{n\to\infty} \int f_k\,d\mu_n\\ &= \int f_k\,d\mu \leqslant \mu(G_k).\end{aligned}$$

Letting $k\to\infty$, we get

$$\overline{\lim}_{n\to\infty} \mu_n(C) \leqslant \mu(C).$$

It is now clear that (iii) and (iv) are equivalent because open sets and closed sets are complements of each other and the whole space has measure unity for all probability measures. We shall now prove that

(iii) and (iv) imply (v). Let $A \in \mathscr{B}_X$, and let A^0 and $\bar{A}$ denote the interior and closure of A respectively. Suppose $\mu(\bar{A}-A^0)=0$. Since $A^0 \subset A \subset \bar{A}$ we have

$$\overline{\lim_{n\to\infty}} \mu_n(A) \leqslant \overline{\lim_{n\to\infty}} \mu_n(\bar{A}) \leqslant \mu(\bar{A}) = \mu(A),$$

$$\underline{\lim_{n\to\infty}} \mu_n(A) \geqslant \underline{\lim_{n\to\infty}} \mu_n(A^0) \geqslant \mu(A^0) = \mu(A).$$

Hence $\lim_{n\to\infty} \mu_n(A) = \mu(A)$.

Now we shall complete the proof by showing that (v) implies (i). Let $f \in C(X)$ and let $\mu_n(A) \to \mu(A)$ as $n \to \infty$ for every borel set A such that $\mu(\bar{A}-A^0)=0$. The distribution μf^{-1} in the real line is concentrated in a bounded interval (a, b). Further μf^{-1} can have at the most a countable number of atoms. Hence we can find for any fixed $\varepsilon>0$ numbers $t_1, t_2, \ldots, t_m$ such that (a) $a=t_0<t_1<\ldots<t_m=b$; (b) $a<f(x)<b$ for all x; (c) $t_j-t_{j-1}<\varepsilon$ for all $j=1, 2, \ldots, m$; (d) $\mu f^{-1}(\{t_j\})=0$ for all $j=1, 2, \ldots, m$. Let

$$A_j = f^{-1}([t_{j-1}, t_j)), \; j=1, 2, \ldots, m.$$

Then $A_1, A_2, \ldots, A_m$ are disjoint borel sets and $X=\bigcup_j A_j$. Further, $\bar{A}_j - A_j^0 \subset f^{-1}(\{t_{j-1}\}) \cup f^{-1}(\{t_j\})$ so that $\mu(\bar{A}_j - A_j^0)=0$. Thus

$$\lim_{n\to\infty} \mu_n(A_j) = \mu(A_j), \; j=1, 2, \ldots, m.$$

Let

$$f^* = \sum_{j=1}^{m} t_{j-1} \chi_{A_j}.$$

Then $|f^*(x) - f(x)| < \varepsilon$ for all x and

$$\begin{aligned} \left|\int f\,d\mu_n - \int f\,d\mu\right| &\leqslant \int |f-f^*|\,d\mu_n + \int |f-f^*|\,d\mu + \left|\int f^*\,d\mu_n - \int f^*\,d\mu\right| \\ &\leqslant 2\varepsilon + \sum_{j=1}^{m} |\mu_n(A_j) - \mu(A_j)|\,|t_{j-1}|. \end{aligned}$$

Letting $n\to\infty$ we have

$$\overline{\lim_{n\to\infty}} \left|\int f\,d\mu_n - \int f\,d\mu\right| \leqslant 2\varepsilon.$$

Letting $\varepsilon \to 0$ we have the required result and the proof is complete.

Corollary. 51.3. Let (Ω, S, P) be any probability space and let $\{f_n\}, \{g_n\}$ be two sequences of X-valued random variables such that

$$\lim_{n\to\infty} d(f_n(\omega), g_n(\omega))=0 \text{ in } P \text{ measure.} \tag{51.2}$$

If $Pf_n^{-1} \Longrightarrow \mu$ in $M_0(X)$, then $Pg_n^{-1} \Longrightarrow \mu$.

Proof. Let ϕ be any bounded real valued uniformly continuous function on X. Then

$$\lim_{n\to\infty}\int \phi(f_n(\omega))\,dP(\omega) = \lim_{n\to\infty}\int \phi\, dPf_n^{-1} = \int \phi\, d\mu. \tag{51.3}$$

Let $\varepsilon > 0$ be arbitrary. By the uniform continuity of ϕ we can choose a $\delta > 0$ such that $|\phi(x) - \phi(y)| < \varepsilon$ whenever $d(x, y) < \delta$. We have

$$P\{|\phi(f_n(\omega)) - \phi(g_n(\omega))| < \varepsilon\} \geqslant P\{d(f_n(\omega), g_n(\omega)) < \delta\}.$$

By Eq. (51.2) the right hand side tends to unity and so does the left hand side of the above inequality. This implies that

$$\lim_{n\to\infty} [\phi(f_n(\omega)) - \phi(g_n(\omega))] = 0 \text{ in } P \text{ measure.}$$

Since ϕ is bounded we have by Lebesgue dominated convergence theorem

$$\lim_{n\to\infty} \int [\phi(f_n(\omega)) - \phi(g_n(\omega))]\, dP(\omega) = 0.$$

Now Eq. (51.3) implies

$$\lim_{n\to\infty}\int \phi\, dPg_n^{-1} = \lim_{n\to\infty}\int \phi(g_n(\omega))\, dP(\omega) = \int \phi\, d\mu.$$

Now criterion (ii) of Proposition 51.2 completes the proof.

Definition. 51.4. For any probability measure μ on X, a set $A \in \mathscr{B}_X$ is called a *μ-continuity set* if the boundary $\bar{A} - A^0$ of A has μ-measure zero.

Exercise. 51.5. Let X be a separable metric space and let $\mu \in M_0(X)$. Then all μ-continuity sets form a boolean algebra.

In the case of real line we can make criterion (v) of Proposition 51.2 much simpler. Indeed, we have the following.

Proposition. 51.6. Let μ_n, μ be probability distributions on R and let F_n, F be their distribution functions respectively, where $n = 1, 2, \ldots$. Then $\mu_n \Longrightarrow \mu$ as $n \to \infty$ if and only if $F_n(x) \to F(x)$ as $n \to \infty$ for every x which is a continuity point of F.

Proof. Let $\mu_n \Longrightarrow \mu$ and let x be a point of continuity of F. Then $(-\infty, x)$ is a continuity set of μ. Hence $F_n(x) \to F(x)$ as $n \to \infty$.

To prove the converse consider any bounded continuous function ϕ on R with $\sup_{x\in R} |\phi(x)| = \alpha$. Let $\varepsilon > 0$ be arbitrary. Since the set of all discontinuity points of F is countable we can choose continuity points, a, b of F such that

$$F(b) - F(a) > 1 - \varepsilon. \tag{51.4}$$

Since $F_n(b) \to F(b)$ and $F_n(a) \to F(a)$ as $n \to \infty$ there exists an integer n_0 such that

$$F_n(b) - F_n(a) > 1 - \varepsilon \text{ for all } n > n_0. \tag{51.5}$$

Since ϕ is uniformly continuous in any bounded interval we can select continuity points $t_0 = a < t_1 < t_2 < \ldots < t_k = b$ of F satisfying

$$\sup_{(a,\, b]} \left| \phi(x) - \sum_{i=0}^{k-1} \phi(t_i)\, \chi_{(t_i,\, t_{i+1}]}(x) \right| < \varepsilon. \tag{51.6}$$

Let

$$\phi_k(x) = \sum_{i=0}^{k-1} \phi(t_i)\, \chi_{(t_i,\, t_{i+1}]}(x).$$

Since t_j's are continuity points of F we have

$$\lim_{n\to\infty} \int_{(a,\, b]} \phi_k \, d\mu_n = \int_{(a,\, b]} \phi_k \, d\mu. \tag{51.7}$$

We have

$$\left| \int \phi d\mu_n - \int \phi \, d\mu \right| \leqslant a \int_{(-\infty,\, a] \cup (b,\, \infty)} 1 \, d\mu_n + a \int_{(-\infty,\, a] \cup (b,\, \infty)} 1 \, d\mu +$$

$$\int_{(a,\, b]} |\phi_k - \phi| \, d\mu_n + \int_{(a,\, b]} |\phi_k - \phi| \, d\mu + \left| \int \phi_k \, d\mu_n - \int \phi_k \, d\mu \right|.$$

Now inequalities (51.4) and (51.5) imply that the first two terms are less than $a\varepsilon$ for $n > n_0$. Inequality (51.6) implies that the next two terms are less than ε. By inequality (51.7) the last term tends to 0 as $n \to \infty$. Hence

$$\overline{\lim_{n\to\infty}} \left| \int \phi \, d\mu_n - \int \phi \, d\mu \right| \leqslant 2\,(a+1)\varepsilon.$$

Since ε is arbitrary we have

$$\lim_{n\to\infty} \int \phi \, d\mu_n = \int \phi \, d\mu.$$

Thus $\mu_n \Longrightarrow \mu$ as $n \to \infty$ and the proof is complete.

As an illustration of the usefulness of this result we shall now deduce the central limit theorem for identically distributed random variables from Proposition 9.3. To this end we need an elementary inequality.

Proposition. 51.7. Let f, g be random variables on any probability space (Ω, S, P) with distribution functions F, G respectively. Suppose $\mathbf{E}\,|f-g|^2 < \varepsilon^2$. Then

$$G(t - \sqrt{\varepsilon}) - \varepsilon \leqslant F(t) \leqslant G(t + \sqrt{\varepsilon}) + \varepsilon \text{ for all } t \in R.$$

Proof. We have by Chebyshev's inequality

$$\begin{aligned} &P(f \leqslant t) - P(g \leqslant t + \sqrt{\varepsilon}) \\ &\leqslant P(f \leqslant t,\ g > t + \sqrt{\varepsilon}) \\ &\leqslant P(|f - g| > \sqrt{\varepsilon}) \leqslant \varepsilon. \end{aligned}$$

Thus $F(t) \leqslant G(t + \sqrt{\varepsilon}) + \varepsilon$. Interchanging F and G and changing t to $t - \sqrt{\varepsilon}$ we obtain $F(t) \geqslant G(t - \sqrt{\varepsilon}) - \varepsilon$. This completes the proof.

Proposition. 51.8. (Central limit theorem.) Let (Ω, S, P) be a probability space and let $f_1, f_2, \ldots$ be independent and identically distributed random variables on Ω with $\mathbf{E}f_i = 0$, $\mathbf{E}f_i^2 = 1$. Then the distribution of $\frac{1}{\sqrt{n}}\,(f_1 + f_2 + \ldots + f_n)$ converges weakly to the standard normal distribution.

Proof. If $f_1, f_2, \ldots$ are simple then the result follows from Remark 9.4 and Proposition 51.6. In the general case we shall approximate by simple random variables. Choose a sequence of non-negative simple functions $\{u_k\}$ on $[0, \infty)$ such that $u_k(t)$ increases to t for every $t \geqslant 0$. Define v_k by

$$\begin{aligned} v_k(t) &= u_k(t) \text{ if } t \geqslant 0, \\ &= -u_k(-t) \text{ if } t < 0. \end{aligned}$$

Then $|v_k(t)| \leqslant |t|$ for all t and $v_k(t) \to t$ as $k \to \infty$. Define the random variables $\{s_{kn}\}$ on Ω by putting

$$\begin{aligned} m_k &= \mathbf{E}(v_k(f_n)); \\ \sigma_k^2 &= \mathbf{V}(v_k(f_n)); \\ s_{kn} &= \frac{v_k(f_n) - m_k}{\sigma_k}. \end{aligned}$$

Then for any fixed k, $s_{k1}, s_{k2}, \ldots$ are independent and identically distributed random variables with mean zero and variance unity. Further by Lebesgue dominated convergence theorem

$$\lim_{k \to \infty} \mathbf{E}\,|s_{kn} - f_n|^2 = 0 \text{ for any } n.$$

Let

$$S_{kn} = \frac{s_{k1} + s_{k2} + \ldots + s_{kn}}{\sqrt{n}},$$

$$S_n = \frac{f_1 + f_2 + \ldots + f_n}{\sqrt{n}}$$

and let F_{kn} and F_n be the distribution functions of S_{kn} and S_n respec tively. We have from the independence of the random variables $s_{k1} - f_1, s_{k2} - f_2, \ldots,$ for any fixed k,

$$\mathbf{E}(S_{kn} - S_n)^2 = \mathbf{E}(s_{k1} - f_1)^2.$$

Let now $\varepsilon > 0$ be arbitrary. Since the right hand side above tends to 0 as $k \to \infty$ we can choose a k_0 so large that

$$\mathbf{E}(S_{k_0 n} - S_n)^2 < \varepsilon \text{ for all } n.$$

By the preceding proposition we have

$$F_{k_0 n}(t - \sqrt{\varepsilon}) - \varepsilon \leqslant F_n(t) \leqslant F_{k_0 n}(t + \sqrt{\varepsilon}) + \varepsilon \text{ for all } t.$$

If $\Phi(t)$ stands for the distribution function of the standard normal distribution we have from the central limit theorem for simple random variables

$$\lim_{n \to \infty} F_{k_0 n}(t) = \Phi(t) \text{ for all } t.$$

Thus

$$\Phi(t - \sqrt{\varepsilon}) - \varepsilon \leqslant \underline{\lim}_{n \to \infty} F_n(t) \leqslant \overline{\lim}_{n \to \infty} F_n(t) \leqslant \Phi(t + \sqrt{\varepsilon}) + \varepsilon.$$

Letting $\varepsilon \to 0$ we have

$$\lim_{n \to \infty} F_n(t) = \Phi(t).$$

Now Proposition 51.6 implies the required weak convergence and completes the proof.

Exercise. 51.9. Let μ_n and μ be probability measures on R with distribution functions F_n and F respectively. If F is continuous everywhere and $\mu_n \Longrightarrow \mu$ as $n \to \infty$ then

$$\lim_{n \to \infty} \sup_{t \in R} |F_n(t) - F(t)| = 0.$$

§52. Prohorov's theorem

The aim of the present section will be to describe a criterion due to Yu. V. Prohorov for a sequence of probability distributions to possess a convergent subsequence.

Proposition. 52.1. Let X be a compact metric space and let $\{\mu_n\}$ be a sequence of probability measures on X. Then $\{\mu_n\}$ possesses a weakly convergent subsequence.

Proof. Consider the Banach space $C(X)$ with the norm

$$\|\phi\| = \sup_{x \in X} |\phi(x)|.$$

Since X is compact $C(X)$ is a separable metric space under the metric induced by this norm. Choose a dense sequence $\{\phi_n\}$ in $C(X)$. By the diagonal procedure we can choose a subsequence $\{\mu_{n_k}\}$ of $\{\mu_n\}$ such that

$$\lim_{k \to \infty} \int \phi_j \, d\mu_{n_k} = a_j, \; j = 1, 2, \ldots$$

exists for all j. Now consider any $\phi \in C(X)$. For any $\phi > 0$, choose a ϕ_i such that $\| \phi - \phi_j \| < \varepsilon$. Then

$$\begin{aligned} &\left| \int \phi \, d\mu_{n_k} - \int \phi \, d\mu_{n_m} \right| \leqslant \\ &\left| \int \phi_j \, d\mu_{n_k} - \int \phi_j \, d\mu_{n_m} \right| + \int | \phi - \phi_j | \, d\mu_{n_k} + \int | \phi - \phi_j | \, d\mu_{n_m}. \end{aligned}$$

The last two terms on the right side are less than ε. The first one tends to zero as k and m tend to ∞. Since ε is arbitrary it is clear that

$$\lim_{k, m \to \infty} \left| \int \phi \, d\mu_{n_k} - \int \phi \, d\mu_{n_m} \right| = 0.$$

Thus $\int \phi \, d\mu_{n_k}$ converges as $k \to \infty$ for every ϕ. Let

$$\Lambda(\phi) = \lim_{k \to \infty} \int \phi \, d\mu_{n_k}, \; \phi \in C(X).$$

Then Λ is a non-negative linear functional on $C(X)$ such that $\Lambda(1) = 1$. By Corollary 33.5 it follows that there exists a $\mu \in M_0(X)$ satisfying

$$\Lambda(\phi) = \int \phi \, d\mu \text{ for all } \phi \in C(X).$$

This shows that $\{\mu_{n_k}\}$ converges weakly to μ as $k \to \infty$.

Definition. 52.2. Let X be a separable metric space and let $\{\mu_n\}$ be a sequence in $M_0(X)$. $\{\mu_n\}$ is said to be *uniformly tight* if for every $\varepsilon > 0$ there exists a compact set $K_\varepsilon \subset X$ such that

$$\mu_n(K_\varepsilon) > 1 - \varepsilon \text{ for all } n = 1, 2, \ldots.$$

Proposition. 52.3. (Prohorov's theorem.) Let X be a separable metric space and let $\{\mu_n\}$ be a sequence in $M_0(X)$ which is uniformly tight. Then $\{\mu_n\}$ has a weakly convergent subsequence.

If X is a complete and separable metric space and $\{\mu_n\}$ converges weakly then it is uniformly tight.

Proof. Let X be a separable metric space. Then by Urysohn's theorem (See p. 125, [12]) X can be considered as a subset of a compact metric space $\widetilde{X}$ with the relative topology. For any $\mu \in M_0(X)$, define $\widetilde{\mu} \in M_0(\widetilde{X})$ by

$$\widetilde{\mu}(A) = \mu(A \cap X), \; A \in \mathscr{B}_{\widetilde{X}}.$$

By Proposition 52.1 we extract a subsequence $\{\tilde{\mu}_{n_k}\}$ from $\{\tilde{\mu}_n\}$ which converges weakly in the space $\tilde{X}$ to a probability measure ν. For each $r = 1, 2, \ldots$ choose a compact set $K_r \subset X$ such that

$$\mu_{n_k}(K_r) \geqslant 1 - \frac{1}{r} \text{ for all } k. \tag{52.1}$$

Since K_r is compact in X, it follows that K_r is compact in $\tilde{X}$ and hence borel in $\tilde{X}$. Further

$$\tilde{\mu}_{n_k}(K_r) = \mu_{n_k}(K_r) \text{ for } r = 1, 2, \ldots; \; k = 1, 2, \ldots.$$

By Proposition 51.2 we have

$$\overline{\lim_{k\to\infty}} \; \tilde{\mu}_{n_k} (K_r) \leqslant \nu(K_r) \text{ for } r = 1, 2, \ldots.$$

Now inequality (52.1) implies that

$$\nu(K_r) \geqslant 1 - \frac{1}{r} \text{ for } r = 1, 2, \ldots.$$

If we write $E_0 = \bigcup_r K_r$ then $E_0 \subset X$, E_0 is borel in $\tilde{X}$ and $\nu(E_0) = 1$. We now claim that there exists a $\mu \in M_0(X)$ such that $\tilde{\mu} = \nu$. Indeed $\mathscr{B}_X = \mathscr{B}_{\tilde{X}} \cap X$. For any $A \in \mathscr{B}_X$ there exists a $B_1 \in \mathscr{B}_{\tilde{X}}$ such that $A = B_1 \cap X$. In such a case define $\mu(A) = \nu(B_1)$. If $B_2 \in \mathscr{B}_{\tilde{X}}$ and $A = B_2 \cap X$ then $B_1 \Delta B_2 \subset X' \subset E_0'$ and $\nu(B_1 \Delta B_2) = 0$. Thus $\nu(B_1) = \nu(B_2)$. In other words $\mu(A)$ is well-defined. Now let $A_i = B_i \cap X$ be a sequence of disjoint sets where $B_i \in \mathscr{B}_{\tilde{X}}$ for all i. Since $B_i \cap E_0 \subset B_i \cap X$ for all i, $B_i \cap E_0$ are disjoint. Thus

$$\mu(\bigcup_i A_i) = \nu(\bigcup_i B_i) = \nu(\bigcup_i (B_i \cap E_0))$$
$$= \sum_i \nu(B_i \cap E_0) = \sum_i \nu(B_i) = \sum_i \mu(A_i).$$

Thus μ is a probability measure such that $\tilde{\mu} = \nu$.

Let C be a closed subset of X. Then there exists a closed subset D of $\tilde{X}$ such that $C = D \cap X$. Since $\tilde{\mu}_{nk} \Longrightarrow \tilde{\mu}$, we have

$$\overline{\lim_{k\to\infty}} \; \mu_n (C) = \overline{\lim_{k\to\infty}} \; \tilde{\mu}_{n_k} (D)$$
$$\leqslant \tilde{\mu}(D) = \mu(C).$$

By Proposition 51.2, $\mu_{n_k} \Longrightarrow \mu$. This proves the first part.

To prove the second part we suppose that X is a complete and separable metric space. Let $\mu_n \Longrightarrow \mu$ as $n \to \infty$ in $M_0(X)$. Since X is separable we can find a sequence of open spheres $S_{n1}, S_{n2}, \ldots$ of radius $1/n$ such that

$$X = \bigcup_{j=1}^{\infty} S_{nj} \text{ for } n = 1, 2, \ldots.$$

Now we claim that for any $\delta > 0$ there exists an integer k_n such that

$$\mu_i\left(\bigcup_{j=1}^{k_n} S_{nj}\right) > 1-\delta \text{ for all } i=1, 2, \ldots .$$

Suppose this is not true. Then there exist a $\delta_0 > 0$ and sequences of integers $i_1 < i_2 < \ldots$ and $k_1 < k_2 < \ldots$ such that

$$\mu_{i_m}\left(\bigcup_{j=1}^{k_m} S_{nj}\right) \leqslant 1-\delta_0 \text{ for } m=1, 2, \ldots .$$

For any fixed r

$$\bigcup_{j=1}^{k_r} S_{nj} \subset \bigcup_{j=1}^{k_m} S_{nj} \text{ for } m \geqslant r.$$

Hence

$$\mu_{i_m}\left(\bigcup_{j=1}^{k_r} S_{nj}\right) \leqslant \mu_{i_m}\left(\bigcup_{j=1}^{k_m} S_{nj}\right) \leqslant 1-\delta_0 \text{ for } m \geqslant r.$$

Since $\mu_{i_m} \Longrightarrow \mu$ as $m \to \infty$ and $\bigcup_{j=1}^{k_r} S_{nj}$ is open we have from Proposition 51.2

$$\mu\left(\bigcup_{j=1}^{k_r} S_{nj}\right) \leqslant \varliminf_{m\to\infty} \mu_{i_m}\left(\bigcup_{j=1}^{k_r} S_{nj}\right) \leqslant 1-\delta_0.$$

Now letting $r \to \infty$ we have $\mu(X) \leqslant 1-\delta_0$, which is a contradiction. Thus our claim holds. We fix n, put $\delta = \varepsilon/2^n$ and choose k_n such that

$$\mu_i\left(\bigcup_{1}^{k_n} S_{nj}\right) > 1 - \frac{\varepsilon}{2^n} \text{ for all } i=1, 2, \ldots . \tag{52.2}$$

We do this for every n and put

$$C_n = \bigcup_{j=1}^{k_n} \bar{S}_{nj};$$

$$K = \bigcap_{n=1}^{\infty} C_n.$$

Since $\mu_i(C_n) > 1 - \frac{\varepsilon}{2^n}$ for all n, we have $\mu_i(K) > 1-\varepsilon$ for all $i=1, 2, \ldots$.

We shall complete the proof by showing that K is compact. Since each C_n is closed it is clear that K is closed. Let now $x_1, x_2, \ldots$ be any sequence of points in K. We shall prove that this sequence has a limit point in K. Since $K \subset C_1$ there exists $n_1 \leqslant k_1$ such that $K \cap \bar{S}_{1n_1} = K_1$ contains infinitely many x_i's. Since $K_1 \subset C_2$ there exists $n_2 \leqslant k_2$ such that $K_1 \cap \bar{S}_{2n_2} = K_2$ has infinitely many x_i's. We repeat this and obtain a sequence $K_1 \supset K_2 \supset \ldots$ such that each K_j has infinitely many x_i's. Since $K_j \subset \bar{S}_{jn_j}$ we have

$$\text{diameter } (K_j) \leqslant \frac{2}{j} \text{ for } j=1, 2, \ldots.$$

The completeness of X now implies that

$$\bigcap_{j=1}^{\infty} K_j = \{x_0\},$$

for some $x_0 \in X$. Further any sphere with x_0 as centre includes a K_j for some large j and hence infinitely many x_i's. In other words x_0 is a limit point of the x_i's. This shows that K is compact and the proof is complete.

Definition. 52.4. Let X be a separable metric space and let $\{\mu_n\}$ be a sequence in $M_0(X)$. $\{\mu_n\}$ is said to be *weakly conditionally compact* or simply *compact* if every subsequence of $\{\mu_n\}$ has a further subsequence which is weakly convergent.

Remark. 52.5. We can now restate Prohorov's theorem as follows: if X is a complete and separable metric space then every compact sequence $\{\mu_n\}$ is uniformly tight. If X is separable then every uniformly tight sequence $\{\mu_n\}$ is compact. It is possible to develop the theory of weak convergence of probability measures very widely by making full use of the power of this result but such a development is outside the scope of this book. The interested reader should consult [1], [17].

Exercise. 52.6. Let $\{\mu_n\}$ be a sequence of probability measures on R^k and let π_j be the map defined by $\pi_j \mathbf{x}=x_j$ for every $j=1, 2, \ldots, k$. Suppose, for every j, the sequence $\{\mu_n \pi_j^{-1}\}$ is compact. Then $\{\mu_n\}$ is compact.

Proposition. 52.7. Let $\{\lambda_n\}$, $\{\mu_n\}$, $\{\nu_n\}$ be three sequences of probability meaures in R^k such that $\lambda_n = \mu_n * \nu_n$ for each n. If $\{\lambda_n\}$ and $\{\mu_n\}$ are compact then $\{\nu_n\}$ is also compact.

Proof. Since $\{\lambda_n\}$ and $\{\mu_n\}$ are compact it follows from Remark 52.5 that for any $\varepsilon > 0$ there exists a compact set K_ε such that

$$\lambda_n(K_\varepsilon) > 1 - \varepsilon, \ \mu_n(K_\varepsilon) > 1 - \varepsilon \text{ for all } n.$$

Then we have

$$\begin{aligned} 1 - \varepsilon < \lambda_n(K_\varepsilon) &= \int \nu_n(K_\varepsilon - \mathbf{x})\, d\mu_n(\mathbf{x}) \\ &\leqslant \int_{K_\varepsilon} \nu_n(K_\varepsilon - \mathbf{x})\, d\mu_n(\mathbf{x}) + \varepsilon, \end{aligned}$$

or

$$\int_{K_\varepsilon} \nu_n(K_\varepsilon - \mathbf{x})\, d\mu_n(\mathbf{x}) > 1 - 2\varepsilon.$$

Hence there exists an $\mathbf{x}_n \in K_\varepsilon$ such that

$$\nu_n(K_\varepsilon - \mathbf{x}_n) > 1 - 3\varepsilon \text{ for all } n.$$

Since $K_\varepsilon - \mathbf{x}_n \subset K_\varepsilon - K_\varepsilon = \{\mathbf{x} - \mathbf{y} : \mathbf{x} \in K_\varepsilon,\ \mathbf{y} \in K_\varepsilon\}$, we have

$$\nu_n(K_\varepsilon - K_\varepsilon) > 1 - 3\varepsilon \text{ for all } n,$$

where $K_\varepsilon - K_\varepsilon$ is a compact set. Thus $\{\nu_n\}$ is uniformly tight and hence compact. This completes the proof.

Proposition. 52.8. Let $\{\lambda_n\}$, $\{\mu_n\}$, $\{\nu_n\}$ be three sequences of probability measures in R^k such that $\lambda_n = \mu_n * \nu_n$ for each n. For any $\mathbf{x} \in R^k$, let $\delta_{\mathbf{x}}$ be the probability measure degenerate at $\mathbf{x}$. Suppose $\{\lambda_n\}$ is compact. Then there exists a sequence $\{\mathbf{x}_n\}$ in R^k such that the sequences $\{\mu_n * \delta_{\mathbf{x}_n}\}$ and $\{\nu_n * \delta_{-\mathbf{x}_n}\}$ are compact.

Proof. Let $\{\varepsilon_n\}$ be a sequence of positive numbers such that $\sum_n \varepsilon_n < \infty$. By Prohorov's theorem we can select a sequence $\{K_r\}$ of compact sets such that

$$\lambda_n(K_r) > 1 - \varepsilon_r \text{ for } r = 1, 2, \ldots .$$

Let $\{\eta_n\}$ be a sequence of positive numbers decreasing to 0 such that $\sum_1^\infty \varepsilon_n \eta_n^{-1} < \frac{1}{2}$. For example we may choose $\varepsilon_n = n^{-2}$ and $\eta_n = c\, n^{-\delta}$ where $0 < \delta < 1$ and c is a suitable constant. Let

$$A_{nr} = \{\mathbf{x} : \mu_n(K_r - \mathbf{x}) > 1 - \eta_r\},$$

$$B_n = \bigcap_{r=1}^{\infty} A_{nr}.$$

Then

$$1 - \varepsilon_r < \lambda_n(K_r) = \int_{A_{nr}} \mu_n(K_r - \mathbf{x})\, d\nu_n(\mathbf{x}) + \int_{A'_{nr}} \mu_n(K_r - \mathbf{x})\, d\nu_n(\mathbf{x})$$

$$\leqslant \nu_n(A_{nr}) + (1 - \eta_r)\, \nu_n(A'_{nr}).$$

Hence

$$\nu_n(A'_{nr}) \leqslant \varepsilon_r\, \eta_r^{-1}.$$

This implies

$$\nu_n(B'_n) \leqslant \sum_{r=1}^{\infty} \varepsilon_r\, \eta_r^{-1} < \tfrac{1}{2}.$$

Hence $B_n \neq \varnothing$. In other words there exists $\mathbf{x}_n \in B_n$ such that

$$\mu_n(K_r - \mathbf{x}_n) > 1 - \eta_r \text{ for all } r \text{ and } n.$$

Equivalently

$$(\mu_n * \delta_{\mathbf{x}_n})\,(K_r) = \mu_n(K_r - \mathbf{x}_n) > 1 - \eta_r \text{ for all } r \text{ and } n.$$

Thus the sequence $\{\mu_n * \delta_{\mathbf{x}_n}\}$ is uniformly tight and hence compact.

We have

$$\lambda_n = \mu_n * \nu_n = (\mu_n * \delta_{\mathbf{x}_n}) * (\nu_n * \delta_{-\mathbf{x}_n}).$$

Proposition 52.7 implies that $\{\nu_n * \delta_{-\mathbf{x}_n}\}$ is also compact. The proof is complete.

Remark. 52.9. Propositions 52.7 and 52.8 are valid when R^k is replaced by a separable B-space. The same proof goes through. For a variety of applications of Proposition 52.8 the reader may refer to [17].

Exercise. 52.10. Let $\{\mu_n\}$ be a sequence of probability measures on R^l such that $\{\mu_n{}^{*l}\}$ is compact for some positive integer l. Then $\{\mu_n\}$ is compact.

§53 Fourier Transforms of Probability Measures in R^k.

One of the basic tools for studying sums of independent random variables and limits of distributions in R^k is the theory of Fourier transforms or characteristic functions. We shall give a brief account of this topic. To this end we introduce some notations and a definition. We shall denote an arbitrary point of R^k by $\mathbf{x}$ which is a column vector with i-th coordinate equal to x_i, $i=1, 2, \ldots, k$. For any two $\mathbf{x}$, $\mathbf{y}$ in R^k we shall write

$$\langle \mathbf{x}, \mathbf{y} \rangle = \sum_{j=1}^{k} x_j \, y_j.$$

For any $\mu \in M_0(R^k)$, we write

$$\hat{\mu}(\mathbf{t}) = \int e^{i\langle \mathbf{t}, \mathbf{x} \rangle} \, d\mu(\mathbf{x}), \; \mathbf{t} \in R^k. \tag{53.1}$$

The complex valued function $\hat{\mu}$ on R^k is called the *Fourier transform* or *characteristic function* of the probability measure or distribution μ. If $\mathbf{f}$ is an R^k valued random variable on a probability space (Ω, S, P) and $\mu = P\mathbf{f}^{-1}$ is the distribution of $\mathbf{f}$, its characteristic function $\hat{\mu}$ is given by

$$\begin{aligned} \hat{\mu}(\mathbf{t}) &= \int e^{i\langle \mathbf{t}, \mathbf{x} \rangle} \, d\mu(\mathbf{x}) \\ &= \int_{\Omega} e^{i\langle \mathbf{t}, \mathbf{f} \rangle} \, dP \\ &= \mathbf{E} \, e^{i\langle \mathbf{t}, \mathbf{f} \rangle} \end{aligned}$$

In this case we also say that $\hat{\mu}$ is the characteristic function of the random variable $\mathbf{f}$.

Proposition. 53.1. For any $\mu \in M_0(R^k)$ the following holds:

(i) $\hat{\mu}(\mathbf{0})=1$;

(ii) $\hat{\mu}(-\mathbf{t})=\bar{\hat{\mu}}(\mathbf{t})$;

(iii) if $a_1, a_2, \ldots, a_n$ are any n complex numbers and $\mathbf{t}_1, \mathbf{t}_2, \ldots, \mathbf{t}_n$ are any n points in R^k, then

$$\Sigma\, a_i\, \bar{a}_j\, \hat{\mu}(\mathbf{t}_i-\mathbf{t}_j) \geqslant 0;$$

(iv) $\hat{\mu}$ is uniformly continuous.

Proof. The first two properties follow immediately from the Eq. (53.1). To prove (iii) we observe that

$$\sum_{r,s} a_r\, \bar{a}_s\, \hat{\mu}(\mathbf{t}_r-\mathbf{t}_s) = \int \Sigma\, a_r\, \bar{a}_s\, e^{i\langle \mathbf{t}_r-\mathbf{t}_s, \mathbf{x}\rangle}\, d\mu(\mathbf{x})$$

$$= \int \left| \Sigma\, a_r\, e^{i\langle \mathbf{t}_r, \mathbf{x}\rangle} \right|^2 d\mu(\mathbf{x}) \geqslant 0.$$

To prove the last part we note that for any $\mathbf{h} \in R^k$,

$$|\hat{\mu}(\mathbf{t}+\mathbf{h})-\hat{\mu}(\mathbf{t})| \leqslant \int |e^{i\langle \mathbf{h}, \mathbf{x}\rangle}-1|\, d\mu(\mathbf{x}).$$

Hence

$$\sup_{\mathbf{t}\in R^k} |\hat{\mu}(\mathbf{t}+\mathbf{h})-\hat{\mu}(\mathbf{t})| \leqslant \int |e^{i\langle \mathbf{h}, \mathbf{x}\rangle}-1|\, d\mu(\mathbf{x}).$$

The integrand on the right side is bounded by the constant 2 and tends to 0 as $\mathbf{h} \to \mathbf{0}$. Hence by Lebesgue's dominated convergence theorem

$$\lim_{|\mathbf{h}|\to 0}\ \sup_{\mathbf{t}\in R^k} |\hat{\mu}(\mathbf{t}+\mathbf{h})-\hat{\mu}(\mathbf{t})| = 0.$$

where $|\mathbf{h}|$ denotes $(\sum_i h_i^2)^{\frac{1}{2}}$. In other words $\hat{\mu}$ is uniformly continuous. This completes the proof.

Remark. 53.2. It is a theorem of S. Bochner that if a complex valued function ϕ on R^k is continuous, $\phi(\mathbf{0})=1$ and for any positive integer n, complex numbers $a_1, a_2, \ldots, a_n$ and points $\mathbf{t}_1, \mathbf{t}_2, \ldots, \mathbf{t}_n$ the inequality $\sum_{i,j} a_i\, \bar{a}_j\, \phi\, (\mathbf{t}_i-\mathbf{t}_j) \geqslant 0$ holds then ϕ is the characteristic function $\hat{\mu}$ of a probability distribution μ in R^k. (For proof the reader may refer to [14].)

Proposition. 53.3. Let $\mu \in M_0(R^k)$ and let $\int |x_j|\, d\mu(\mathbf{x}) < \infty$, for some j. Then $\hat{\mu}$ is partially differentiable in t_j and

$$\frac{\partial \hat{\mu}}{\partial t_j}(\mathbf{t}) = i \int x_j\, e^{i\langle \mathbf{t}, \mathbf{x}\rangle}\, d\mu(\mathbf{x}). \tag{53.2}$$

The above derivative is uniformly continuous in $\mathbf{t}$.

Proof. Let $\mathbf{e}_j$ be the vector with jth coordinate unity and rest of the coordinates 0. Then

$$\frac{\hat{\mu}(\mathbf{t}+h\,\mathbf{e}_j)-\hat{\mu}(\mathbf{t})}{h}=\int e^{i<\mathbf{t},\,\mathbf{x}>}\left(\frac{e^{ihx_j}-1}{h}\right)d\mu(x).$$

The absolute value of the integrand on the right hand side is less than or equal to $|x_j|$. Letting $h\to 0$ and using dominated convergence theorem we obtain Eq. (53.2). The second part is proved exactly in the same manner as the last part of Proposition 53.1. This completes the proof.

Exercise. 53.4. Let $\mu\in M(R^k)$ be such that

$$\int |x_1^{r_1}x_2^{r_2}\ldots x_k^{r_k}|\,d\mu(\mathbf{x})<\infty.$$

Then $\dfrac{\partial^{r_1+\ldots+r_k}}{\partial t_1^{r_1}\,\partial_2^{r_2}\ldots\,\partial t_k^{r_k}}(\hat{\mu}(\mathbf{t}))$ exists as a uniformly continuous function and

$$\frac{\partial^{j_1+\ldots+j_k}}{\partial t_1^{j_1}\partial t_2^{j_2}\ldots\partial t_k^{j_k}}\hat{\mu}(\mathbf{t})=i^{j_1+\ldots+j_k}\int x_1^{j_1}x_2^{j_2}\ldots x_k^{j_k}\,e^{i<\mathbf{t},\,\mathbf{x}>}\,d\mu\,(\mathbf{x})$$

for all $j_1\ \ r_1, j_2\leqslant r_2, \ldots j_k\leqslant r_k$.

Conversely, if $\hat{\mu}\,(\mathbf{t})$ is differentiable k times in the variable t_1, where k is an even integer, then $\int x_1^k\,d\mu\,(\mathbf{x})<\infty$.

In Section 36 we defined the convolution of probability measures in R^k. If $\mathbf{f}$ and $\mathbf{g}$ are independent random variables with values in R^k and distributions μ, ν respectively then $\mu*\nu$ is the distribution of $\mathbf{f}+\mathbf{g}$. Hence the characteristic function of $\mu*\nu$ is

$$\mathbf{E}\,e^{i<\mathbf{t},\,\mathbf{f}+\mathbf{g}>}=\mathbf{E}\,e^{i<\mathbf{t},\,\mathbf{f}>}\,e^{i<\mathbf{t},\,\mathbf{g}>}=\hat{\mu}(\mathbf{t})\,\hat{\nu}(\mathbf{t}).$$

Thus we have the following.

Proposition. 53.5. For any $\mu,\nu\in M_0\,(R^k)$,

$$(\mu*\nu)^\wedge\,(\mathbf{t})=\hat{\mu}(\mathbf{t})\,\hat{\nu}(\mathbf{t}).$$

Exercise. 53.6. Let $\mathbf{f}$ be a random variable with values in R^k and distribution μ. If $\mathbf{A}$ is a $k\times k$ real matrix and $\mathbf{a}\in R^k$, then the characteristic function of $\mathbf{A}\,\mathbf{f}+\mathbf{a}$ is $e^{i<\mathbf{t},\,\mathbf{a}>}\,\hat{\mu}(\mathbf{A}'\mathbf{t})$, where $\mathbf{A}'$ is the transpose of $\mathbf{A}$.

Exercise. 53.7. (i) The binomial distribution of the number of successes in n independent trials with probability p for success has characteristic function $(pe^{it}+q)^n$.

(ii) The Poisson distribution with parameter λ has characteristic function $\exp\left\{\lambda(e^{it}-1)\right\}$.

(iii) The gamma distribution μ on R defined by

$$\mu(E) = \frac{a^\alpha}{\Gamma(\alpha)} \int_{E \cap (0,\infty)} e^{-ax} x^{\alpha-1}\, dx,\ a>0,\ \alpha>0$$

has characteristic function $\left(1 - \frac{it}{a}\right)^{-\alpha}$.

Proposition. 53.8. The multivariate normal distribution in R^k with mean vector $\mathbf{m}$ and covariance matrix Σ has characteristic function $\exp i \langle \mathbf{t}, \mathbf{m} \rangle - \frac{1}{2} \mathbf{t}' \Sigma \mathbf{t}$.

Proof. Let

$$\psi(z) = \frac{1}{\sqrt{2\pi}} \int_{-\infty}^{+\infty} e^{zx} e^{-\frac{1}{2}x^2}\, dx.$$

We note that ψ is an analytic function defined in the entire complex plane. If $z=t$ is real a routine integration shows that

$$\psi(t) = e^{\frac{1}{2}t^2}.$$

Thus the analytic function $\psi(z)$ and $\exp \frac{1}{2}z^2$ agree on the entire real axis. Hence they agree everywhere. Thus

$$\psi(it) = \exp\left\{-\frac{1}{2}t^2\right\}.$$

In other words the standard normal distribution on R has characteristic function $\exp\left\{-\frac{1}{2}t^2\right\}$.

Now consider the multivariate normal distribution with mean vector $\mathbf{o}$ and covariance matrix I. Then the coordinates are independent and identically distributed with standard normal distribution. Hence it has characteristic function

$$\exp\{-\tfrac{1}{2}(t_1^2+t_2^2+\ldots+t_k^2)\} = \exp\{-\tfrac{1}{2}\langle \mathbf{t}, \mathbf{t}\rangle\}.$$

Let $\boldsymbol{\xi}$ be a random variable with $\xi_1, \xi_2, \ldots, \xi_k$ being independently distributed with standard normal distribution. Let $\boldsymbol{\eta} = \mathbf{A}\boldsymbol{\xi} + \mathbf{m}$ where $\mathbf{A}$ is any $k \times k$ matrix and $\mathbf{m}$ a constant vector. Then by Exercise 53.6 the characteristic function of $\boldsymbol{\eta}$ is

$$e^{i\langle \mathbf{t}, \mathbf{m}\rangle - \frac{1}{2}\langle \mathbf{A}'\mathbf{t}, \mathbf{A}'\mathbf{t}\rangle}$$

$$= e^{i\langle \mathbf{t}, \mathbf{m}\rangle - \frac{1}{2}\mathbf{t}'\mathbf{A}\mathbf{A}'\mathbf{t}}$$

But the distribution of $\boldsymbol{\eta}$ is multivariate normal with mean $\mathbf{m}$ and covariance matrix $\mathbf{AA}'$. But any positive semi-definite matrix Σ is of the form $\mathbf{AA}'$ for some $\mathbf{A}$. Hence the proof is complete.

Proposition. 53.9. (Inversion theorem) Let $\mu, \nu \in M_0(R^k)$ and let $\hat{\mu}(\mathbf{t})=\hat{\nu}(\mathbf{t})$ for all $\mathbf{t}$. Then $\mu=\nu$.

Proof. We have

$$\frac{1}{(2\pi)^k}\int e^{i\langle \mathbf{x}-\mathbf{u},\,\mathbf{y}\rangle-\frac{1}{2}\sigma^2\langle \mathbf{y},\mathbf{y}\rangle}\,d\mathbf{y}$$

$$=\frac{1}{(2\pi)^{k/2}\sigma^k}\,e^{-\frac{1}{2\sigma^2}\langle \mathbf{x}-\mathbf{u},\,\mathbf{x}-\mathbf{u}\rangle} \text{ for all } \mathbf{x},\mathbf{u}\in R^k,$$

where $d\mathbf{y}$ denotes integration with respect to Lebesgue measure. Integrating both sides of the above equation with respect to μ in the variable $\mathbf{x}$ and using Fubini's theorem we get

$$\frac{1}{(2\pi)^k}\int \hat{\mu}(\mathbf{y})\,e^{-i\langle \mathbf{u},\mathbf{y}\rangle-\frac{1}{2}\sigma^2\langle \mathbf{y},\mathbf{y}\rangle}\,d\mathbf{y}$$

$$=\frac{1}{\sigma^k(\sqrt{2\pi})^k}\int e^{-\frac{1}{2\sigma^2}\langle \mathbf{x}-\mathbf{u},\,\mathbf{x}-\mathbf{u}\rangle}\,d\mu(\mathbf{x}).$$

Now integrating both sides with respect to $\mathbf{u}$ in the region

$$A=\{\mathbf{u}: -\infty<u_i<a_i,\ i=1,2,\ldots,k\} \tag{53.3}$$

and using Fubini's theorem for the right hand side only we get

$$\int_A\left[\frac{1}{(2\pi)^k}\int \mu(\mathbf{y})\,e^{-i\langle \mathbf{u},\mathbf{y}\rangle-\frac{1}{2}\sigma^2\langle \mathbf{y},\mathbf{y}\rangle}\,d\mathbf{y}\right]d\mathbf{u}$$

$$=\int\left[\prod_{j=1}^{k}\frac{1}{\sqrt{2\pi}\,\sigma}\int_{-\infty}^{a_j} e^{-\frac{1}{2\sigma^2}(u_j-x_j)^2}\,du_j\right]d\mu(\mathbf{x})$$

$$=\int\left[\prod_{j=1}^{k}\frac{1}{\sqrt{2\pi}}\int_{-\infty}^{\frac{a_j-x_j}{\sigma}} e^{-\frac{1}{2}u^2}\,du\right]d\mu(\mathbf{x}). \tag{53.4}$$

Now observe that the function within square brackets on the right hand side above is always less than or equal to 1. As $\sigma\to 0$ it converges to the limit function

$$\begin{aligned}\rho(\mathbf{x}) &= 0 \text{ if } x_j > a_j \text{ for some } j,\\ &= 1 \text{ if } x_j < a_j \text{ for all } j,\\ &= \frac{1}{2^{k-l}} \text{ if } x_j < a_j \text{ for } l \text{ values of } j \text{ and}\\ &\qquad x_j = a_j \text{ for } k-l \text{ values of } j.\end{aligned}$$

Letting $\sigma \to 0$ in (53.4) and applying dominated convergence theorem we get

$$\lim_{\sigma \to 0} \int_A \left[\frac{1}{(2\pi)^k} \int \hat{\mu}(y)\, e^{-i\langle \mathbf{u}, \mathbf{y}\rangle - \frac{\sigma^2}{2}\langle \mathbf{y}, \mathbf{y}\rangle} d\mathbf{y} \right] d\mathbf{u}$$

$$= \sum_{r=0}^{k} \frac{1}{2^r} \sum_{i_1 < i_2 < \ldots < i_r} \mu\{\mathbf{x}: x_{i_j} = a_{i_j}, j = 1, 2, \ldots, r;$$

$$x_i < a_i \text{ for } i \notin (i_1, i_2, \ldots i_r)\} \tag{53.5}$$

where A is defined by Eq. (53.3). This equation holds for every probability measure μ. If $\hat{\mu} = \hat{\nu}$ and $\mathbf{a}$ is any point in R^k such that

$$\mu\{\mathbf{x} : x_i = a_i\} = \nu\{\mathbf{x} : x_i = a_i\} = 0 \text{ for all } i,$$

then Eq. (53.5) implies that

$$\mu\{\mathbf{x} : x_j \leqslant a_j \text{ for all } j = 1, 2, \ldots, k\} =$$
$$\nu\{\mathbf{x} : x_j \leqslant a_j \text{ for all } j = 1, 2, \ldots, k\}.$$

Hence μ and ν agree on finite disjoint unions of differences of such sets. In other words μ and ν agree on a boolean algebra of sets generating the borel σ-algebra of R^k. Thus $\mu = \nu$. This completes the proof.

Remark. 53.10. Equation (53.5) may be called the inversion formula for characteristic functions. It expresses the measure μ in terms of the characteristic function $\hat{\mu}$. In the case of real line it assumes the following simple form:

$$\lim_{\sigma \to 0} \frac{1}{2\pi} \int_{-\infty}^{x} \left[\int_{-\infty}^{+\infty} \hat{\mu}(y)\, e^{-iuy - \frac{1}{2}\sigma^2 y^2}\, dy \right] du$$

$$= \mu((-\infty, x)) + \tfrac{1}{2}\mu(\{x\}).$$

Exercise. 53.11. Let $\mu \in M_0(R)$ and let $\hat{\mu}$ be its characteristic function. Then

(i) $$\lim_{T \to \infty} \frac{1}{\pi} \int_{-T}^{+T} \hat{\mu}(t)\, e^{-ita} \frac{\sin th}{t}\, dt$$

$$= \mu((a-h, a+h)) + \tfrac{1}{2}[\mu(\{a-h\}) + \mu(\{a+h\})];$$

(ii) $$\lim_{T \to \infty} \frac{1}{2\pi T} \int_{-T}^{+T} \hat{\mu}(t)\, e^{-ita}\, dt = \mu(\{a\});$$

(iii) $$\lim_{T\to\infty} \frac{1}{2\pi T} \int_{-T}^{+T} |\hat{\mu}(t)|^2 \, dt = \sum_{a:\mu(\{a\})>0} \mu(\{a\})^2.$$

(Hint: Use (ii) and the fact that $|\hat{\mu}(t)|^2$ is the characteristic function of $\mu * \tilde{\mu}$ where $\tilde{\mu}(E)=\mu(-E)$ for all borel sets E.)

(iv) The distribution function of μ is continuous if and only if

$$\lim_{T\to\infty} \frac{1}{T} \int_{-T}^{T} |\hat{\mu}(t)|^2 \, dt = 0.$$

In particular, if $\lim_{t\to\infty} \hat{\mu}(t)=0$, then its distribution function is continuous.

Exercise. 53.12. If $\mu \in M_0(R)$ and $\hat{\mu}(h)=1$ for some $h \neq 0$ then

$$\mu\left\{x : x = \frac{2\pi n}{h} \text{ for some } n=0, \pm 1, \pm 2, \ldots\right\} = 1.$$

In particular, if $|\hat{\mu}(h_1)| = |\hat{\mu}(h_2)| = 1$ and $\frac{h_1}{h_2}$ is not rational then μ is degenerate at some point.

Exercise. 53.13. If $\mu \in M_0(R^k)$ and $\hat{\mu} \in L_1(R^k)$ then μ is absolutely continuous with respect to the Lebesgue measure in R^k and μ has density function $f(\mathbf{x})$ given by

$$f(\mathbf{x}) = \frac{1}{(2\pi)^k} \int_{-\infty}^{+\infty} \hat{\mu}(\mathbf{y}) \, e^{-i\langle \mathbf{x}, \mathbf{y}\rangle} \, d\mathbf{y}$$

Exercise. 53.14. Let $A \subset R^k$ be a bounded borel set of positive Lebesgue measure. Suppose μ and ν are two probability measures on R^k such that $\mu(A+\mathbf{x})=\nu(A+\mathbf{x})$ for all $\mathbf{x}$ in R^k. Then $\mu=\nu$. (Hint: Let λ be the probability measure defined by

$$\lambda(F)=L(A)^{-1} \int_F \chi_A(\mathbf{x}) \, d\mathbf{x},$$ where $L(A)$ is the Lebesgue measure of A. Then $\mu * \lambda = \nu * \lambda$. Now equate the characteristic functions of both sides and observe that the set $\{\mathbf{t} : \hat{\lambda}(\mathbf{t})=0\}$ is of Lebesgue measure zero. Indeed, $\int \exp(x_1 z_1 + \ldots + x_k z_k) \, d\lambda(\mathbf{x})$ is an entire function in the complex variables $z_1, z_2, \ldots, z_k$.)

We shall now prove an inequality concerning characteristic functions.

Proposition. 53.15. Let $\mu \in M_0(R^k)$ and let $\hat{\mu}$ be its characteristic function. Then

$$\mu\left\{\mathbf{x}: \max_{1 \leqslant j \leqslant k} |x_j| > \frac{2}{\tau}\right\}$$

$$\leqslant 2\left[1 - \left|\frac{1}{(2\tau)^k} \int_{\max_i |t_i| \leqslant \tau} \hat{\mu}(\mathbf{t})\, d\mathbf{t}\right|\right] \text{ for all } \tau > 0.$$

where $d\mathbf{t}$ denotes integration with respect to Lebesgue measure.

Proof. We have by Fubini's theorem

$$\left|\frac{1}{(2\tau)^k} \int_{\max |t_i| \leqslant \tau} \hat{\mu}(\mathbf{t})\, d\mathbf{t}\right| = \left|\frac{1}{(2\tau)^k} \int_{\max |t_i| \leqslant \tau} \int e^{i\langle \mathbf{t}, \mathbf{x}\rangle} d\mu(\mathbf{x})\, d\mathbf{t}\right|$$

$$= \left|\int \prod_{j=1}^{k} \frac{\sin \tau x_j}{\tau x_j}\, d\mu(\mathbf{x})\right|$$

$$\leqslant \int_{\max_j |x_j| \leqslant a} d\mu(\mathbf{x}) + \int_{\max_j |x_j| > a} \prod_{j=1}^{k} \left|\frac{\sin \tau x_j}{\tau x_j}\right| d\mu(\mathbf{x})$$

$$\leqslant \mu\{\mathbf{x}: \max |x_j| \leqslant a\} + \frac{1}{\tau a}\, \mu\left\{\mathbf{x}: \max_j |x_j| > a\right\}$$

$$= 1 - \left(1 - \frac{1}{\tau a}\right) \mu\{\mathbf{x}: \max |x_j| > a\}$$

for all $a > 0$. Putting $a = 2/\tau$ we get the required inequality. This completes the proof.

Corollary. 53.16. Let $\{\mu_n\}$ be a sequence of probability measures on R^k. Then $\{\mu_n\}$ is compact if the sequence of characteristic functions $\{\hat{\mu}_n\}$ is equicontinuous at 0, i.e.,

$$\lim_{\mathbf{t}\to\mathbf{0}} \sup_n |\hat{\mu}_n(\mathbf{t}) - 1| = 0.$$

Proof. Let $\{\hat{\mu}_n\}$ be equicontinuous at $\mathbf{0}$. Let $\varepsilon > 0$ be arbitrary. Then there exists a $\tau > 0$ such that

$$|\hat{\mu}_n(\mathbf{t}) - 1| < \varepsilon \text{ for all } n,$$

if $\max_j |t_j| \leqslant \tau$. Then

$$1 - \left|\frac{1}{(2\tau)^k} \int_{\max |t_j| \leqslant \tau} \hat{\mu}_n(\mathbf{t})\, d\mathbf{t}\right| \leqslant \int_{\max_j |t_j| \leqslant \tau} |1 - \hat{\mu}_n(\mathbf{t})|\, d\mathbf{t} \leqslant \varepsilon$$

for all n.

Hence by the preceding proposition

$$\mu_n \left\{ \mathbf{x}: \max_j |x_j| > \frac{2}{\tau} \right\} < 2\varepsilon \text{ for all } n.$$

In other words $\{\mu_n\}$ is uniformly tight. Hence by Prohorov's theorem $\{\mu_n\}$ is compact.

Proposition. 53.17. (Levy-Cramer continuity theorem.) Let $\{\mu_n\}$ be a sequence of probability measures on R^k. Suppose the sequence $\{\hat{\mu}_n\}$ of characteristic functions converges in Lebesgue measure over every bounded rectangle to a function ϕ, where $\phi(\mathbf{0})=1$ and ϕ is continuous at the origin $\mathbf{0}$. Then there exists a probability measure μ such that $\hat{\mu}(\mathbf{t})=\phi(\mathbf{t})$ a.e. $\mathbf{t}$ (Lebesgue measure) and $\mu_n \Longrightarrow \mu$ as $n\to\infty$.

Conversely if $\mu_n \Longrightarrow \mu$ as $n\to\infty$ then the sequence $\{\hat{\mu}_n\}$ of characteristic functions converges pointwise to the characteristic function $\hat{\mu}$.

Proof. Since $|\hat{\mu}_n(\mathbf{t})| \leqslant 1$ for all n and all $\mathbf{t}$ we may assume without loss of generality that $|\phi(\mathbf{t})| \leqslant 1$ for all $\mathbf{t}$. The continuity of ϕ at the origin implies that for any given $\varepsilon>0$ we can find a $\tau>0$ such that

$$|1-\phi(\underline{t})| < \frac{\varepsilon}{4}$$

for all $\mathbf{t}$ such that $\max_j |t_j| < \tau$. Hence

$$1 - \left| \frac{1}{(2\tau)^k} \int_{\max_j |t_j| \leqslant \tau} \phi(\mathbf{t})\, d\mathbf{t} \right| \leqslant \frac{\varepsilon}{4}. \tag{53.6}$$

Since $\{\hat{\mu}_n\}$ is uniformly bounded and converges in Lebesgue measure to ϕ we have by Lebesgue's dominated convergence theorem

$$\lim_{n\to\infty} \frac{1}{(2\tau)^k} \int_{\max_j |t_j| \leqslant \tau} \hat{\mu}_n(\mathbf{t})\, d\mathbf{t} = \frac{1}{(2\tau)^k} \int_{\max_j |t_j| \leqslant \tau} \phi(\mathbf{t})\, d\mathbf{t}. \tag{53.7}$$

Combining Eqs. (53.6) and (53.7) we observe that there exists a positive integer n_0 such that

$$1 - \left| \frac{1}{(2\tau)^k} \int_{\max_j |t_j| \leqslant \tau} \hat{\mu}_n(\mathbf{t})\, d\mathbf{t} \right| < \frac{\varepsilon}{2} \text{ for all } n > n_0.$$

By proposition 53.15, we have

$$\mu_n \left\{ \mathbf{x} : \max_j |x_j| > \frac{2}{\tau} \right\} < \varepsilon \text{ for all } n > n_0.$$

In other words the sequence $\{\mu_n\}$ is uniformly tight. By Prohorov's theorem there exists a subsequence $\{\mu_{n_k}\}$ converging weakly to a probability measure μ. Since $e^{\mathbf{i}\langle \mathbf{t}, \mathbf{x} \rangle}$ is continuous in $\mathbf{x}$ for every fixed $\mathbf{t}$ we have by the definition of weak convergence

$$\lim_{k \to \infty} \hat{\mu}_{n_k}(\mathbf{t}) = \hat{\mu}(\mathbf{t}) \text{ for every } \mathbf{t}.$$

Hence $\hat{\mu}(\mathbf{t}) = \phi(\mathbf{t})$ a.e. $\mathbf{t}$ (Lebesgue measure). If ν is any other limit of $\{\mu_n\}$ then $\hat{\nu}(\mathbf{t}) = \phi(\mathbf{t}) = \hat{\mu}(\mathbf{t})$ a.e. $(\mathbf{t})$. Since $\hat{\mu}$ and $\hat{\nu}$ are continuous we have $\hat{\mu}(\mathbf{t}) = \hat{\nu}(\mathbf{t})$ for all $\mathbf{t}$. Hence by inversion theorem $\mu = \nu$. Thus $\{\mu_n\}$ is compact and all its limits are equal to μ. In other words $\mu_n \Longrightarrow \mu$ as $n \to \infty$. This completes the proof of the first part. The second part follows trivially from the definition of weak convergence and the proof is complete.

Exercise. 53.18. If $\mu_n \Longrightarrow \mu$ as $n \to \infty$ in $M_0(R^k)$ then $\hat{\mu}_n$ converges uniformly over every bounded interval to $\hat{\mu}$. (Hint: Establish the equicontinuity of the sequence $\{\hat{\mu}_n\}$ and apply Ascoli's theorem (of topology). Hence there exists a metric d in $M_0(R^k)$ which makes $M_0(R^k)$ a complete and separable metric space and $\mu_n \Longrightarrow \mu$ as $n \to \infty$ if and only if $d(\mu_n, \mu) \to 0$ as $n \to \infty$.

As an application of the Levy-Cramer continuity theorem we shall prove some of the elementary results of probability theory.

Proposition. 53.19. Let μ be a probability measure on the real line R such that $\int |x| \, d\mu(x) < \infty$. Suppose $\int x \, d\mu(x) = 0$. Then

$$\lim_{n \to \infty} \left[\hat{\mu}\left(\frac{t}{n}\right) \right]^n = 1.$$

If further $\int x^2 \, d\mu(x) = 1$ then

$$\lim_{n \to \infty} \left[\hat{\mu}\left(\frac{t}{\sqrt{n}}\right) \right]^n = e^{-\frac{1}{2}t^2}.$$

Proof. Suppose $\int |x| \, d\mu(x) < \infty$. By Proposition 53.3, $\hat{\mu}(t)$ is differentiable and the derivative is continuous. If $\int x \, d\mu(x) = 0$, then $\hat{\mu}'(0) = 0$. By mean value theorem we have

$$\hat{\mu}(t) = \hat{\mu}(0) + t \, \hat{\mu}'(\theta t) \text{ for all } |t| \leqslant 1,$$

where $\theta = \theta(t) \in (0, 1)$. Hence

$$\hat{\mu}\left(\frac{t}{n}\right) = 1 + \frac{t}{n}\,\hat{\mu}'\left(\theta_n \frac{t}{n}\right)$$
$$= 1 + t\,\alpha_n(t), \text{ say,}$$

where

$$\lim_{n\to\infty} n\,\alpha_n(t) = \hat{\mu}'(0) = 0 \text{ for every } t.$$

Hence

$$\lim_{n\to\infty}\left[\hat{\mu}\left(\frac{t}{n}\right)\right]^n = 1 \text{ for all } t.$$

This proves the first part. Now suppose $\int x^2 d\mu(x) = 1$. Then by Exercise 53.4, $\hat{\mu}$ is twice differentiable, $\hat{\mu}''$ is continuous and $\hat{\mu}'(0)=0$, $\hat{\mu}''(0)=-1$. Hence by Taylor's theorem we have

$$\hat{\mu}(t) = \hat{\mu}(0) + t\,\hat{\mu}'(0) + \tfrac{1}{2}t^2\,\hat{\mu}''(\theta t) \text{ for all } |t| \leqslant 1,$$

where $\theta = \theta(t) \in (0, 1)$. Thus

$$\hat{\mu}\left(\frac{t}{\sqrt{n}}\right) = \hat{\mu}(0) + \frac{t^2}{2n}\,\hat{\mu}''\left(\theta_n \frac{t}{\sqrt{n}}\right)$$
$$= 1 + \beta_n(t), \text{ say,}$$

where

$$\lim_{n\to\infty} n\beta_n(t) = \tfrac{1}{2}t^2\,\hat{\mu}''(0) = -\tfrac{1}{2}t^2 \text{ for all } t.$$

Thus

$$\lim_{n\to\infty}\left[\hat{\mu}\left(\frac{t}{\sqrt{n}}\right)\right]^n = e^{-\frac{1}{2}t^2}.$$

This completes the proof.

Corollary. 53.20. (Weak law of large numbers.) Let (Ω, S, P) be a probability space and let $f_1, f_2, \ldots.$ be a sequence of independent and identically distributed random variables with mean m. Then the random variable $\frac{1}{n}(f_1+f_2+\ldots+f_n)$ converges in P measure to the constant random variable m.

Proof. We can write

$$\tilde{f}_n = \frac{1}{n}(f_1+f_2+\ldots+f_n) - m$$
$$= \frac{1}{n}((f_1-m)+(f_2-m)+\ldots+(f_n-m)).$$

Let the common distribution of $f_i - m$ be μ. Then the characteristic function of $\tilde{f}_n$ is $\left[\hat{\mu}\left(\frac{t}{n}\right)\right]^n$. By Proposition 53.19 this converges to the constant function 1 which is the characteristic function of the distribution degenerate at 0. Hence by the Levy-Cramer continuity theorem the distribution of $\tilde{f}_n$ converges weakly to the distribution degenerate at 0. This completes the proof.

Exercise. 53.21. Using Proposition 53.19 deduce the central limit theorem of Proposition 51.8.

We shall now state and prove a version of central limit theorem for not necessarily identically distributed random variables.

Proposition. 53.22. (Liapunov's central limit theorem.) Let (Ω, S, P) be a probability space and let $f_1, f_2, \ldots$ be a sequence of independent random variables on Ω. Let

$$\mathbf{E}f_i = m_i;\ \mathbf{E}(f_i - m_i)^2 = \sigma_i^2;\ \mathbf{E}\,|f_i - m_i|^3 = \beta_i;$$

$$A_n^2 = \sum_1^n \sigma_i^2,\ B_n = \sum_1^n \beta_i.$$

Suppose

$$\lim_{n\to\infty} B_n A_n^{-3} = 0;\ \lim_{n\to\infty} A_n = \infty. \tag{53.8}$$

Then the distribution of the random variable

$$\tilde{f}_n = A_n^{-1}\,[f_1 + f_2 + \ldots + f_n - (m_1 + m_2 + \ldots + m_n)\,]$$

converges weakly to the standard normal distribution as $n \to \infty$.

Proof. For $|z| < 1$ define $\log(1+z)$ by the power series

$$\log(1+z) = z - \frac{z^2}{2} + \frac{z^3}{3} - \ldots$$

Then

$$|\log(1+z) - z| \leqslant |z|^2 \text{ if } |z| < \tfrac{1}{2}. \tag{53.9}$$

Let the characteristic function of $f_i - m_i$ be $\phi_i(t)$. Then ϕ_i is three times differentiable and ϕ_i''' is a continuous function (see Exercise 53.4). By Taylor's theorem we have

$$\phi_i(t) = 1 + t\,\phi_i'(0) + \frac{t^2}{2}\,\phi_i''(0) + \frac{t^3}{3!}\,\phi_i'''(\theta_i t) \text{ if } |t| \leqslant 1,$$

where $\theta_i = \theta_i(t) \in (0, 1)$. We have once again by Exercise 53.4,

$$|\phi_i'''(t)| \leqslant \beta_i \text{ for all } t \text{ and all } i = 1, 2, \ldots.$$

Thus

$$\phi_i(t)=1-\frac{t^2}{2}\sigma_i^2+\frac{t^3}{6}\alpha_i(t) \text{ for } |t|\leqslant 1,$$

where

$$|\alpha_i(t)|\leqslant \beta_i \text{ for all } t \text{ and all } i. \tag{53.10}$$

The characteristic function $\psi_n(t)$ of $\tilde{f}_n$ is given by

$$\psi_n(t)=\prod_{i=1}^{n}\phi_i\left(\frac{t}{A_n}\right).$$

Let t be fixed. Since $A_n\to\infty$, $\left|\frac{t}{A_n}\right|\leqslant 1$ for all large n. Hence

$$\log\phi_i\left(\frac{t}{A_n}\right)=\log\left(1-\frac{t^2}{2A_n^2}\sigma_i^2+\frac{t^3}{6A_n^3}\alpha_i\left(\frac{t}{A_n}\right)\right)$$

for all large n. Put

$$z_{in}=-\frac{t^2}{2A_n^2}\sigma_i^2+\frac{t^3}{6A_n^3}\alpha_i\left(\frac{t}{A_n}\right). \tag{53.11}$$

Since $\sigma_i\leqslant\beta_i^{1/3}$ (see Proposition 34.13)

$$\left(\frac{\sigma_i}{A_n}\right)^3\leqslant\frac{\beta_i}{A_n^3}\leqslant\frac{B_n}{A_n^3} \text{ for } i\leqslant n, \tag{53.12}$$

and inequality (53.10) implies

$$\frac{|\alpha_i(t/A_n)|}{A_n^3}\leqslant\frac{B_n}{A_n^3} \tag{53.13}$$

Hence Eqs. (53.8), (53.11) to (53.13) imply

$$\lim_{n\to\infty}\sup_{i\leqslant n}|z_{in}|=0.$$

By inequality (53.9)

$$\left|\log\phi_i\left(\frac{t}{A_n}\right)-z_{in}\right|\leqslant|z_{in}|^2 \text{ for all } i\leqslant n,$$

if n is sufficiently large. By inequality (53.10)

$$\left|\left[\sum_{i=1}^{n}\log\phi_i\left(\frac{t}{A_n}\right)\right]+\tfrac{1}{2}t^2\right|\leqslant\sum_{i=1}^{n}\left|z_{in}\right|^2+\frac{|t|^3}{6A_n^3}B_n. \tag{53.14}$$

Since $|\alpha+\beta|^2\leqslant 2(|\alpha|^2+|\beta|^2)$ for any two complex numbers α, β we have from Eq. (53.11)

$$\left|z_{in}\right|^2\leqslant\frac{t^4\sigma_i^4}{2A_n^4}+\frac{t^6}{18A_n^6}\beta_i^2,\ i\leqslant n. \tag{53.15}$$

But

$$\frac{\sum_{i=1}^{n} \sigma_i^4}{A_n^4} \leqslant \frac{\max_j \sigma_j}{A_n} \frac{\sum_{i=1}^{n} \sigma_i^3}{A_n^3}$$

$$\leqslant \frac{\sum_{i=1}^{n} \beta_i}{A_n^3} = \frac{B_n}{A_n^3}, \tag{53.16}$$

and

$$\frac{\sum_{i=1}^{n} \beta_i^2}{A_n^6} \leqslant \frac{(\sum_{i=1}^{n} \beta_i)^2}{A_n^6} = \left(\frac{B_n}{A_n^3}\right)^2. \tag{53.17}$$

Inequalitites (53.14) to (53.17) imply

$$\left|\left[\sum_i \log \phi_i\left(\frac{t}{A_n}\right)\right] + \frac{t^2}{2}\right| \leqslant \left(\frac{|t|^3}{6} + \frac{t^4}{2}\right)\frac{B_n}{A_n^3} + \frac{t^6}{18}\frac{B_n^2}{A_n^6}.$$

By Eq. (53.8) we have

$$\lim_{n\to\infty} \sum_{i=1}^{n} \log \phi_i\left(\frac{t}{A_n}\right) = -\tfrac{1}{2}t^2.$$

Taking exponentials we have

$$\lim_{n\to\infty} \prod_{i=1}^{n} \phi_i\left(\frac{t}{A_n}\right) = e^{-\frac{1}{2}t^2}.$$

Now an application of Levy-Cramer continuity theorem completes the proof.

Remark. 53.23. The proof of the preceding proposition gives a flavour of the subject of limit theorems for sums of independent random variables. For a much deeper study of this subject the reader should refer to [1], [6], and [17].

Exercise. 53.24. Consider a χ_n^2 random variable. Then the distribution of $\frac{\chi_n^2 - n}{\sqrt{2n}}$ converges weakly to the standard normal distribution.

Exercise. 53.25. Let (Ω, S, P) be any probability space and let $\Omega = \bigcup_{i=1}^{k} A_i$ be a partition of Ω into disjoint sets $A_i \in S$. Let

$p_i = P(A_i) \neq 0$, $i = 1, 2, \ldots, k$. Repeat n independent trials of the experiment (Ω, S, P). If $\omega_1, \omega_2, \ldots, \omega_n$ are the outcomes let

$$f_{in} = \chi_{A_i}(\omega_1) + \chi_{A_i}(\omega_2) + \ldots + \chi_{A_i}(\omega_n), \; i = 1, 2, \ldots, k.$$

Let

$$\chi^2 = \sum_{i=1}^{k} \frac{(f_{in} - np_i)^2}{np_i}.$$

Then the distribution of χ^2 converges weakly to the χ^2_{k-1} distribution (with $k-1$ degrees of freedom) as $n \to \infty$.

(Hint: Use Proposition 9.3 and Exercise 37.11). It may be noted that the χ^2 test in statistics for fitting a distribution model is based on this limit theorem of probability theory.

CHAPTER EIGHT

Invariant Measures on Groups

§54. Haar Measure

Consider the space R^k with the Lebesgue measure L on the borel σ-algebra. We have seen earlier that

$$L(E) = L(E + \mathbf{a}) \text{ for all } \mathbf{a} \in R^k, \; E \in \mathscr{B}_{R^k}.$$

If we replace R^k by a group X with a borel structure $\mathscr{B}$, where the map $(x, y) \to xy^{-1}$ from $(X \times X, \mathscr{B} \times \mathscr{B})$ onto $(X, \mathscr{B})$ is a borel map we can ask the following question: does there exist a non-zero σ-finite measure μ on $\mathscr{B}$ such that

$$\mu(Ex) = \mu(E) \text{ for all } E \in \mathscr{B}, \; x \in X,$$

where

$$Ex = \{yx, \; y \in E\}?$$

The aim of the present chapter is to investigate this problem in some detail. Further this topic is a testing ground for understanding the interplay of measure, topology and groups. This is also the starting point of the theory of group representations.

Throughout this section we shall assume that X is a locally compact second countable metric space which is also a group with identity element e, where multiplication between two elements x and y is written as xy and the inverse of any element x is written as x^{-1}. We also assume that the map $(x, y) \to xy^{-1}$ from $X \times X$ onto X is continuous. We say briefly that X is a locally compact second countable metric group. For any two subsets A, B of X and any $z \in X$ we write

$$\begin{aligned} AB &= \{xy : x \in A, y \in B\}; \\ A^{-1} &= \{x^{-1}, x \in A\}; \\ Az &= \{xz, x \in A\}; \\ zA &= \{zx, x \in A\}. \end{aligned}$$

Remark. 54.1. The space R^k under the usual topology and addition operation is a locally compact second countable metric group; The space $GL(n, R)$ of all $n \times n$ non-singular real matrices with the relative topology of R^{n^2} and multiplication as the group operation is another example of such a group. Closed subgroups of $GL(n, R)$ are the most common examples of such groups met with in practice.

Proposition. 54.2. (Haar measure theorem.) Let X be a locally compact second countable metric group. Then there exists a σ-finite measure μ on $\mathscr{B}_X$ with the following properties:

(i) $\mu(K) < \infty$ for every compact subset K;

(ii) $\mu(G) > 0$ for every non-empty open subset G;

(iii) for any $A \in \mathscr{B}_X$, $x \in X$ the set $xA \in \mathscr{B}_X$ and $\mu(xA) = \mu(A)$.

Proof. Let d be the metric in X. Let $S_\varepsilon = \{x : d(x, e) < \varepsilon\}$ denote the open sphere of radius ε and centre e for every $\varepsilon > 0$. Without loss of generality we may assume that the closed spheres $\bar{S}_\varepsilon$ are compact for all $\varepsilon \leqslant 1$. For any compact set K and any $\varepsilon > 0$ the family $\{xS_\varepsilon, x \in X\}$ is an open covering of K. Hence it admits finite subcoverings. Let $m(\varepsilon, K)$ be the smallest number of sets of the form xS_ε required to cover K. Put

$$\lambda_\varepsilon(K) = \frac{m(\varepsilon, K)}{m(\varepsilon, \bar{S}_1)}, \quad K \in \mathscr{K}_X,$$

where $\mathscr{K}_X$ is the class of all compact subsets of X. Then we have

$$\begin{aligned} &\lambda_\varepsilon(\bar{S}_1) = 1; \ \lambda_\varepsilon(K) \leqslant m(1, K); \\ &\lambda_\varepsilon(xK) = \lambda_\varepsilon(K) \text{ for all } x; \\ &\lambda_\varepsilon(K_1 \cup K_2) \leqslant \lambda_\varepsilon(K_1) + \lambda_\varepsilon(K_2); \\ &\lambda_\varepsilon(K_1) \leqslant \lambda_\varepsilon(K_2) \text{ if } K_1 \subset K_2, \end{aligned} \tag{54.1}$$

Now let K_1, K_2 be two disjoint compact sets. Then there exists a positive δ depending on K_1 and K_2 such that

$$K_1 S_\varepsilon \cap K_2 S_\varepsilon = \emptyset \text{ for all } \varepsilon \leqslant \delta.$$

Thus

$$\lambda_\varepsilon(K_1 \cup K_2) = \lambda_\varepsilon(K_1) + \lambda_\varepsilon(K_2) \text{ for all } \varepsilon \leqslant \delta. \tag{54.2}$$

Now consider the compact topological space

$$\Gamma = \prod_{K \in \mathscr{K}_X} [0, m(1, K)].$$

By Eq. (54.1) $\lambda_\varepsilon \in \Gamma$ for all $\varepsilon \leqslant 1$. Further $\{\lambda_\varepsilon\}$ is indexed by the directed set (0, 1] under the reverse ordering of the real line. The compactness of Γ implies that as $\varepsilon \to 0$ we can extract a convergent subnet of $\{\lambda_\varepsilon\}$ with the limit λ. Because of Eqs. (54.1) and (54.2) it follows that λ is a compact content on $\mathscr{K}_X$. By Corollary 20.12 we construct a measure μ satisfying properties (i) and (ii) of the same corollary. This shows that

$$\begin{aligned} &\lambda(K) \leqslant \mu(K) < \infty, \\ &\mu(xK) = \mu(K) \text{ for all } K \in \mathscr{K}_X \text{ and } x \in X. \end{aligned}$$

The regularity of μ in every compact set shows that $\mu(xA)=\mu(A)$ for all $A \in \mathscr{B}_X$ and $x \in X$.

Now suppose that G is an open set with $\mu(G)=0$. Then $\{xG, x\in X\}$ is an open covering of X. Hence there exists a countable subcovering. Since $\mu(xG)=\mu(G)=0$ for all x we have $\mu(X)=0$. But $\mu(\bar{S}_1)\geqslant\lambda(\bar{S}_1)=1$; which is a contradiction. In other words $\mu(G)>0$ for every non-empty open set. This completes the proof.

Remark. 54.3. The measure μ of the above proposition is called a *left invariant Haar measure* of X. If we put $\tilde{\mu}(E)=\mu(E^{-1})$ then $\tilde{\mu}(K)<\infty$ for every compact K and $\tilde{\mu}(Ex)=\tilde{\mu}(E)$ for all $E\in\mathscr{B}_X$; $x\in X$. Further $\tilde{\mu}$ is a measure. This may be called a *right invariant Haar measure* of X.

In the actual construction of invariant measures the following elementary proposition is very useful.

Proposition. 54.4. Let $\Omega\subset R^k$ be an open set and let T be a homeomorphism of Ω such that

$$T(\mathbf{x}) = \begin{pmatrix} t_1(\mathbf{x}) \\ t_2(\mathbf{x}) \\ \cdot \\ \cdot \\ \cdot \\ t_k(\mathbf{x}) \end{pmatrix} \qquad \text{for all } \mathbf{x}\in\Omega.$$

Suppose each t_i has continuous partial derivatives of first order. Let there exist a non-negative function ρ on Ω which is integrable with respect to the Lebesgue measure L over every compact subset of Ω and which satisfies the equation

$$\left|\det\left(\left(\frac{\partial t_i}{\partial x_j}\right)\right)\right| = \rho(\mathbf{x})\ \rho(T\mathbf{x})^{-1} \text{ a.e. } \mathbf{x}(L).$$

Then the measure μ defined by

$$\mu(E) = \int_E \rho(\mathbf{x})\, d\mathbf{x}. \quad E\in\mathscr{B}_\Omega$$

is invariant under T.

Proof. Let ϕ be any non-negative continuous function on Ω. Then by the change of variable formula

$$\begin{aligned}\int \phi(T\mathbf{x})\, d\mu(\mathbf{x}) &= \int \phi(T\mathbf{x})\ \rho(\mathbf{x})\, d\mathbf{x} \\ &= \int \phi(\mathbf{y})\ \rho(T^{-1}\mathbf{y})\ [\rho(T^{-1}\mathbf{y})\ \rho(\mathbf{y})^{-1}]^{-1}\, d\mathbf{y} \\ &= \int \phi(\mathbf{y})\ \rho(\mathbf{y})\, d\mathbf{y} \\ &= \int \phi\, d\mu.\end{aligned}$$

Hence $\mu=\mu T^{-1}$ and the proof is complete.

54.5. Example. Let X be the multiplicative group of all 2×2 matrices of the form

$$\begin{pmatrix} x & y \\ 0 & 1 \end{pmatrix}, \text{ where } x>0, \ -\infty<y<\infty.$$

For any fixed $\begin{pmatrix} a & b \\ 0 & 1 \end{pmatrix}$, we have

$$\begin{pmatrix} x & y \\ 0 & 1 \end{pmatrix} \begin{pmatrix} a & b \\ 0 & 1 \end{pmatrix} = \begin{pmatrix} xa & xb+y \\ 0 & 1 \end{pmatrix},$$

$$\begin{pmatrix} a & b \\ 0 & 1 \end{pmatrix} \begin{pmatrix} x & y \\ 0 & 1 \end{pmatrix} = \begin{pmatrix} ax & ay+b \\ 0 & 1 \end{pmatrix}.$$

We can identify X with the set $\Omega \subset R^2$ defined by

$$\Omega = \{(x, y): x>0, \ -\infty<y<\infty\}.$$

Right and left translations by (a, b) are the maps

$$(x, y) \to (ax, bx+y),$$
$$(x, y) \to (ax, ay+b)$$

and their Jacobians are a and a^2 respectively. If we put

$$\rho_r(x, y) = \frac{1}{x},$$

$$\rho_l(x, y) = \frac{1}{x^2},$$

then by the preceding proposition the measures μ_r and μ_l defined by

$$\mu_r(E) = \int_E \frac{1}{x}\, dx,$$

$$\mu_l(E) = \int_E \frac{1}{x^2}\, dx, \quad E \in \mathscr{B}_\Omega,$$

are right and left invariant Haar measures on X.

Exercise. 54.6. Let X be the multiplicative group of all upper triangular $n\times n$ real matrices of the form

$$\xi = \begin{pmatrix} x_{11} & x_{12} & x_{13} & \dots & x_{1n} \\ 0 & x_{22} & x_{23} & \dots & x_{2n} \\ 0 & 0 & x_{33} & \dots & x_{3n} \\ \dots & \dots & \dots & \dots & \dots \\ 0 & 0 & 0 & \dots \ 0 & x_{nn} \end{pmatrix}$$

where the diagonal entries x_{jj} are all positive. Then the right and left invariant Haar measures μ_r and μ_l are given by

$$d\mu_r(\xi) = \left[x_{11}\ x_{22}^2\ x_{33}^3\ \cdots\ x_{nn}^n\right]^{-1} \prod_{1 \leqslant i \leqslant j \leqslant} dx_{ij},$$

$$d\mu_l(\xi) = \left[x_{11}^n\ x_{22}^{n-1}\ x_{33}^{n-2}\ \cdots\ x_{nn}\right]^{-1} \prod_{1 \leqslant i \leqslant j \leqslant n} dx_{ij},$$

respectively, where $\prod_{1 \leqslant i \leqslant j \leqslant n} dx_{ij}$ indicates Lebesgue measure in the Euclidean space of dimension $\dfrac{n(n+1)}{2}$.

Definition. 54.7. Let X be a locally compact second countable metric group. *A* σ-finite measure μ on X is said to be *right* (left) *quasi invariant* if $\mu(Ex)=0$ ($\mu(xE)=0$) for all x whenever $\mu(E)=0$, $E \in \mathscr{B}_X$.

Proposition. 54.8. Let X be a locally compact second countable metric group and let λ be a right or left quasi invariant measure. Let μ_l and μ_r be respectively left and right invariant Haar measures on X. Then $\lambda \equiv \mu_l \equiv \mu_r$.

Proof. Let λ be a right quasi invariant σ-finite measure. Let $\lambda(E)=0$ for some $E \in \mathscr{B}_X$. By Fubini's theorem we have

$$\begin{aligned} 0 &= \int \lambda(Ey^{-1})\, d\mu_l(y) = \int\int \chi_E(xy)\, d\lambda(x)\, d\mu_l(y) \\ &= \int \left[\, \int \chi_E(xy) d\mu_l(y)\, \right] d\lambda(x) = \int \mu_l(x^{-1} E) d\lambda(x) \\ &= \int \mu_l(E)\, d\lambda(x). \end{aligned}$$

This implies that $\mu_l(E)=0$. Conversely, if $\mu_l(E)=0$, going through the above equations in the reverse order we have

$$\int \lambda(Ey^{-1}) d\mu_l(y) = 0.$$

Hence for some $y_0 \in X$, $\lambda(Ey_0^{-1})=0$. Since λ is right quasi invariant we have $\lambda(E)=0$. Thus $\lambda \equiv \mu_l$. In particular $\mu_r \equiv \mu_l$. The proof is complete.

Corollary. 54.9. (Uniqueness theorem.) Let μ, μ' be two left (right) invariant Haar measures on a locally compact second countable metric group X. Then there exists a constant c such that $\mu'(E)=c\mu(E)$ for all $E \in \mathscr{B}_X$.

Proof. By the preceding proposition $\mu' \equiv \mu$. Let $f = \dfrac{d\mu'}{d\mu}$. Then

$$\begin{aligned} \int_E f\, d\mu &= \mu'(E) = \mu'(z^{-1} E) \\ &= \int_{z^{-1}E} f\, d\mu = \int_E f(zx)\, d\mu(x). \end{aligned}$$

for all $E \in \mathscr{B}_X$ and $z \in X$. Hence

$$f(x) = f(zx) \text{ a.e. } x\ (\mu),$$

for every z. By Fubini's theorem (See Corollary 35.16) there exists x_0 such that

$$f(x_0) = f(zx_0) \text{ a.e. } z\ (\mu).$$

Since μ is right quasi invariant $f(z) = c$, a constant a.e. z (μ). Hence $\mu'(E) = c\mu(E)$ for all $E \in \mathscr{B}_X$. This completes the proof.

Remark. 54.10. Suppose μ is a left invariant Haar measure on X, For any element x the set function $\mu(Ex)$, $E \in \mathscr{B}_X$ is again a left invariant Haar measure. By the above corollary there exists a constant $\Delta(x)$ such that

$$\mu(Ex) = \Delta(x)\mu(E) \text{ for all } E \in \mathscr{B}_X. \tag{54.3}$$

It is clear that $\Delta(xy) = \Delta(x)\Delta(y)$ for all x, y. It is easy to see that Δ is a continuous function on X with values in $(0, \infty)$. It is called the *modular function* of the group X.

If a group X has no non-trivial continuous homomorphisms into $(0, \infty)$ considered as a multiplicative group then $\Delta(x) \equiv 1$, and the left invariant Haar measure is automatically right invariant. Such groups are called *unimodular*. For example every compact metric group is unimodular.

If Δ is the modular function, put

$$\nu(E) = \int_E \Delta(x)^{-1}\, d\mu(x).$$

Then Eq. (54.3) implies the ν is right invariant. Thus the left and right invariant measures μ and ν respectively can always be chosen so that

$$\frac{d\mu}{d\nu} = \Delta.$$

Exercise. 54.11. The group $SL(n, R)$ of all $n \times n$ real matrices of determinant unity is unimodular.

Exercise. 54.12. Let μ be a left invariant Haar measure on X. For any borel set $E \subset X$ such that $\mu(E) < \infty$, the function $\mu(xE \Delta E)$ is continuous in x. If $\mu(E) > 0$ then $E^{-1}E$ is a neighbourhood of e. (Hint: Proceed exactly as in the proof of Proposition 21.2).

Exercise. 54.13. Let μ be as in the preceding exercise. Consider the B-space $L_p(\mu)$ for any $p \geqslant 1$. Let L_z be the map

$$(L_z f)(x) = f(z^{-1}x), f \in L_p(\mu), z \in X.$$

Then L_z is an operator on $L_p(\mu)$ and $\| L_z f \|_p = \| f \|_p$. Further $L_{z_1} L_{z_2} = L_{z_1 z_2}$ for all $z_1, z_2 \in X$ and

$$\lim_{z \to e} \| L_z f - f \| = 0 \text{ for all } f \in L_p(\mu).$$

(Hint: Proceed exactly as in the proof of Corollary 39.2).

Exercise. 54.14. Let $X_0 \subset X$ be a countable dense subgroup. Let $D \subset X$ be a subset such that $D \cap (zX_0)$ is a single point set for every $z \in X$. If μ is the left invariant Haar measure then D is not μ^*-measurable.

Let X/X_0 be the space of all left cosets of the form zX_0, $z \in X$. Let π be the cononical map $z \to zX_0$ from X onto X/X_0. Declare a set $E \subset X/X_0$ to be measurable if $\pi^{-1}(E)$ is measurable in X. Choose a probability measure $\lambda \equiv \mu$. On the σ-algebra $\mathscr{B}_0$ of all measurable sets in X/X_0 define $\lambda_0 = \lambda \pi^{-1}$. Then $(X/X_0, \mathscr{B}_0, \lambda_0)$ is a non-standard probability space. (Hint: Proceed exactly as in the proof of Proposition 21.5).

§55. Quasi Invariant Measures on Homogeneous Spaces

Let X be a locally compact second countable metric group and let T be a locally compact metric space with a continuous map $(x, t) \to xt$ from $X \times T$ onto T with the following properties:

(i) $y(xt) = (yx)t$ for all $y, x \in X$ and $t \in T$;

(ii) for each fixed $x \in X$, the map $t \to xt$ is a homeomorphism of T;

(iii) for any two points $t_1, t_2 \in T$ there exists an element $x \in X$ such that $xt_1 = t_2$.

(iv) $xt = t$ for all $t \in T$ if and only if $x = e$.

Then T is called a *homogeneous space* under the group X. We shall consider only such homogeneous spaces.

Let $t_0 \in T$ be fixed. Then the set $\{x : xt_0 = t_0\}$ is a closed subgroup of X. It is called the *little group* at t_0 or the *isotropy subgroup* at t_0.

We shall write π for the cononical map $x \to xt_0$ from X into T. By condition (iii) π is actually an onto map. For any borel set $E \subset T$ and any $x \in X$, we write xE for the set $\{xt, t \in E\}$. Then xE is also a borel set.

We shall choose and fix left and right invariant Haar measures μ_l and μ_r in X such that

$$\frac{d\mu_l}{d\mu_r} = \Delta, \tag{55.1}$$

where Δ is the modular function on X. On the little group X_0 at t_0 we choose and fix left and right invariant measures ν_l and ν_r such that

$$\frac{d\nu_l}{d\nu_r} = \delta, \tag{55.2}$$

where δ is the modular function on X_0.

With these notations we shall investigate the problem of existence of invariant measures on T. To this end we introduce the following definition.

Definition. 55.1. A σ-finite measure λ on T is said to be *quasi invariant* under the group X if for any borel set $E \subset T$, $\lambda(xE) = 0$ for all $x \in X$ whenever $\lambda(E) = 0$. λ is said to be *invariant* under the group X if $\lambda(xE) = \lambda(E)$ for all $x \in X$ and $E \in \mathscr{B}_T$.

Proposition. 55.2. Let T be a homogeneous space under the action of the group X. Then there exists a quasi invariant probability measure λ on $\mathscr{B}_T$. If λ_1 is any quasi invariant σ-finite measure on T then $\lambda \equiv \lambda_1$.

Proof. Choose a strictly positive function ϕ on X such that $\int \phi \, d\mu_l = 1$. Define the probability measure μ by

$$\mu(A) = \int_A \phi \, d\mu_l, \; A \in \mathscr{B}_X.$$

Put $\lambda = \mu\pi^{-1}$, where π is the canonical map described earlier. Then λ is a probability measure on T. Let $E \in \mathscr{B}_T$ be such that $\lambda(E) = 0$. For any x we have

$$\lambda(xE) = \mu(\pi^{-1}(xE)) = \mu(x\pi^{-1}(E))$$

because $\pi(xy) = x\pi(y)$ for all $x, y \in X$. If $\lambda(E) = 0$, then $\mu\pi^{-1}(E) = 0$. Since $\mu \equiv \mu_l$ it follows that $\mu(x\pi^{-1}(E)) = 0$. Thus λ is quasi invariant.

Let now λ_1 be any σ-finite quasi invariant measure. Then

$$\begin{aligned} \lambda_1(E) = 0 &\Longrightarrow \lambda_1(x^{-1}E) = 0 \text{ for all } x \\ &\Longrightarrow \int \lambda_1(x^{-1}E)\, d\mu(x) = 0 \\ &\Longrightarrow \int\int \chi_E(xt)\, d\mu(x)\, d\lambda_1(t) = 0 \\ &\Longrightarrow \mu\{x : xt \in E\} = 0 \text{ for some } t. \end{aligned}$$

Choose an x_0 such that $x_0 t = t_0$. Then the quasi invariance of μ under right translation by x_0 implies that

$$\mu\{x : x\, t_0 \in E\} = \mu\pi^{-1}(E) = \lambda(E) = 0.$$

Thus $\lambda \ll \lambda_1$. Going through the above implications in the reverse order we see that $\lambda_1 \ll \lambda$. This completes the proof.

Exercise. 55.3. Let λ be a quasi invariant σ-finite measure on T under X. For any $x \in X$ let λ^x be the measure defined by $\lambda^x(E) = \lambda(xE)$ for all $E \in \mathscr{B}_T$. In $L_2(\lambda)$ define the operator U_x by

$$(U_x f)(t) = \sqrt{\frac{d\lambda}{d\lambda^x}}(x^{-1} t)\, f(x^{-1} t).$$

Then U_x is an operator on $L_2(\lambda)$ with the properties:

$$(U_x f_1, U_x f_2) = (f_1, f_2) \text{ for } x \in X;\ f_1, f_2 \in L_2(\lambda),$$
$$U_x U_y = U_{xy} \text{ for all } x, y \in X,$$

where $(.,.)$ stands for the inner product in $L_2(\lambda)$. (See Exercise 48.9)

Exercise. 55.4. Let μ be a probability measure on X such that

$$\mu(A) = \int_A \phi(x)\, d\mu_l(x),\ A \in \mathscr{B}_X,$$

where $\phi(x) > 0$ for all x. Let μ^x be the measure defined by

$$\mu^x(A) = \mu(xA) \text{ for all } x \text{ and } A \in \mathscr{B}_X. \text{ Then}$$

$$\frac{d\mu^{x_1}}{d\mu}(x) = \phi(x_1 x)\, \phi(x)^{-1} \text{ a.e. } x(\mu_l) \text{ for all } x_1.$$

Proposition. 55.5. Let T be a homogeneous space under X and let $t_0 \in T$. For any compact set $K \subset X$ the function

$$\alpha(x) = \int_{X_0} \chi_K(x\xi)\, d\nu_l(\xi)$$

where X_0 is the little group at t_0 and ν_l is its left invariant Haar measure is a bounded function of x wihch is right invariant under X_0, i.e., $\alpha(x\xi) = \alpha(x)$ for all $\xi \in X_0$.

Proof. The right invariance of α is trivial. It is clear that

$$\alpha(x) = \nu_l(x^{-1} K \cap X_0).$$

If $x^{-1} K \cap X_0 \neq \varnothing$ then there exists a $k(x) \in K$ such that $x^{-1} k(x) = h(x) \in X_0$. Thus $x^{-1} = h(x)\, k(x)^{-1}$ and

$$\begin{aligned}\nu_l(x^{-1} K \cap X_0) &= \nu_l(h(x)\, k(x)^{-1} K \cap X_0)\\ &= \nu_l(k(x)^{-1} K \cap X_0)\\ &\leqslant \nu_l(K^{-1} K \cap X_0) < \infty.\end{aligned}$$

This completes the proof.

Proposition. 55.6. Let T be a homogeneous space under the group X and let X_0 be the little group at $t_0 \in T$. Let μ_l and ν_l be the left invariant Haar measures of X and X_0 respectively. Let $\mathscr{B}_0 \subset \mathscr{B}_X$ be the sub σ-algebra $\{A: A \in \mathscr{B}_X, A\xi = A \text{ for all } \xi \in X_0\}$. Let μ be a probability measure equivalent to μ_l and let $\alpha(x) = \dfrac{d\mu}{d\mu_l}(x)$. For any

$x \in X$ let μ^x be the probability measure defined by $\mu^x(A) = \mu(xA)$ for all $A \in \mathscr{B}_X$. Then

$$\mathbf{E}\left(\frac{d\mu^x}{d\mu}\,\middle|\, \mathscr{B}_0\right)(y) = \beta(xy)\,\beta(y)^{-1} \text{ a.e. } y(\mu), \tag{55.3}$$

where

$$\beta(y) = \int_{X_0} \alpha(y\,\xi^{-1})\,\Delta(\xi)^{-1}\,d\nu_l(\xi), \tag{55.4}$$

and Δ is the modular function of X. Further

$$\beta(y\xi) = \beta(y)\,\frac{\delta(\xi)}{\Delta(\xi)} \text{ a.e. } y\,(\mu) \tag{55.5}$$

for every $\xi \in X_0$, where δ is the modular function of X_0. The space T admits a σ-finite invariant measure if and only if $\delta(\xi) = \Delta(\xi)$ for all $\xi \in X_0$. In such a case the invariant measure is unique.

Proof. Let $\gamma(x, y) = \mathbf{E}\left(\frac{d\mu^x}{d\mu}\,\middle|\, \mathscr{B}_0\right)(y)$. Since $\gamma(x.)$ is $\mathscr{B}_0$ measurable we have

$$\gamma(x, y\xi) = \gamma(x, y) \text{ a.e. } y(\mu), \tag{55.6}$$

for each fixed $x \in X$, $\xi \in X_0$. By the preceding proposition for any compact K the function

$$\int_{X_0} \chi_K(y\,\xi)\,d\nu_l(\xi)$$

is right invariant under X_0 and bounded. Hence it is $\mathscr{B}_0$ measurable. By the definition of conditional expectation we have

$$\begin{aligned} &\int_X \left(\int_{X_0} \chi_K(y\,\xi)\,d\nu_l(\xi)\right) \gamma(x, y)\,\alpha(y)\,d\mu_l(y) \\ &= \int_X \left(\int_{X_0} \chi_K(y\,\xi)\,d\nu_l(\xi)\right) d\mu^x(y) \\ &= \int_X \left(\int_{X_0} \chi_K(y\,\xi)\,d\nu_l(\xi)\right) \alpha(xy)\,d\mu_l(y). \end{aligned}$$

Put $y\xi = y'$ on both sides. By Fubini's theorem, Eqs. (55.1) and (55.6) we get

$$\begin{aligned} &\int \chi_K(y')\,\gamma(x, y')\left[\int_{X_0} \alpha(y'\,\xi^{-1})\,\Delta(\xi)^{-1}\,d\nu_l(\xi)\right] d\mu_l(y') \\ &= \int \chi_K(y')\left[\int \alpha(xy'\,\xi^{-1})\,\Delta(\xi)^{-1}\,d\nu_l(\xi)\right] d\mu_l(y') \end{aligned}$$

for every compact K. Hence

$$\gamma(x, y) = \frac{\int \alpha(xy\xi^{-1})\,\Delta(\xi)^{-1}\,d\nu_l(\xi)}{\int \alpha(y\xi^{-1})\,\Delta(\xi)^{-1}\,d\nu_l(\xi)} \quad \text{a.e. } y(\mu)$$

which proves Eq. (55.3). For any $\eta \in X_0$ we have

$$\begin{aligned} \beta(y\eta) &= \int \alpha(y\eta\xi^{-1})\,\Delta(\xi)^{-1}\,d\nu_l(\xi) \\ &= \int \alpha(y\eta\xi^{-1})\,\Delta(\xi)^{-1}\,\delta(\xi)\,d\nu_r(\xi). \end{aligned}$$

Changing the variable ξ to $\xi\eta$ we get Eq. (55.5).

Now suppose there exists a σ-finite measure λ on T which is invariant under X action. By Proposition 55.2, $\lambda \equiv \mu\pi^{-1}$. Let

$$\rho(t) = \frac{d\lambda}{d\mu\pi^{-1}}(t).$$

Then

$$\lambda(E) = \int_E \rho(t)\, d\mu\pi^{-1}(t) = \int_{\pi^{-1}(E)} \rho(\pi(y))\, d\mu(y).$$

On the other hand

$$\begin{aligned}\lambda(E) = \lambda(xE) &= \int_{\pi^{-1}(xE)} \rho(\pi(y))\, d\mu\,(y)\\ &= \int_{x\pi^{-1}(E)} \rho(\pi(y))\, d\mu\,(y)\\ &= \int_{\pi^{-1}(E)} \rho(\pi(xy))\, \frac{d\mu^x}{d\mu}\,(y)\, d\mu\,(y)\\ &= \int_{\pi^{-1}(E)} \rho(\pi(xy))\, \mathbf{E}\left(\frac{d\mu^x}{d\mu}\,\middle|\, \mathscr{B}_0\right)(y)\, d\mu(y).\end{aligned}$$

Hence

$$\mathbf{E}\left(\frac{d\mu^x}{d\mu}\,\middle|\, \mathscr{B}_0\right)(y) = \rho(\pi(y))\, \rho(\pi(xy))^{-1} \text{ a.e. } y\ (\mu).$$

By Eq. (55.3) we have

$$\beta(xy)\, \beta(y)^{-1} = \rho(\pi(y))\, \rho(\pi(xy))^{-1} \text{ a.e. } y\ (\mu)$$

for each x. Thus

$$\rho(\pi(y))\beta(y) = \rho(\pi(xy))\, \beta(xy) \text{ a.e. } y\ (\mu)$$

for each x. By Fubini's theorem

$$\rho(\pi(y)), \beta(y) = c, \text{ a constant a.e. } y\ (\mu).$$

But $\rho(\pi(y\xi)) = \rho(\pi(y))$ for all y whenever $\xi \in X_0$. Hence $\beta(y\xi) = \beta(y)$ a.e. y for each $\xi \in X_0$. Hence Eq. (55.5) implies that $\delta(\xi) = \mathbf{\Delta}(\xi)$ for each $\xi \in X_0$. Conversely, if $\delta(\xi) = \mathbf{\Delta}(\xi)$ for all $\xi \in X_0$ then $\beta(y)$ defined by Eq. (55.4) is right invariant under X_0. Then there exists a function $\rho(t)$ on T such that

$$\rho(\pi(y)) = \beta(y)^{-1}.$$

Putting

$$\lambda(E) = \int_E \rho(t)\, d\mu\, \pi^{-1}\,(t)$$

we get a σ-finite invariant measure. If λ' is another σ-finite invariant measure on T then $\lambda' \equiv \lambda$. Let

$$\frac{d\lambda'}{d\lambda}(t) = a(t),\ t \in T.$$

The invariance of λ and λ' imply

$$a(xt)=a(t) \text{ a.e. } t(\lambda) \text{ for each } x\in X.$$

By Fubini's theorem

$$a(xt_1)=a(t_1) \text{ a.e. } x(\mu_l)$$

for some t_1. There exists an x_1 such that $x_1t_1=t_0$. Hence

$$a(xt_0)=\text{constant a.e. } x(\mu_l).$$

Since $\mu\equiv\mu_l$ and $\mu\pi^{-1}\equiv\lambda$ it follows that

$$a(t)=\text{constant a.e. } t\,(\lambda).$$

Thus $\lambda'=c\lambda$ for some constant c. This completes the proof.

Example. 55.7. We shall now study an example which we come across in physics. Consider the space R^4 and denote an arbitrary point of R^4 by

$$\mathbf{x}=\begin{pmatrix} x_0 \\ x_1 \\ x_2 \\ x_3 \end{pmatrix}.$$

For any two points $\mathbf{x}$, $\mathbf{y}$, let

$$\langle \mathbf{x}, \mathbf{y}\rangle=x_0y_0-x_1y_1-x_2y_2-x_3y_3$$

or, equivalently,

$$\langle \mathbf{x}, \mathbf{y}\rangle=\mathbf{x}'\Lambda\mathbf{y},$$

$$\Lambda=\begin{pmatrix} 1 & 0 & 0 & 0 \\ 0 & -1 & 0 & 0 \\ 0 & 0 & -1 & 0 \\ 0 & 0 & 0 & -1 \end{pmatrix}.$$

Let

$$SO(1, 3)=\{A : A=((a_{ij}))0\leqslant i,j\leqslant 3, \det A=1,$$
$$\langle A\mathbf{x}, A\mathbf{y}\rangle=\langle \mathbf{x}, \mathbf{y}\rangle \text{ for all } \mathbf{x}, \mathbf{y} \in R^4\}.$$

Then $SO\ (1, 3)$ is a group. It is clear that

$$SO(1, 3)=\{A : A'\,\Lambda A=\Lambda, \det A=1\} \tag{55.7}$$

where A' denotes the transpose of A. Let

$$T=\{\mathbf{x} : \langle \mathbf{x}, \mathbf{x}\rangle=m^2\}$$

where m is a fixed positive constant. The map $(A, \mathbf{x})\to A\mathbf{x}$ makes T a homogeneous space under the group $SO\ (1, 3)$. Let

$$\mathbf{p}=\begin{pmatrix} m \\ 0 \\ 0 \\ 0 \end{pmatrix}.$$

The little group at $\mathbf{p}$ consists of all matrices $((a_{ij}))$ in $SO\ (1, 3)$ which have the property

$$a_{00}=1,\ a_{01}=a_{02}=a_{03}=a_{10}=a_{20}=a_{30}=0.$$

(It is isomorphic to the group of orthogonal 3×3 matrices of determinant unity and hence compact.) Both $SO\ (1, 3)$ and the little group at $\mathbf{p}$ have modular function unity. Hence T has an invariant σ-finite measure. We shall now compute it. We have

$$T=\{\mathbf{x} : x_0>0,\ x_0=(m^2+x_1^{\ 2}+x_2^{\ 2}+x_3^{\ 2})^{\frac{1}{2}}\} \cup$$
$$\{\mathbf{x} : x_0<0,\ x_0=-(m^2+x_1^{\ 2}+x_2^{\ 2}+x_3^{\ 2})^{\frac{1}{2}}\}=T_+ \cup T_- \text{ say.}$$

Thus T is the union of two connected sets T_+ and T_- each of which is homeomorphic to R^3. For any $A=((a_{ij}))$, the transformation

$$\mathbf{x}\rightarrow A\mathbf{x}$$

considered as a transformation in T has the form

$$\begin{pmatrix} x_1 \\ x_2 \\ x_3 \end{pmatrix} \rightarrow \begin{pmatrix} a_{10}x_0+a_{11}x_1+a_{12}x_2+a_{13}x_3 \\ a_{20}x_0+a_{21}x_1+a_{22}x_2+a_{23}x_3 \\ a_{30}x_0+a_{31}x_1+a_{32}x_2+a_{33}x_3 \end{pmatrix}$$

in T_+ and T_-, where $x_0=\pm(m^2+x_1^{\ 2}+x_2^{\ 2}+x_3^{\ 2})^{\frac{1}{2}}$. Hence the modulus of the Jacobian is the absolute value of

$$\det \begin{pmatrix} a_{10}\dfrac{x_1}{x_0}+a_{11} & a_{10}\dfrac{x_2}{x_0}+a_{12} & a_{10}\dfrac{x_3}{x_0}+a_{13} \\ a_{20}\dfrac{x_1}{x_0}+a_{21} & a_{20}\dfrac{x_2}{x_0}+a_{22} & a_{20}\dfrac{x_3}{x_0}+a_{23} \\ a_{30}\dfrac{x_1}{x_0}+a_{31} & a_{30}\dfrac{x_2}{x_0}+a_{32} & a_{30}\dfrac{x_3}{x_0}+a_{33} \end{pmatrix}.$$

But the above expression is same as

$$\frac{1}{x_0}\det \begin{pmatrix} x_0 & -x_1 & -x_2 & -x_3 \\ a_{10} & a_{11} & a_{12} & a_{13} \\ a_{20} & a_{21} & a_{22} & a_{23} \\ a_{30} & a_{31} & a_{32} & a_{33} \end{pmatrix}.$$

But Eq. (55.7) implies that this is x_0'/x_0 where

$$x'_0=a_{00}x_0+a_{01}x_1+a_{02}x_2+a_{03}x_3$$

Thus the modulus of the Jacobian is of the form $\rho(\mathbf{x})\ \rho(\mathbf{Ax})^{-1}$ where $\rho(\mathbf{x})=\dfrac{1}{|x_0|}$. By Proposition 54.4 it follows that the invariant measure λ on T is given by

$$d\lambda(x)=\frac{dx_1\ dx_2\ dx_3}{(m^2+x^2_{\ 1}+x^2_{\ 2}+x^2_{\ 3})^{\frac{1}{2}}}$$

Exercise. 55.8. Let $GL(n, R)=X$ be the group of all non-singular $n\times n$ matrices of the form $((x_{ij}))_{1\leq i, j\leq n}$. Then the left and right invariant Haar measures μ_l and μ_r are given by

$$d\mu_r=d\mu_l=|\det((x_{ij}))|^{-n} \prod_{1\leq i, j\leq n} dx_{ij}.$$

The modular function of the subgroup X_0 of all upper triangular matrices of the form $((x_{ij}))$ where $x_{ij}=0$ if $j<i$, is given by

$$\delta\left(\begin{pmatrix} x_{11} & x_{12} & \dots & x_{1n} \\ 0 & x_{22} & \dots & x_{2n} \\ 0 & 0 & \dots 0 & x_{nn} \end{pmatrix}\right)=\left|\prod_{j=1}^{n} x_{jj}{}^{-(n-2j+1)}\right|$$

Hence there does not exist an X-invariant measure on the quotient space X/X_0.

§56. Mackey-Weil Theorem

We shall now establish that a standard borel group $(X, \mathscr{B})$ with a quasi invariant measure μ has a locally compact topology whose borel σ-algebra is $\mathscr{B}$ and μ is equivalent to the Haar measure.

Let $(X, \mathscr{B})$ be a standard borel space which is also a group such that the map $(x, y)\to xy^{-1}$ from $(X\times X, \mathscr{B}\times\mathscr{B})$ onto $(X, \mathscr{B})$ is borel. Then $(X, \mathscr{B})$ is called a *standard group*. We shall adopt the notations of Section 54. For any $A\in\mathscr{B}$, consider the set $E=\{(x, y): xy^{-1}\in A\}$. Since $A^{-1}=\{y: (e, y)\in E\}$ and $Ay=\{x: (x, y)\in E\}$ are sections of measurable sets they are themselves measurable. Combining the two we see that yA is also measurable. Thus inversion and right and left translations are borel automorphisms of $(X, \mathscr{B})$. A σ-finite measure μ on $(X, \mathscr{B})$ is said to be *left quasi invariant* if for any $A\in\mathscr{B}$, $\mu(xA)=0$ for all x whenever $\mu(A)=0$.

Let $(X, \mathscr{B}, \mu)$ be a standard group with a left quasi invariant measure and let $L_2(\mu)$ be the real Hilbert space of square integrable functions with respect to μ. For any $x\in X$ we introduce the operator L_x defined by

$$(L_x f)(y)=\sqrt{\frac{d\mu}{d\mu^x}}(x^{-1}y)\, f(x^{-1}y), f\in L_2(\mu) \tag{56.1}$$

where μ^x is the measure defined by $\mu^x(A)=\mu(xA)$ for all $A\in\mathscr{B}$. Then L_x has the following properties:

$$L_{x_1} L_{x_2}=L_{x_1x_2} \text{ for all } x_1, x_2\in X;$$
$$(L_x f_1, L_x f_2)=(f_1, f_2) \text{ for all } f_1, f_2\in L_2(\mu),$$

where (f_1, f_2) denotes the inner product in $L_2(\mu)$. If

$$\alpha(x, y) = \frac{d\mu}{d\mu^x}(y)$$

then $\alpha(e, y) = 1$ and $\alpha(x_1 x_2, y) = \alpha(x_1, x_2 y)\, \alpha(x_2, y)$ a.e. y (μ) for every $x_1, x_2 \in X$. (See Exercise 48.9). In particular,

$$\frac{d\mu}{d\mu^x}(x^{-1} y) = \frac{d\mu^{x^{-1}}}{d\mu}(y).$$

Let $\mathcal{U}$ denote the set of all operators U on $L_2(\mu)$ such that U is onto and $(Uf_1, Uf_2) = (f_1, f_2)$ for all $f_1, f_2 \in L_2(\mu)$. Then $\mathcal{U}$ is a group. Give $\mathcal{U}$ the smallest topology which makes every map

$$U \to (Uf, g)$$

continuous for all fixed pairs f, g of elements in $L_2(\mu)$. Since $L_2(\mu)$ is a complete and separable metric space it follows that $\mathcal{U}$ is also a complete and separable metric space where the map $(U, V) \to UV^{-1}$ from $\mathcal{U} \times \mathcal{U}$ onto $\mathcal{U}$ is continuous. $\mathcal{U}$ is called the *unitary group of* $L_2(\mu)$ and any element of $\mathcal{U}$ is called a *unitary operator*.

Proposition. 56.1. Let $(X, \mathcal{B}, \mu)$ be a standard group with a left quasi invariant probability measure μ. The map $x \to L_x$ defined by Eq. (56.1) from X into $\mathcal{U}$ is borel.

Proof. It is enough to prove that for any A, B in $\mathcal{B}$ the function

$$\phi_{A,B}(x) = \int \left(\frac{d\mu}{d\mu^x}\right)^{\frac{1}{2}} (x^{-1} y)\, \chi_A(x^{-1} y)\, \chi_B(y)\, d\mu(y) \tag{56.2}$$

is borel. Let λ be the measure on $\mathcal{B} \times \mathcal{B}$ defined by

$$\lambda(C) = \int \chi_C(x, xy)\, d\mu \times \mu\, (x, y),\ C \in \mathcal{B} \times \mathcal{B}. \tag{56.3}$$

Since μ is left quasi invariant it follows that $\mu \times \mu$ is quasi invariant under the transformation $(x, y) \to (x, xy)$. Hence $\lambda \equiv \mu \times \mu$. Let

$$\psi(x, y) = \frac{d\lambda}{d\mu \times \mu}(x, y).$$

Then

$$\lambda(A \times B) = \int \left[\int \psi(x, y)\, \chi_B(y)\, d\mu(y)\right] \chi_A(x)\, d\mu(x). \tag{56.4}$$

By Eq. (56.3)

$$\begin{aligned}
\lambda(A \times B) &= \int \left[\int \chi_A(x)\, \chi_B(xy)\, d\mu(y)\right] d\mu(x) \\
&= \int \chi_A(x) \left[\int \chi_B(y) \frac{d\mu^{x^{-1}}}{d\mu}(y)\, d\mu(y)\right] d\mu(x) \\
&= \int \chi_A(x) \left[\int \chi_B(y) \frac{d\mu}{d\mu^x}(x^{-1} y)\, d\mu(y)\right] d\mu(x).
\end{aligned} \tag{56.5}$$

Comparing Eqs. (56.4) and (56.5) we see that

$$\frac{d\mu}{d\mu^x}(x^{-1} y) = \psi(x, y) \text{ a.e. } y(\mu) \text{ a.e. } x(\mu).$$

Hence Eq. (56.2) can be rewritten as

$$\phi_{A,B}(x) = \int \psi(x,y)^{\frac{1}{2}} \chi_A(x^{-1}y)\, \chi_B(y)\, d\mu(y).$$

Since the integrand is measurable on $X \times X$ Fubini's theorem implies that $\phi_{A,B}$ is borel. This completes the proof.

Proposition. 56.2. The mapping $x \to L_x$ from X into $\mathscr{U}$ is one-one.

Proof. If for some $x \neq y$ $L_x = L_y$ then $L_{xy^{-1}} = L_x L_y^{-1} = I$, where I is the identity operator. Hence it is enough to prove that $x = e$ if $L_x = I$. Suppose for some x_0, $L_{x_0} = I$. Then the constant function 1 is in $L_2(\mu)$ and $L_{x_0}1 = 1$. This implies $\frac{d\mu}{d\mu^{x_0}}(x_0^{-1}y) = 1$ a.e. $y(\mu)$. Since μ is left quasi invariant we have $\frac{d\mu}{d\mu^{x_0}}(y) = 1$ a.e. $y(\mu)$ or equivalently, $\mu^{x_0} = \mu$. Since $L_{x_0}\chi_A = \chi_A$ for all $A \in \mathscr{B}$ we have $\chi_A(x_0^{-1}y) = \chi_A(y)$ a.e. $y(\mu)$ and $\mu(x_0 A \,\Delta\, A) = 0$ for all $A \in \mathscr{B}$. Since X has a separable metric topology whose borel sets constitute $\mathscr{B}$ we can choose a metric d and a countable dense set $D = \{z_1, z_2, \ldots\}$ in X such that the spheres under d with centres at z_j, $j = 1, 2, \ldots$ belong to $\mathscr{B}$. Let for every positive rational r and every positive integer j, $S_{jr} = \{x : d(x, z_j) < r\}$. Then

$$\mu\{\bigcup_{j,r} (x_0 S_{jr} \,\Delta\, S_{jr})\} = 0.$$

But the set within brackets above is the same as $\{x : x_0^{-1}x \neq x\}$. Hence $x_0 = e$ and the proof is complete.

Proposition. 56.3. Let $A \in \mathscr{B}$ and $\mu(A) > 0$. Then

$$\{x : \int |L_x \chi_A - \chi_A|^2 d\mu < \mu(A)\} \subset AA^{-1}. \tag{56.6}$$

If $\Upsilon = \{L_x, x \in X\}$ then Υ is a locally compact subgroup of the unitary group $\mathscr{U}$.

Proof. We have by Schwarz's inequality

$$\begin{aligned}\int |L_x \chi_A - \chi_A|^2 d\mu &= 2\mu(A) - 2\int \chi_A L_x \chi_A \, d\mu \\ &= 2\mu(A) - 2\int \left[\frac{d\mu^{x^{-1}}}{d\mu}(y)\right]^{\frac{1}{2}} \chi_{xA \cap A}(y)\, d\mu(y) \\ &\geqslant 2\mu(A) - 2\mu(A \cap x^{-1}A)^{\frac{1}{2}} \mu(xA \cap A)^{\frac{1}{2}}.\end{aligned}$$

Let B denote the left hand side of (56.6). If $x \in B$, then

$$\mu(xA \cap A)^{\frac{1}{2}} \mu(A \cap x^{-1}A)^{\frac{1}{2}} \geqslant \tfrac{1}{2}\mu(A) > 0.$$

Hence $xA \cap A \neq \emptyset$ and $x \in AA^{-1}$. This proves Eq. (56.6).

Since $L: x \to L_x$ is a borel mapping from the standard space $(X, \mathscr{B})$ into the standard space $\mathscr{U}$, Lusin's theorem implies that there exists a set $A_0 \in \mathscr{B}$ such that $L(A_0)$ is a borel set in $\mathscr{U}$ and $\mu(A_0) = 1$. Since

μL^{-1} is a probability measure in $\mathscr{U}$ we can find a compact set $K \subset L(A_0)$ such that $\mu L^{-1}(K) > 0$. In Eq. (56.6) put $A = L^{-1}(K)$. Let Γ be the neighbourhood of the identity operator in $\mathscr{U}$ defined by

$$\Gamma = \{U : \|U\chi_A - \chi_A\|^2 < \mu(A)\}.$$

Then Eq. (56.6) can be written as

$$\Gamma \cap \Upsilon \subset KK^{-1} \subset \Upsilon.$$

Since K is compact in $\mathscr{U}$, KK^{-1} is also compact in $\mathscr{U}$. Thus a neighbourhood of I in Υ is contained in a compact subset of Υ. In other words Υ is a locally compact subgroup of $\mathscr{U}$. This completes the proof.

Proposition. 56.4. Let $(X, \mathscr{B}, \mu)$ be a standard group with a σ-finite left quasi invariant measure μ. Then there exists a topology on X which makes it a locally compact second countable metric group with the following properties: (i) the borel σ-algebra $\mathscr{B}_0$ of the topology is contained in $\mathscr{B}$; (ii) μ is equivalent to the left invariant (and hence right invariant) Haar measure on $\mathscr{B}_0$; (iii) the completion of $\mathscr{B}_0$ under the Haar measure includes $\mathscr{B}$.

Proof. Without loss of generality we may assume that μ is a probability measure (otherwise we can construct an equivalent probability measure ν and replace μ by ν). Now construct the group $\Upsilon \subset \mathscr{U}$ as in Proposition 56.3. Let $\mathscr{B}_\Upsilon$ be its borel σ-algebra. Put $\mathscr{B}_0 = \{L^{-1}(E), E \in \mathscr{B}_\Upsilon\}$, where L denotes the map $x \to L_x$. Give the smallest topology to X which makes L continuous. Since L is one-one and Υ is locally compact it follows that under this topology X is a locally compact second countable metric group whose borel σ-algebra is $\mathscr{B}_0$. Since L is a borel map, $\mathscr{B}_0 \subset \mathscr{B}$. Since μ on $\mathscr{B}_0$ is left quasi invariant, Proposition 54.7 implies that $\mu \equiv \lambda$ where λ is a left invariant Haar measure on $(X, \mathscr{B}_0)$. Thus we have proved (i) and (ii).

Since $(X, \mathscr{B})$ and $(\mathscr{U}, \mathscr{B}_U)$ are standard borel spaces and L is a one-one borel map from X into $\mathscr{U}$ we can find a set $A \in \mathscr{B}$ such that $L(A) \in \mathscr{B}_U$ and $\mu(A) = 1$. This implies that $A \in \mathscr{B}_0$ and $\lambda(A') = 0$. Thus (iii) holds and the proof is complete.

Remark. 56.5. If we make use of Kuratowski's theorem (see Remark 24.27) we can conclude that L is a borel isomorphism between X and Υ and hence $\mathscr{B}_0 = \mathscr{B}$. Thus every standard group $(X, \mathscr{B})$ with a left quasi invariant σ-finite measure μ has a topology which makes X a locally compact second countable metric group with borel σ-algebra $\mathscr{B}$. Further the Haar measure on X is equivalent to μ. This is precisely the Mackey-Weil theorem.

References

1. Billingsley, P., *Convergence of Probability Measures*, New York: Wiley, 1968.
2. Doob, J. L., *Stochastic Processes*, New York: Wiley, 1953.
3. Dunford, N. and Schwarz, J. T., *Linear Operators*, Vol. I, New York: Interscience, 1958.
4. Feller, W., *An Introduction to Probability Theory and its Applications*, Vols. I and II, New Delhi: Wiley Eastern, 1972.
5. Friedman, N. A., *Introduction to Ergodic Theory*, New York: Van Nostrand Reinhold Company, 1970.
6. Gnedenko, B. V. and Kolmogorov, A. N., *Limit Distributions for Sums of Independent Random Variables*, (Translated from Russian) Cambridge, Massachussets: Addison-Wesley, 1954.
7. Halmos, P. R., *Measure Theory*, Princeton, New Jersey: Van Nostrand, 1962.
8. ———, *Introduction to Hilbert Space and the Theory of Spectral Multiplicity*, New York: Chelsea, 1951.
9. ———, *Lectures on Ergodic Theory*, The Mathematical Society of Japan, 1956.
10. Hardy, G. H., *A Course of Pure Mathematics*, London: Cambridge University Press, 1952.
11. Hormander, L., *Linear Partial Differential Operators*, Berlin: Springer-Verlag, 1969.
12. Kelley, J. L., *General Topology*, New York: Van Nostrand, 1961.
13. Kingman, J. F. C. and Taylor, S. J., *Introduction to Measure and Probability*, London: Cambridge University Press, 1966.
14. Lukacs, E., *Characteristic Functions*, London: Griffin, 1970.
15. Mackey, G. W., *Mathematical Foundations of Quantum Mechanics*, New York: Benjamin, 1962.
16. Neumark, M. A., *Normed Rings*, (Translated from Russian), Groningen: Noordhoff, 1959.

17. Parthasarthy, K. R., *Probability Measures on Metric Spaces*, New York: Academic Press, 1967.

18. ———, 'Probability Theory on the Closed Subspaces of a Hilbert Space', *Les Probabilités sur les Structures Algebriques*, Paris: C.N.R.S., 1970, pp. 265-92.

19. Schwarz, L., *Theorie des distributions*, Paris: Hermann, 1966.

20. Sobolev, S. L., *Applications of Functional Analysis in Mathematical Physics*, Translations of Mathematical Monographs, Vol. 7, Providence: Amer. Math. Soc., 1963.

21. Szego, G., Orthogonal Polynomials, Providence: Colloquium Publications, Amer. Math. Soc., 1948.

22. Varadarajan, V. S., *Geometry of Quantum Theory*, Vol. I and II, Princeton, New Jersey: Van Nostrand, 1968, 1970.

23. Zygmund, A., *Trigonometrical Series*, New York: Dover, 1955.

Index